The Imperial College Lectures in
PETROLEUM ENGINEERING

Reservoir Engineering

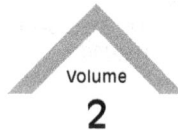

Volume
2

Martin J Blunt
Imperial College London, UK

World Scientific

NEW JERSEY · LONDON · SINGAPORE · BEIJING · SHANGHAI · HONG KONG · TAIPEI · CHENNAI · TOKYO

Published by

World Scientific Publishing Europe Ltd.

57 Shelton Street, Covent Garden, London WC2H 9HE

Head office: 5 Toh Tuck Link, Singapore 596224

USA office: 27 Warren Street, Suite 401-402, Hackensack, NJ 07601

Library of Congress Cataloging-in-Publication Data

Names: Blunt, Martin J., author.

Title: Reservoir engineering / Martin Blunt, (Imperial College London, UK).

Description: [Hackensack] New Jersey : World Scientific, [2017] |

   Series: The Imperial College lectures in petroleum engineering ; volume 2 | Includes index.

Identifiers: LCCN 2016049016 | ISBN 9781786342096 (hc : alk. paper)

Subjects: LCSH: Oil reservoir engineering.

Classification: LCC TN870.57 .B58 2017 | DDC 553.2/8--dc23

LC record available at https://lccn.loc.gov/2016049016

British Library Cataloguing-in-Publication Data

A catalogue record for this book is available from the British Library.

Desk Editors: Dipasri Sardar/Mary Simpson/Shi Ying Koe

Typeset by Stallion Press
Email: enquiries@stallionpress.com

Printed in Singapore

Preface

Access to sustainable, affordable water, energy and food are some of the major technological challenges of the 21st century. How do we manage precious and scarce water resources, while preventing pollution? How do we extract the most of our remaining supplies of conventional oil and gas? How can we extract, safely, shale gas and oil? Can we collect and store carbon dioxide in the subsurface to prevent atmospheric emissions and help avoid dangerous climate change? Can we manage to provide sufficient energy for a growing world's population with an aspiration for improved prosperity? An understanding of these challenges involves multiphase flow in porous media — the flow of water, oil and gas with associated pollutants — underground in geological formations.

The subject of multiphase flow in porous media is undergoing a revolution, not just as a result of its many important applications, but because of developments in our quantitative understanding of how fluids are arranged and moved, combined with the ability to image fluids at the micron scale inside rocks.

This book will apply concepts of multiphase flow in porous media to understand and design recovery from oil and gas reservoirs. This is one of the major challenges referred to above — at present we recover only around one third of the oil from fields we have discovered. How can we improve this to the 50–60% now achievable with the best engineering methods, and beyond?

This course will describe how different methods can be used to:

- assess the development potential of oil and gas reservoirs;
- identify the principal displacement mechanisms controlling performance;
- predict recovery and oil in place;
- understand reservoir simulation methods;
- understand single and multiphase flow in porous media.

It is assumed that you already know about hydrocarbon phase behaviour, reservoir simulation and the principal reservoir drive mechanisms. You will also need to know Darcy's law and the meaning of relative permeability and fractional flow, although these are described again in these notes.

The main audience for this book are students and practitioners of reservoir engineering, but this work is also of value to those interested in some of the problems mentioned above, namely carbon dioxide storage and contaminant transport. The material is based on lectures given to MSc students in Petroleum Engineering at Imperial College London and at Politecnico di Milano.

The emphasis will be on learning fundamentals with some time taken to cover basic concepts. I will not repeat details that are well covered in other textbooks, or which are not strictly relevant. Furthermore, these notes will not illustrate the concepts with field examples: this is better left to project work or indeed industrial experience. This is not a manual for reservoir engineers, but a teaching tool to establish the fundamentals.

About the Author

Martin Blunt joined Imperial in June 1999 as a Professor of Petroleum Engineering. He is also a visiting professor at the Politecnico di Milano. He served as Head of the Department of Earth Science and Engineering from 2006 to 2011. Previous to this he was Associate Professor of Petroleum Engineering at Stanford University in California. Before joining Stanford in 1992, he was a research reservoir engineer with BP in Sunbury-on-Thames. He holds MA and PhD (1988) degrees in theoretical physics from Cambridge University.

Professor Blunt's research interests are in multiphase flow in porous media with applications to geological carbon storage, oil and gas recovery, and contaminant transport and clean-up in polluted aquifers. He performs experimental, theoretical and numerical research into many aspects of flow and transport in porous systems, including pore-scale modelling of displacement processes, and large-scale simulation using streamline-based methods.

He has taught reservoir engineering to thousands of students and professionals worldwide for over 20 years and has been given teaching awards at both Stanford University and Imperial College London.

Contents

Chapter 1

Introduction to Reservoir Engineering

The main aim of this work is to understand how oil, water and gas flow deep underground with application to hydrocarbon recovery.

1.1. The Three Main Concepts: Material Balance, Darcy's Law and Data Integration

Before I present any details, there are three main points that need to be understood by any good reservoir engineer. In the end, everything can be expressed with reference to one of these three fundamental concepts.

1. **Material balance.** Mass is conserved; what leaves a reservoir (is produced) minus what is injected is the change of mass in the sub-surface. For every field, under every circumstance, a reservoir engineer needs to check material balance — ideally by hand — to understand and interpret production data. This will be the basic principle on which I will base the analysis of fields under primary production. Furthermore, it lies at the heart of the derivation of the flow equations used to predict flow performance. However, for this we also need an equation for flow — point 2 below.

2. **Darcy's law for fluid flow.** Fluid flows in response to a pressure gradient. The linear relationship between the gradient of pressure (or, more generally, the potential) and flow rate is Darcy's law. It is the basis for any understanding and prediction of flow.

3. **Look at all the data and have a coherent, consistent understanding of the field.** A reservoir engineer assesses different information from several sources: geological interpretations, seismic surveys, log analysis, core analysis and fluid properties combined with production (rate and pressure) data. All of this data needs to be incorporated into a model of the reservoir to predict future performance and design production. A model in this context is not solely a complicated computer realisation of what the field might be like, but more a conceptual understanding of the field that includes the type of fluids present, the geological structure and the production mechanism. Too frequently, the time-consuming yet intellectually mundane task of operating reservoir simulation software overwhelms the effort to understand the field rationally; what are the major uncertainties in the understanding of the field, what data is needed to remove or reduce these uncertainties, what is happening now, what controls production and, physically, what are the consequences of alternative production strategies? The essence of good reservoir engineering is combining data, identifying uncertainty and describing production mechanisms. It is not playing computer games with sophisticated software as a smokescreen for a poor understanding of the basic mechanisms by which oil is produced.

1.2. What is a Reservoir and What is a Porous Medium?

Figure 1.1 is a schematic of an oil field, which also contains gas, contained underneath impermeable cap rock. The diagram is reasonable, but rather underestimates the typical depth of the field. Usually, the oil is several kilometres below ground, while the depth of the column of oil itself is often less than 100 m. The areal extent is generally several square kilometres; later we will discuss some of the world's larger oil fields, but the total volume of oil-bearing rock is typically around 10^9 m^3, with, of course, a huge variation.

The gas and oil are held in the pore spaces of the rock at high temperatures and pressures. It is possible to estimate these values

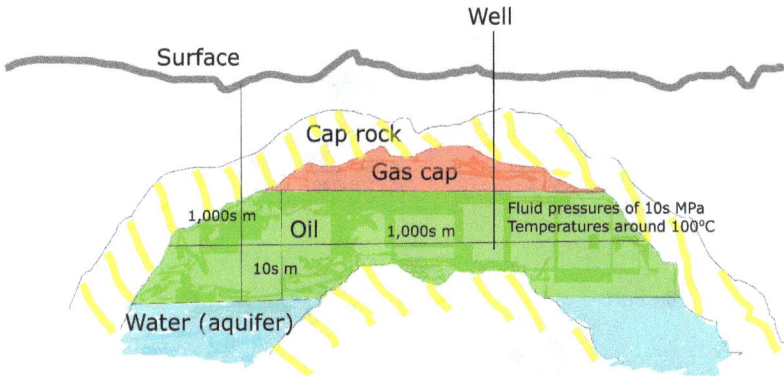

Figure 1.1. A schematic of an oil reservoir. The picture is reasonable, but the oil is generally found several kilometres below ground, while the water, oil and gas are all contained in porous rock.

from the known depth and the geothermal gradient, as well as the pressure gradient. A typical geothermal gradient is $30°\mathrm{C/km}$, giving temperatures of around $100°\mathrm{C}$ for reservoirs a few kilometres deep.

The oil and gas are held in a porous rock. What does this mean? Soils, sand, gravel, sedimentary rock and fractured rock all have some void space — i.e. gaps between the solid, as shown in Fig. 1.2. These systems are all porous media. If this space is continuous, in however a tortuous a fashion, it is possible for a fluid that occupies the voids to flow through the system — the material is said to be permeable. Soil, sand and gravel consist of small solid particles packed together. Consolidated rock is normally found deep underground where the individual particles have fused together. Volcanic rock that does not naturally contain any void space can still be permeable if it has a continuous pathway of fractures.

1.3. Fluid Pressures

The fluid pressure can be estimated from the weight of fluid above it in the pore space. Pressure increases with depth as

$$P = P_o + \rho g h, \qquad (1.1)$$

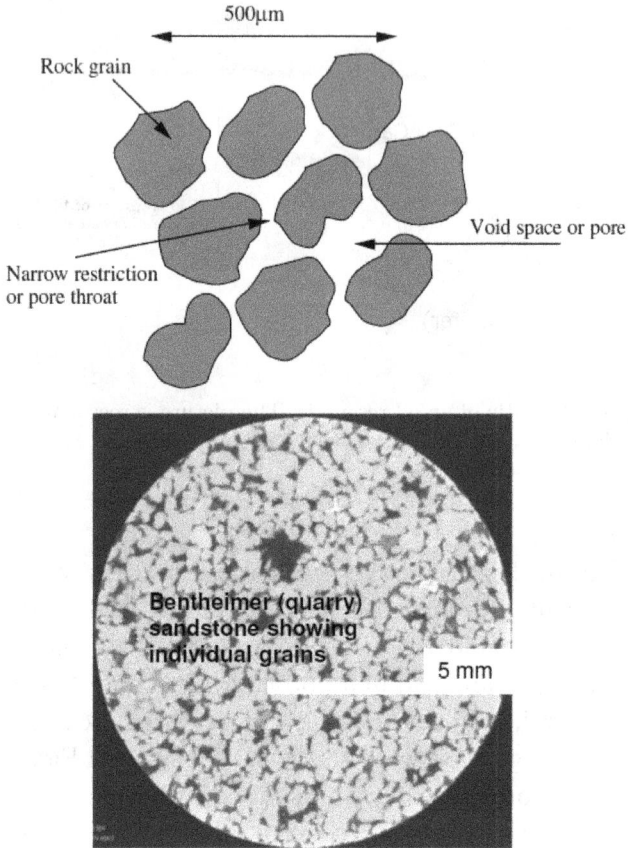

Figure 1.2. Top, a schematic two-dimensional (2D) cross-section through a porous rock; bottom, a 2D cross-section of a 3D image of a sandstone showing individual grains. Approximately one quarter of the rock volume is void space. A porous medium contains void space — in reservoir engineering this void space may contain oil, gas and water.

where P_o is a reference pressure, ρ is the fluid density and g is the acceleration due to gravity $= 9.81 \text{ ms}^{-2}$.

Putting in representative values of depth and (water) density yields pressures of several tens of megapascals (MPa),[1] or hundreds

---

[1]This volume will use, where possible, SI units. The US oil industry sticks doggedly to its peculiar and non-sensical system that is often curiously described as "British". In places I will have to use them, in order to understand the

of times atmospheric pressure (which is approximately 0.1 MPa). We will use this equation later when it is employed to determine the depths of oil–water and gas–oil contacts.

In modern petroleum engineering, oil fields are detected through seismic imaging, where sound waves are sent through the rock; the returning waves detect changes in the acoustic properties of the rock and can be used to detect possible traps where hydrocarbons could accumulate. It is also possible in some cases to infer directly the likely presence of hydrocarbons.

Then an exploration well is drilled. You can never be sure that you have an oil field until you have drilled a well and oil is produced; the seismic image may have been wrongly interpreted, or the field might contain oil, but the flow rate is so slow as to make production uneconomic. When the well is drilled, fluid and rock samples can be collected and brought to the surface for further analysis.

1.4. Oil Initially in Place

The first consideration is to estimate how much oil is contained in the field. This quantity is called stock tank oil initially in place (STOIIP) and is computed as follows:

$$N = \phi S_o V / B_o, \tag{1.2}$$

where N is the STOIIP, ϕ is the porosity, S_o is the oil saturation, V is the gross rock volume and B_o is the oil formation volume factor. Let's go through each of the terms. The seismic image, and the thickness of the field (or the thickness of oil-bearing rock) directly contacted by the well, give a good inference of the extent of the field; i.e. the volume of porous rock that contains oil. This is the *gross rock volume, V*.

---

current literature and practice. However, this is not an excuse — often used — to employ them yourselves in serious engineering calculations when this is not strictly necessary. Certainly the use of unit conversions in equations is absolutely ridiculous and should never be contemplated.

1.4.1. *Definition of Porosity and Saturation*

However, the oil field is not an underground lake, or cavern full of oil. The oil resides in porous rock. Only a fraction of that rock contains void space.

The *porosity*, ϕ, is the fraction of the volume of the porous medium occupied by void space. This means that the porosity is the volume of void space in a soil or rock divided by the total volume of the soil or rock (including void spaces). More strictly speaking, we mean the *effective porosity*, or the volume fraction of the porous medium containing connected void spaces through which fluids may flow; it excludes regions of void space entirely enclosed by solid material. For most soils and unconsolidated rock the effective porosity and void fraction are the same, but they may be different for some rocks, such as carbonates and highly porous soils. From now on when we mention porosity, we mean the effective porosity.

The porosity is around 35%–40% for, say, sand on a beach, see Table 1.1, but is much lower deep underground, where the grains comprising the rock have been fused together at high temperatures and pressures. Typical porosities lie in the range 10%–25%. The porosity can be measured directly on core samples (centimetre-long samples taken while drilling the well) or estimated from so-called log or down-hole measurements.

Furthermore, not all the void space is full of oil. Initially, the rock is saturated with (salty) water. Oil formed from the

Table 1.1. The porosity of natural soils reservoir rocks are generally consolidated and have lower porosities typically in the range 15%–30%.

Description	Porosity (%)
Uniform sand, loose	46
Uniform sand, dense	34
Glacial till, very mixed-grain	20
Soft glacial clay	55
Stiff glacial clay	37
Soft very organic clay	75
Soft bentonite clay	84

chemical transformation of organic matter (generally shallow marine organisms) in deep sediments rises slowly upwards over geological time. The oil collects under traps, from which it cannot escape, displacing the water that was initially present. However, not all the water can be squeezed out of the rock — there is always some water initially present. *Saturation* is defined as the fraction of the void space occupied by a given phase.

The *water saturation*, S_w, is the fraction of the void space of the soil occupied by water. The volume of water per unit volume of soil or rock is ϕS_w. ϕS_w is called the moisture content, θ, in the groundwater literature. However, here we are more concerned with systems containing oil and natural gas.

Water saturation can, again, be measured from extracted core samples and from log measurements of electrical resistivity (water conducts much better than gas or oil). Generally, we see a water saturation of between 10% and 40% initially in oil fields.

In oil reservoirs the void spaces may contain water, oil and gas. The *oil saturation* is the fraction of the void space occupied by oil and the *gas saturation* is the fraction of the void space occupied by gas. The sum of the saturations of all phases is one (why?).

1.4.2. *Conversion From Reservoir to Surface Volumes*

We now can calculate the volume of oil in the reservoir. However, when the oil is brought to the surface — where it is sold — its volume changes. Hence, it is the universal practice in petroleum engineering always to refer to oil volumes at so-called *stock tank conditions*. This is the oil volume at a standard temperature (60°F or around 18°C) and pressure (atmospheric). The ratio of reservoir volume to surface volume is called the *oil formation volume factor*. This ratio generally lies between 1 and 2. The oil shrinks when it is brought to the surface. At first sight, this seems counterintuitive, since you would expect a fluid to expand as the pressure drops. However, as Fig. 1.3 explains, the oil contains dissolved gas (as we discuss later); the exsolution of this gas as the pressure drops means that the oil contains fewer molecules and overall its volume decreases. The oil formation volume factor, B_o, is measured on fluid samples taken from the well.

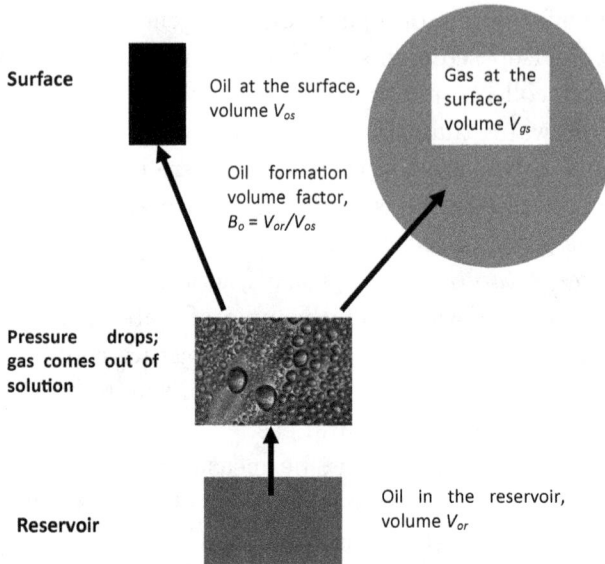

Figure 1.3. When oil flows up to the surface, its pressure drops. Bubbles of gas exsolve from the oil. At the surface, both oil and gas are produced. The volume of oil at the surface is lower than that in the reservoir, because gas has come out of solution. The oil formation volume factor is the ratio of the reservoir volume of oil to the volume at standard or stock tank conditions ($60°$F and atmospheric pressure).[2]

So, now we know how much oil we have underground — from all the terms in Eq. (1.2) — but how do we produce it?

1.5.   Oil Production

Oil fields are produced by drilling wells through the reservoir and allowing the oil to flow up through the well to be collected at the surface. This process is called primary production; this is the first process that occurs and uses the reservoir energy — essentially the pressure of the rock and fluids — to drive out the oil. The problem with this approach to production is twofold. First, once the pressure has dropped sufficiently, the field will stop producing even though it is still full of oil, so it is extremely inefficient. Secondly, and related to this, is that oil at high pressures and temperatures is a mix of

---

[2]Image from dehaanservices.ca.

hundreds of chemical constituents and some of these — principally the gaseous fractions methane, ethane, propane and butane — come out of solution when the oil is brought to the surface and the pressure drops. The pressure at which gas first exsolves from the oil is called the bubble point. Gas has a lower viscosity than oil and so flows much more readily; as a consequence, the gas is produced preferentially to the oil, leaving the oil behind. Hence, in general, production is designed to maintain the reservoir pressure above the bubble point.

How do we recover more oil, while preventing the preferential production of gas? If the well, or wells, only produce oil, then the pressure will, inevitably, fall over time. Hence, we need to apply so-called secondary production, where another fluid — normally gas or water — is injected into the reservoir through injection wells. This process serves two purposes. Firstly, it helps maintain the reservoir pressure (above the bubble point) and keeps a high driving force to keep the oil flowing. Secondly, the water (or gas) displaces the oil from the pore space of the rock, leading — potentially — to high recoveries.

The final process in a field life is tertiary (third) recovery: when another fluid is injected instead of, say, just water to remove more of the oil. Sometimes the expressions *improved oil recovery*, or *enhanced oil recovery* are used.[3] In general, these terms refer to the injection of something other than just water to recover as much oil as possible from the field, and are not strictly related to the time sequence. Enhanced oil recovery can, in theory, be either a secondary or tertiary process. Enhanced oil recovery can include the injection of gases (natural gas or carbon dioxide), the use of foams, polymers and surfactants, and thermal methods, such as steam injection, where the reservoir oil is heated, lowering its viscosity and aiding flow. The injection of low-salinity water is another example of an improved (or perhaps even enhanced) recovery process.

---

[3]Generally, enhanced oil recovery and improved oil recovery mean the same. However, the word "enhanced" sounds stronger than "improved" and so is strictly reserved for the injection of something other than normal water, while improved oil recovery is sometimes used to cover a range of different reservoir engineering practices that the engineer considers to be non-routine.

Overall, a sophisticated suite of techniques can be applied to recover the oil. However, the huge volumes of fluid involved and the oil price do severely limit the technologies that can be employed economically. While the oil price from 2008 to 2014 of around $100/barrel was historically high, 1 barrel is around 160 litres, so the price is only around 60 cents per litre.

As touched upon previously, around two-thirds of the oil that has been discovered is left underground when the field is abandoned. Improvements in technology are now routinely seeing recoveries of 50% or better, thanks to better seismic imaging of the rock (and seeing how the images change over time), targeted drilling and the use of horizontal and slanted wells, better simulation technologies to model the likely movement of fluids, and generally a much better understanding of flow in porous media.

This work is concerned with this better understanding, laying a rigorous foundation for improving oil recovery around the world. To help place this in perspective, Fig. 1.4 shows a modern simulation model, where a structurally complex reservoir is described with spatially varying properties (porosity and permeability, defined later). We will go through the concepts that underlie the construction of a model of this type and the equations that are used to solve for fluid flow. What this book is *not* about is how to run simulators or other computer codes — that you can do when you are working in the industry. What you learn here are the underlying concepts that will allow you to understand and interpret the complex and fascinating behaviour of hydrocarbon fields.

1.6. The World's Largest Oil Fields

Table 1.2 (taken from Wikipedia)[4] lists some of the world's largest oil fields. While we are still discovering oil around the world — principally deep offshore Angola, Brazil and the Gulf of Mexico — the vast

---

[4]In general, I would recommend not using Wikipedia, or indeed other internet sources blindly, but rather going back to the original source material. However, in some circumstances — such as this — the information is extremely valuable (and convenient) for illustrative purposes.

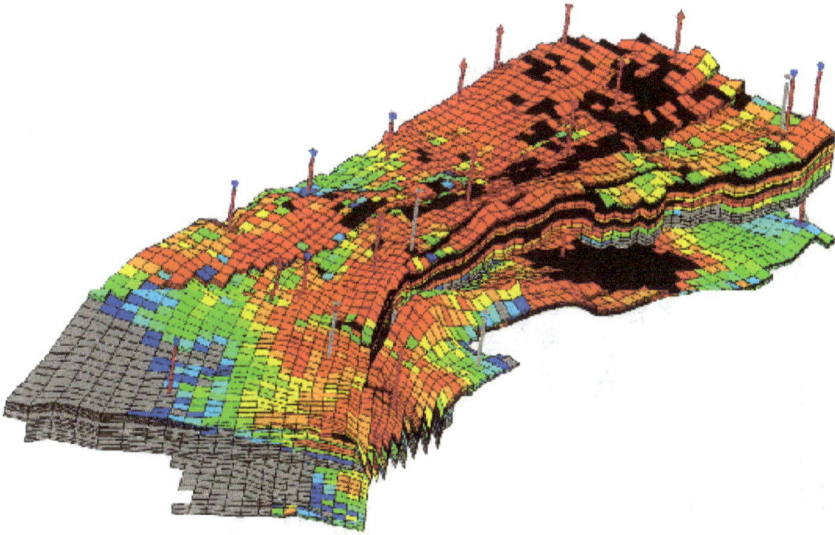

Figure 1.4. A modern simulation model, showing a representation of a structurally complex reservoir and the locations of wells (for both injection and production). The colours indicate porosity.

majority of the world's largest fields were discovered many decades ago. To maintain oil production, and indeed to increase it in order to supply oil to the developing world's population, who have a legitimate aspiration to share in the prosperity enjoyed by the Western world, we need to recover the oil from the fields that we have already found as efficiently as possible; we cannot rely on finding new conventional oil. In addition, we can look for new sources, such as shale oil (oil contained in shale or source rock), oil shale (immature source rock where the organic material has to be heated to produce oil), or other resources. We currently produce around 30 billion barrels of oil each year,[5] while we discover at most half that in new fields.

The world's largest oil field is in Saudi Arabia — Ghawar — which is undergoing the world's largest water injection project. It is a huge carbonate field, with fractures and zones of high permeability

---

[5] A great source of public domain information on oil (and other energy) production is the BP statistical review of world energy: http://www.bp.com/en/global/corporate/energy-economics/statistical-review-of-world-energy.html

Table 1.2. Oil fields greater than 1 billion barrels (160×10^6 m^3).

Field	Location	Discovered	Started Production	Recoverable Oil (Billion Barrels)	Production (Million Barrels/ Day)
Ghawar Field	Saudi Arabia	1948	1951	75-83	5
Burgan Field	Kuwait	1937	1948	66-72	1.7
Ferdows/Mound/ Zagheh Field	Iran	2003		7-9 (38 Gb resource)	
Sugar Loaf field	Brazil	2007		possibly 25-40	
Cantarell Field	Mexico	1976	1981	18	0.408
Bolivar Coastal Field	Venezuela	1917	1922	30-32	2.6-3
Azadegan field	Iran	2004		9	
Lula Field	Brazil, Santos Basin	2007		5-8	
Safaniya–Khafji Field	Saudi Arabia/ Neutral Zone	1951		30	
Esfandiar Field	Iran			30	
Rumaila Field	Iraq	1953		17	1.3
Tengiz Field	Kazakhstan	1979	1993	26-40	0.53
Ahwaz Field	Iran	1958		10.1	0.700
Kirkuk Field	Iraq	1927	1934	8.5	0.480
Shaybah Field	Saudi Arabia			15	
Agha Jari Field	Iran	1937		8.7	0.200
Majnoon Field	Iraq	1975		11-20	0.5
Samotlor Field	Russia, West Siberia	1965	1969	14-16	0.844
Romashkino Field	Russia Volga-Ural	1948	1949	16-17	0.301 (2006)
Prudhoe Bay	United States, Alaska	1969		13	0.9

Field	Location				
Sarir Field	Libya	1961	1961	12 (6.5 billion recoverable)	
Priobskoye field	Russia, West Siberia	1982	2000	13	0.680 (2008)
Lyantorskoye field	Russia, West Siberia	1966	1979	13	0.168 (2004)
Abqaiq Field	Saudi Arabia			12	0.43
Chicontepec Field	Mexico	1926		6.5 (19 certified)	
Berri Field	Saudi Arabia	1965		12	
Zakum Field	Abu Dhabi, UAE		1967	12	
West Qurna Field	Iraq	1973		15–21	0.18–0.25 (potential)
Manifa Field	Saudi Arabia	1971		11	
Fyodorovskoye Field	Russia, West Siberia	1971	1974	11	1.9
East Baghdad Field	Iraq	1976		8	0–0.05 (potential)
Faroozan-Marjan Field	Saudi Arabia/Iran			10	
Marlim Field	Brazil, Campos Basin			10–14	
Awali	Bahrain			1	
Aghajari Field	Iran	1999		14	
Azadegan Field	Iran			5.2	
Gachsaran Field	Iran	1927		15	
Marun Field	Iran			16	
Mesopotamian Foredeep Basin	Kuwait			66–72	
Minagish	Kuwait			2	

(Continued)

Table 1.2. (*Continued*)

Field	Location	Discovered	Started Production	Recoverable Oil (Billion Barrels)	Production (Million Barrels/Day)
Raudhatain	Kuwait			11	
Sabriya	Kuwait			3.8–4	
Yibal	Oman			1	
Dukhan Field	Qatar			2.2	
Halfaya Field	Iraq			4.1	
Az Zubayr Field	Iraq			6	
Nahr Umr Field	Iraq			6	
Abu-Sa'fah field	Saudi Arabia			6.1	
Hassi Messaoud	Algeria			9	
Kizomba Complex	Angola			2	
Dalia (oil field)	Angola			1	
Belayim	Angola			>1	
Zafiro	Angola			1	
Zelten oil field	Libya			2.5	
Agbami Field	Nigeria			0.8–1.2	
Bonga Field	Nigeria			1.4	
Azeri-Chirag-Guneshli	Azerbaijan	1985	1997	5.4	
Karachaganak Field	Kazakhstan	1972		2.5	
Kashagan Field	Kazakhstan	2000		30	
Kurmangazy Field	Kazakhstan			6–7	
Darkhan Field	Kazakhstan			9.5	
Zhanazhol Field	Kazakhstan			3	

Field	Country				
Uzen Field	Kazakhstan			7	
Kalamkas Field	Kazakhstan			3.2	
Zhetybay Field	Kazakhstan			2.1	
Nursultan Field	Kazakhstan			4.5	
Ekofisk oil field	Norway			3.3	
Troll Vest	Norway			1.4	
Statfjord	Norway			3.4	
Gullfaks	Norway			2.1	
Oseberg	Norway	1979	1988	2.2	3.78
Srorre	Norway			1.5	
Mamontovskoye Field	Russia			8	
Russkoye Field	Russia			2.5	
Kamennoe Field	Russia			1.9	
Vankor Field	Russia	1983	2009	3.8	
Vatyeganskoye Field	Russia			1.4	
Tevlinsko-Russkinskoye Field	Russia			1.3	
Sutorminskoye Field	Russia			1.3	
Urengoy group	Russia			1	
Ust-Balykskoe Field	Russia			>1	
Tuymazinskoe Field]	Russia			3	
Arlanskoye Field	Russia			>2	
South-Hilchuy Field	Russia			3.1	

(Continued)

Table 1.2. (*Continued*)

Field	Location	Discovered	Started Production	Recoverable Oil (Billion Barrels)	Production (Million Barrels/Day)
North-Dolginskoye Field	Russia			2.2	
Nizhne-Chutinskoe Field	Russia			1.7	
South-Dolginskoye Field	Russia			1.6	
Prirazlomnoye Field	Russia			1.4	
West-Matveevskoye Field	Russia			1.1	
Sakhalin Islands	Russia			14	
Odoptu	Russia			1	
Arukutun-Dagi	Russia			1	
Piltum-Astokhskoye Field	Russia			1	
Ayash Field East-Odoptu Field	Russia			4.5	
Verhne-Chonskoye Field	Russia			1.3	
Talakan Field	Russia			1.3	
North-Caucasus Basin	Russia			1.7	
Clair oil field	United Kingdom	1977		1.75	
Forties oil field	United Kingdom	1970		5	

Jupiter field	Brazil			7
Cupiagua/Cusiana	Colombia			1
Boscán Field, Venezuela	Venezuela			1.6
Pembina	Canada	1953	1953	
Swan Hills	Canada			
Rainbow Lake	Canada			
Hibernia	Canada	1979	1997	3
Terra Nova Field	Canada	1984	2002	1.0
Kelly-Snyder / SACROC	United States, Texas			1.5
Yates Oil Field	United States, Texas	1926	1926	3.0 (2.0 billion recovered; 1.0 reserve remaining)
Kuparuk oil field	United States, Alaska	1969		6
Alpine	United States, Alaska			0.4–1
East Texas Oil Field	United States, Texas	1930		6
Spraberry Trend	United States, Texas	1943		10
Wilmington Oil Field	United States, California	1932		3

(*Continued*)

Table 1.2. (*Continued*)

Field	Location	Discovered	Started Production	Recoverable Oil (Billion Barrels)	Production (Million Barrels/ Day)
South Belridge Oil Field	United States, California	1911		2	
Coalinga Oil Field	United States, California	1887		1	
Elk Hills	United States, California	1911		1.5	
Kern River	United States, California	1899		2.5	
Midway-Sunset Field	United States, California	1894		3.4	
Thunder Horse Oil Field	United States, Gulf of Mexico			>1	
Kingfish	Australia			1.2	
Halibut	Australia			1	
Daqing Field	China	1959	1960	16	

(fast flow). The second largest field — Burgan — is mainly sandstone and situated in Kuwait. Historically, the most oil of any country has been produced in the US; note, however, that its largest oil field — Prudhoe Bay off the North coast of Alaska — is quite some way down the list. At present, Saudi Arabia and Russia are the two largest oil producers, but the US is catching up fast, thanks to new discoveries and shale oil.

The recoverable oil is the amount of oil that can be extracted from the field using current technology; the amount of oil underground is typically (as I said above) some three times (or more) larger. If you consider that the oil price is $100/barrel, then a 1-billion-barrel oil field represents $100 billion of potential value. Imagine that a sensible application of the ideas in this book could improve overall recovery of, say, 1% for a single field; this represents a considerable amount of money!

1.7. Fluid Pressure Regimes

Fluid mass above a point increases with depth and thus pressure increases with depth as given by Eq. (1.1). This can be written in terms of a pressure gradient with z referring to depth:

$$\frac{\partial P}{\partial z} = \rho g. \tag{1.3}$$

Pressure increases fastest for water with depth, then oil, then gas.

In oil field operations the fluid pressure can be measured down-hole using a repeat formation tester (RFT). Consider Fig. 1.5, a schematic diagram of pressure as a function of depth. Pressure measurements can be used to identify the locations of the contacts between oil and water, and gas and oil, even if these are not detected directly by logs.

If the pressure profiles as a function of depth from different wells do not superimpose, then this can imply the presence of non-communicating regions (shales and faults), indicating reservoir compartmentalisation.

As Fig. 1.5 indicates, the pressure in the hydrocarbon-bearing zones is *higher* than in the surrounding aquifer (water-saturated

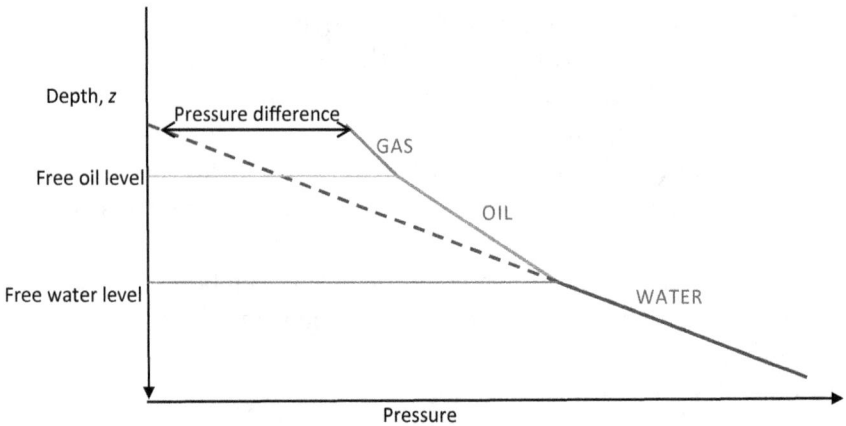

Figure 1.5. A schematic of the pressure profile with depth for an oil field with a gas cap. The oil and water pressures are equal at the free water level. The pressure difference — where the gas pressure is higher than that in the surrounding aquifer — is indicated at the top of the reservoir.

rock). This has nothing to do with compressibility, but is caused simply by the density differences between the fluids. The water and oil have the same pressure at the free water level; the oil pressure decreases less rapidly with height (decreasing depth) than water, and so the oil pressure is higher than that of the water above the free water level. The same is true for the gas column. There is an analogy with keeping a balloon underwater in the bath — you need to force the balloon down to prevent it rising as it is buoyant. This is equivalent to the fluid pressure in the cap rock necessary to stop the oil and gas escaping upwards.

The higher pressure in the hydrocarbon-bearing zone compared to a water column at the same depth poses a potential hazard during drilling. A water-filled drill hole, when it encountered oil or gas, would allow a blowout; the higher-pressure hydrocarbon would flow into the hole and rise uncontrolled to the surface. This is the principal reason for the use of *drilling mud* — dense fluid mixed with water or oil — which ensures a higher pressure in the well than the formation. The mud also cools and lubricates the drill bit and helps carry cuttings to the surface.

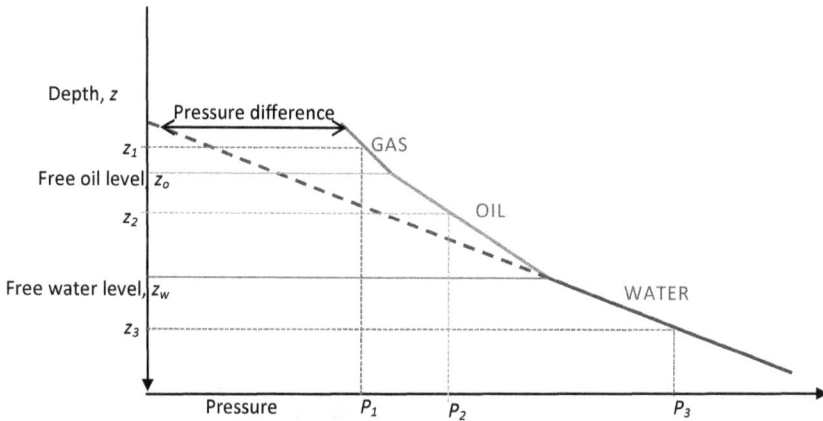

Figure 1.6. A schematic of the pressure profile with depth for an oil field with a gas cap, where specific pressure measurements at the indicated depths are shown. From this the depths of the free water and oil levels can be computed.

Figure 1.6 shows a similar diagram, but where specific pressure measurements have been made at the depths indicated. If we know the densities of the fluids (these can be obtained from measurements on down-hole samples, or — more simply — from the slopes of measured pressure as a function of depth), then the depths of the free water and oil levels can be computed.

Then using the measurements together with Eqs. (1.1) and (1.3), we can write for the gas, oil and water pressures, respectively:

$$P_g = P_1 + (z - z_1)\rho_g g, \tag{1.4}$$

$$P_o = P_2 + (z - z_2)\rho_o g, \tag{1.5}$$

$$P_w = P_3 + (z - z_3)\rho_w g. \tag{1.6}$$

The free water level is — by definition — when the oil and water pressures are the same. So, equating Eqs. (1.5) and (1.6) we find[6]

$$z_w = \frac{P_2 - P_3 + z_3\rho_w g - z_2\rho_o g}{(\rho_w - \rho_o)g}. \tag{1.7}$$

---

[6]One tip for the calculation: get your units straight. If depth is in m and density in $kg \cdot m^{-3}$, then you need to do the calculation in Pa (not MPa or anything else, such as — horrors — psi, a senseless unit which should be forever banned from petroleum engineering).

Similarly, for the free oil level where the oil and gas pressures are the same,

$$z_o = \frac{P_1 - P_2 + z_2\rho_o g - z_1\rho_g g}{(\rho_o - \rho_g)g}. \tag{1.8}$$

The height of the oil column — used to determine the initial oil in place — is simply $z_w - z_o$.

The other way to determine the contacts between phases and the height of the oil column is directly from log measurements. Here, down-hole readings of resistivity (to distinguish between water and hydrocarbons) and density (to distinguish between oil and gas) can be used to locate the depths of the contacts. In practice, however, the well may not penetrate the contact directly, or the readings may be ambiguous or open to different interpretations. The essence of reservoir engineering is to consider all the data together — there is never one overriding DNA-type test that trumps all others — so you must always assess all the evidence carefully and arrive at a determination that is consistent with it.

There is a difference anyway between the free water level determined from the pressure and the contact determined from logs — this difference is the *capillary pressure*. In a reservoir that is initially water-wet, a pressure difference between oil and water (that is, a higher pressure in the oil) is necessary for oil to enter the porous medium during primary oil migration. This means that the true contact — where the oil saturation is significant — generally lies above the point where the oil and water pressures are the same — where the capillary pressure (the difference between the oil and water pressures) is equal to the capillary entry pressure. In general — for permeable sandstones — this difference is less than 1 m, but it may be significant in lower permeability systems. A full understanding of this requires a knowledge of capillary pressure, provided in the following sections.

We conclude this analysis with some definitions.

A normally pressured reservoir. The *water* pressure is as expected for its depth — there is a continuous pathway of water in the pore space to the surface. If z is the depth from either the water table (onshore) or the sea surface (offshore), then the water pressure is given by

$$P_w = P_{\text{atm}} + z\rho_w g, \tag{1.9}$$

where P_{atm} is atmospheric pressure and we have assumed a constant water (brine) density with depth. Note that we only consider the water pressure — the oil and gas pressures are higher than water for a given depth, as discussed above.

Over-pressured. The reservoir pressure higher than expected — i.e. the real water pressure is higher than you would expect for a normally pressured reservoir. This means that the reservoir has been uplifted since filling. Note that this is the pressure relative to that expected for depth — it is not an absolute measure. In an over-pressured reservoir production may be rapid, but during drilling there is a risk of blowouts, as the reservoir pressure could be higher than the mud pressure down-hole.

Under-pressured. This is when the water pressure lower than expected. The reservoir has been downthrown over geological time — more sediment has been deposited over the field since it was charged with oil and gas.

1.8. Reservoir Fluids

We have already defined the oil formation volume factor. In this section, we define some other terms necessary for our analysis of oil production. In Fig. 1.7, we extend the diagram we showed previously in Fig. 1.3 to include the more general case where both oil and gas are produced from the reservoir — that is, oil and gas are both present in the reservoir itself.

The various terms are easy to define using equations based on Fig. 1.8, but require great care when described in words.

So, for completeness, here are the definitions. The *oil formation volume factor*, B_o, is the ratio of the reservoir volume of oil to the

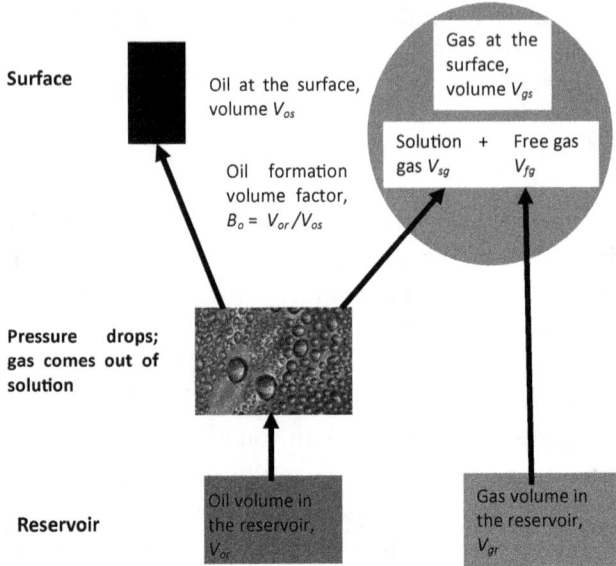

Figure 1.7. A schematic showing the production of oil and gas from a reservoir with the corresponding volumes of oil and gas at the surface. The oil volume shrinks because of the exsolution of dissolved gas; the gas produced is both solution gas (from the oil) and free gas (which was gas in the reservoir as well).

surface volume of oil,

$$B_o = \frac{V_{or}}{V_{os}}. \tag{1.10}$$

Traditionally, oil volumes are measured in barrels. To emphasise the nature of the conversion, oil volumes measured at the surface are called stock tank barrels (stb), while volumes in the reservoir are measured in reservoir barrels (rb). Hence the common unit for B_o — while it is strictly dimensionless — is rb/stb. A typical range is from 1 (heavy oils with little or no associated gas) to more than 2 for highly volatile oils.

A *saturated oil* is defined as one that cannot hold any more gas. This means oil at, or below, the bubble point, when gas first comes out of solution. The volume of the oil *decreases* as the pressure drops.

An *undersaturated oil* is one that can dissolve more gas if it is available. The volume of oil *increases* as the pressure drops, until the bubble point is reached (the oil becomes saturated).

The terms saturated and undersaturated are commonly used, but I find them confusing. It shows more clarity and confidence to refer simply to an oil that is above or below the bubble point.

The *gas formation volume factor*, B_g, is the reservoir volume of free gas divided by the surface volume of that free gas (the solution gas is not included):

$$B_g = \frac{V_{gr}}{V_{fg}}.$$
(1.11)

Again the units here should be dimensionless and indeed they are in SI units (rm^3/sm^3 where the r and s stand for reservoir and surface, respectively); in field units the situation is a little more confused. Reservoir volumes of both oil and gas are measured in rb, while gas volumes at the surface are measured in cubic feet (scf). So the units of B_g are traditionally rb/scf.[7]

R is the *producing gas/oil ratio*, or the ratio of the total gas production rate to the oil production rate, with both volumes measured at surface conditions:

$$R = \frac{V_{gs}}{V_{os}} = \frac{V_{sg} + V_{fg}}{V_{os}}.$$
(1.12)

The field units for R are scf/stb. Note that strictly this is a ratio of rates — or volumes — produced over a fixed, short, time period, such as one day. It is not a ratio of total volumes since the field was produced — this is defined and used later.

R_s is the *solution gas/oil ratio* and is the volume of gas (measured at standard conditions) that will dissolve in unit stock tank volume

---

[7]At this point you may think that some conversion factors may be helpful. When I need them I simply look them up on the internet — this is one occasion where this is appropriate and convenient.

of oil at reservoir conditions:

$$R_s = \frac{V_{sg}}{V_{os}}. \tag{1.13}$$

For gases, we often do not quote the formation volume factor directly, but invoke a gas law. The ideal gas law — with which I hope you are familiar — is

$$PV = nRT, \tag{1.14}$$

where P is pressure, V is (gas) volume, n is the number of moles and T is the absolute temperature (kelvins — I don't even know what silly absolute temperature unit has been invented for the Fahrenheit scale). R in this context is the *universal gas constant*, whose value in SI units is $8.314 \, \text{J} \cdot \text{K}^{-1} \cdot \text{mol}^{-1}$ (here I simply refuse to contemplate its value in inconsistent, should-have-died-out-with-the-dinosaurs units).

The ideal gas law assumes that the molecules in the gas behave as point particles with no mutual interaction. This is a good approximation at atmospheric temperatures and pressures (stock tank conditions) but the nature of the "gas" encountered in the reservoir is very different. At pressures that are hundreds of times atmospheric, the gas has a density comparable to oil and there is significant interaction between the gas molecules. Indeed the distinction between gas and liquid is not obvious — strictly we refer to gas as the less dense phase and oil as the more dense phase when two hydrocarbon phases are present in the reservoir.

In these cases a non-ideal gas law is used:

$$PV = ZnRT, \tag{1.15}$$

where Z is an empirical factor, sometimes called a compressibility factor, which is a function of temperature, pressure and composition. It can either be measured directly on gas samples in the laboratory, or estimated from correlations tuned to match experimental data.

It is possible to determine B_g from Z using Eq. (1.15). If we assume that $Z = 1$ at surface conditions (denoted by the subscript s) then, since when the gas expands from reservoir to surface conditions the number of moles is the same, we have

$$\frac{P_s V_s}{T_s} = \frac{PV}{ZT}, \tag{1.16}$$

and so using Eq. (1.11) and the nomenclature of Eq. (1.16)

$$B_g = \frac{V}{V_s} = Z \frac{P_s T}{P T_s}. \tag{1.17}$$

Figure 1.8 shows values of B_g, B_o and R_s as a function of pressure for a typical North Sea oil.

Note that B_o reaches its maximum value at the bubble point. Above this pressure, oil expands as the pressure drops (like a normal fluid) and so B_o rises as pressure drops. Below this pressure, however, the exsolution of gas causes the oil to shrink as the pressure drops, and B_o correspondingly falls.

The range of R_s in hydrocarbon fields is considerable, from infinite (for a dry gas — a gas that produces no liquid at the surface) to zero (a heavy oil with no associated gas). It is useful to have some concept of whether or not a field is producing more oil or gas. From a volume perspective, the gas wins, since it is much less dense than oil at the surface. This is further confused by the baffling and unhelpful units of scf/stb (or MScf[8]/stb). So let's consider this on a mass basis. Imagine that the gas is largely methane; then we can estimate the density at standard conditions from the ideal gas law, Eq. (1.14). The molecular mass, M, of methane is $0.016 \ \mathrm{kg \cdot mol^{-1}}$ (note the units — don't try some sort of idiotic compromise over units and muddle yourself with a molecular "weight" in g/mol — this won't work and you know it won't work) and so the density is given by

$$\rho = \frac{nM}{V} = \frac{PM}{RT}. \tag{1.18}$$

---

[8]Another stupidity with field units. M stands for million (10^6) of course — or at least in SI units. In field units M stands for the French word "mille" or thousand — hardly "British" units! Million is represented by MM. Confusing — stick to the M and MM convention for field units, but don't get into a silly muddle and do the same when SI units are (grudgingly) applied.

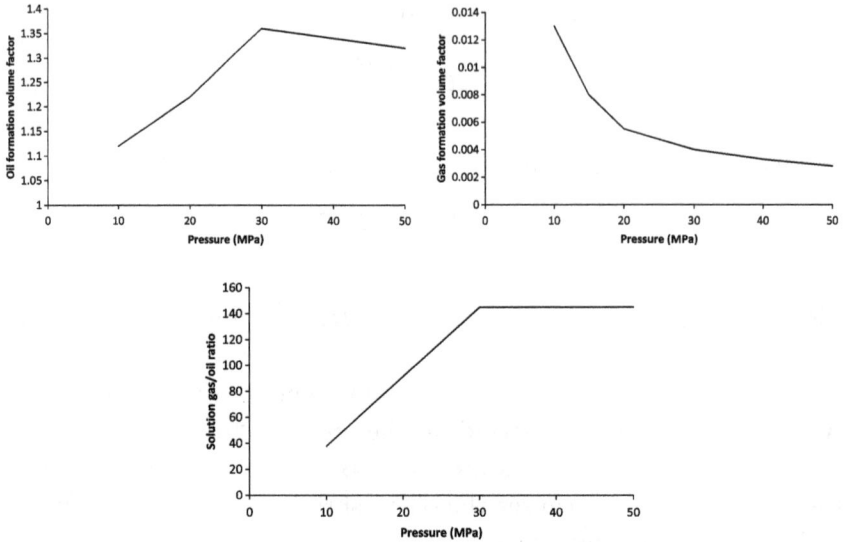

Figure 1.8. Values of the oil formation volume factor, the gas formation volume factor and the solution gas/oil ratio for a typical North Sea oil. The oil formation factor is maximum at the bubble point — this is also the pressure below which the gas solubility falls. In these graphs I use SI units: the formation factors and gas/oil ratio are dimensionless.

Atmospheric pressure is approximately 1.01×10^5 Pa and standard temperature is around $288\,\mathrm{K}$, so using the value of R quoted previously, we obtain a density of $0.67\,\mathrm{kg \cdot m^{-3}}$. This is more than a thousand times less dense than water and explains why — intuitively — we consider gases and liquids as being very different, even though their properties may be similar at reservoir conditions.

We now compare this with typical oil densities (at surface conditions). I will take a representative value of $800\,\mathrm{kg \cdot m^{-3}}$.[9]

---

[9] In petroleum engineering it is common to categorize an oil by its density. Fine, since this does tend to correlate with composition and viscosity as well. However, the unit is bizarre and essentially unusable in any engineering calculation: API gravity (API stands for the American Petroleum Institute, an organisation, in my opinion, of dubious utility). Needless to say, if you have got this far in the notes and want to use API gravity, it is time to move to another textbook. API gravity will not feature again here.

If a hydrocarbon field produces the same mass of gas as oil, then what is the value of R_s? In SI units this would be $800/0.67 = 1,200\,\text{m}^3/\text{m}^3$. In field units, since $1\,\text{m}^3 = 6.2898\,\text{stb}$ and $1\,\text{ft} = 0.3048\,\text{m}$, we have a ratio of $1.2/(6.2898 \times 0.3048^3) = 1\,\text{stb} = 6.7\,\text{Mscf}/\text{stb}$.

So, a field with R_s of greater than around 6 Mscf/stb–7 Mscf/stb is gas-like, while one with a smaller gas/oil ratio is oil-like. In the next section, we provide strict thermodynamic definitions, but this is a useful guide. Wet gases (that produce some liquid at the surface) generally have R_s around 50 Mscf/stb or larger; gas condensates are in the intermediate range 5 Mscf/stb–30 Mscf/stb; volatile oils 2 Mscf/stb–3 Mscf/stb; and black oils (we define these terms later) 0.1 Mscf/stb–2 Mscf/stb. I will repeat that these are simply rough estimates and should never be used to define a producing field.

1.9. Phase Behaviour

Figure 1.9 provides a strict thermodynamic definition of the different types of hydrocarbon that are produced from conventional oil and

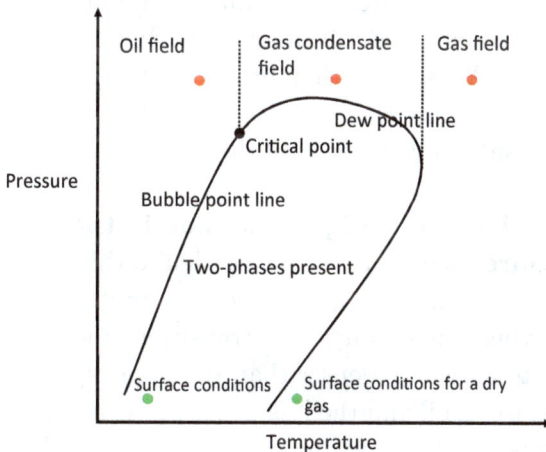

Figure 1.9. A schematic phase diagram for a hydrocarbon mixture for different pressures and temperatures. The mixture separates into two phases — oil and gas — in the region indicated. The red spots mark the initial reservoir conditions for different types of hydrocarbon field; the green spots label the surface conditions.

gas fields. The diagram shows the phases present for a given hydrocarbon composition as a function of pressure (y-axis) and temperature (x-axis). Within the region indicated, two hydrocarbon phases (oil and gas) are present; outside this region only one phase is present.

This phase diagram is not known *a priori*; it needs to be computed based on experimental measurements on fluid samples taken from the reservoir. Normally, measurements are used to tune an equation of state model, from which this phase diagram can be determined; further discussion lies outside the scope of these notes.

Now consider a field that has just been discovered. It has a high pressure initially; what happens as the field is produced and the pressure drops? To a very good approximation we assume that the reservoir temperature stays constant (it can only change the temperature if we inject a considerable quantity of liquid that is hotter, say steam, or cooler, say water, than the reservoir, or induce thermal reaction, such as *in situ* combustion). **In a gas field — in the reservoir itself — the fluid expands without a phase transition.** This is the definition of a gas field — no phase change in the reservoir as the pressure drops. Now, at the surface — at a lower temperature — some liquid (oil) may be produced (it condenses from the gas stream). A gas field that produces liquid oil at the surface is called a wet gas (note that this has absolutely nothing to do with water production), while a gas field that produces no liquid is called a dry gas.

A gas condensate field produces oil in the reservoir itself as the pressure drops. How can we define this as a gas field, as opposed to an oil field? Not on some *ad hoc* limit on the value of R_s. No, because when there is a phase transition, the first drop of the second phase to appear is denser than the majority phase; we define this denser phase as oil and the less dense phase as gas. The pressure at which this occurs is called the dew point. Remember that we are dealing with high pressures, and the densities of the oil and gas may be similar.

An oil field produces gas as the pressure is dropped in the reservoir. We know that this is gas, as the first bubbles of the second phase are less dense than the original oil phase. The pressure when this occurs is the bubble point.

The critical point marks the transition from an oil field to a gas condensate and is the point where the oil and gas properties are indistinguishable. To the left of the critical point on the diagram we have an oil field; to the right we have a gas condensate field.

The fluid properties have an impact on production mechanisms. In a gas field, the preferred method of production is simply to drop the pressure. The gas expands in the field with no phase separation in the subsurface; if you drop the pressure to close to atmospheric, recoveries of 90%–95% are possible (see the section on Material Balance later). In contrast, in an oil field, when the pressure drops below the bubble point, gas is formed. This gas is less viscous than the surrounding oil and — once it is connected in the pore space — will flow preferentially to the oil. Thus you produce gas and leave behind the (more valuable) oil. Even if you drop the pressure to near atmospheric, you still have a field full of oil and, since the oil is not very compressible, this results in rather poor recovery factors (typically around 20%). So, for an oil field, it is necessary to maintain the pressure above the bubble point, normally through the injection of water.

A gas condensate is more complex. The lowest pressures are near the well. If the pressure drops below the dew point, oil is formed. Once again the more valuable fraction of the hydrocarbon — the liquids — are left behind in the reservoir. Moreover, the liquid blocks the pore space and reduces the productivity of the well. There are two options. One is to maintain pressure, as in an oil field, or produce more slowly, although this has an economic cost. The other is to drop the pressure further until we are back in the single-phase (gas) region — see Fig. 1.10. Then the reservoir can be produced as a normal gas field.

The phase diagram is often used to describe different types of oil field. A black oil is not specifically a reference to the colour of crude oil (which indeed is black) but to how it may be described thermodynamically. Most fields are characterised as having an oil with dissolved gas with a solution gas/oil ratio, R_s, defined as a

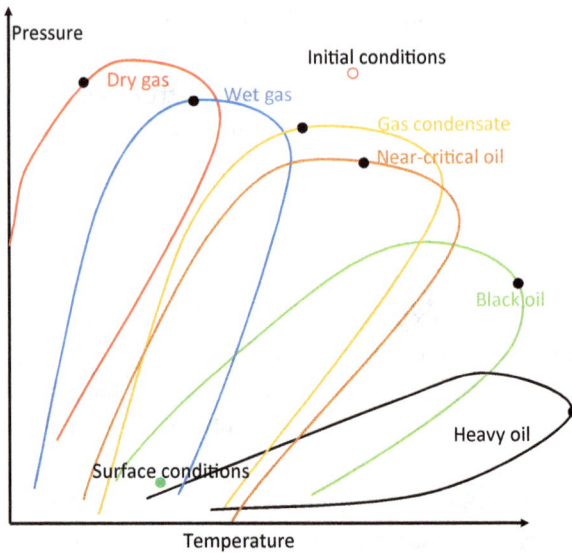

Figure 1.10. Schematic phase diagrams for different types of hydrocarbon reservoir. The open circle marks the initial reservoir conditions; the green spot labels surface conditions. The black spots mark the critical points for the different types of field.

function of temperature and pressure. The gas and oil also have properties that vary only with temperature and pressure. This seems obvious, but in reality, the exsolved gas has a different composition, dependent on temperature and pressure and production process — after all the overall hydrocarbon composition of the reservoir changes if gas is, for instance, preferentially produced compared to oil. For most fields it is a reasonable approximation to assume that the gas has a fixed composition, independent of production path. However, for fields with a temperature near the critical temperature, this is not so accurate. These can be described as volatile or near-critical oil fields and often a more sophisticated compositional characterisation of their behaviour is required. Furthermore, for gas injection, we need also to account for gases that have distinct properties — the injected gas will not normally have a composition identical to exsolved gas. In contrast, a heavy oil may have little or no dissolved gas present.

Figure 1.9 assumes a fixed hydrocarbon composition in the reservoir and explores the behaviour for different initial temperatures. In reality, the variation in initial reservoir temperature is less dramatic than changes in composition — which can vary from pure methane (a dry gas) to virtual solids (oils with a composition dominated by molecules with chains of C_{30} — that is, 30 carbon atoms — or longer).

Figure 1.10 assumes that the initial and surface conditions are fixed and shows the effect of composition on the phase diagram. The figure should be self-explanatory and is consistent with the definitions provided previously. Note that the phase diagrams tend to shift downwards (two-phase conditions at lower pressures) and to the right (two-phase conditions at higher temperatures) as the composition varies. Typically, a dry gas is almost entirely methane (C_1), wet gas contains significant quantities of C_2–C_6, gas condensates and near-critical oils contain a range of hydrocarbons with still large quantities of C_1–C_4 and some heavier components, a black oil has a typical composition with C_6–C_{12} being most common, while heavy oils are predominately C_{10} and heavier.

This concludes a simple overview of hydrocarbon phase behaviour — for further reading the classic work in this area is *The Properties of Petroleum Fluids* by W.D. McCain, PenWell Books, 2nd edition, (1990). What we have described is sufficient though for the discussion of material balance in the next section.

Chapter 2

Material Balance

As mentioned previously, material balance is one of the fundamental principles of reservoir engineering: keeping track of mass, and being able to convert readily from reservoir to surface conditions. In this section, we will encounter this concept several times. This is an essential, standard and powerful tool for using production data to predict recovery and to determine the principal reservoir drive mechanisms. It is used mainly to analyse primary production, but can — and should — be used as a check for any producing field.

As an introduction, consider Fig. 2.1, which gives a schematic overview of the changes that occur during production.

In this example, the gross rock volume is V and the average porosity is ϕ. The initial reservoir volume of oil is then $V N_G \phi S_{oi}$. The subscript i refers to initial conditions. We have added a new concept here: N_G or the *net-to-gross*. Traditionally, the porosity and oil saturation are only defined for the portions of the gross rock volume (a volume that encompasses all the oil present in the reservoir) that contain producible hydrocarbon (i.e. oil that can reasonably flow to a well). The fraction of the gross rock volume that has sufficient porosity, oil saturation and permeability to allow significant production is called the net-to-gross. What exactly is meant by "sufficient" porosity, etc.? This is normally empirically defined by petrophysicists using log analysis and experience, and is open to interpretation. However, a discussion of this topic lies outside the scope of this chapter.

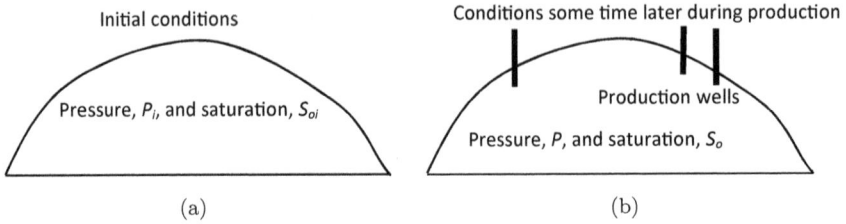

Figure 2.1. Schematic diagram showing an oil field at initial pressure and saturation (a), and at some time later (b), during production, when the oil saturation and pressure have decreased. The bold vertical lines represent production wells.

Therefore, converting to surface conditions, the stock tank oil initially in place (STOIIP), N, is given by an extension to Eq. (2.1):

$$N = \phi N_G S_{oi} V / B_{oi}. \tag{2.1}$$

During production the pressure drops as oil is produced. Moreover, the oil saturation may decrease, as gas comes out of solution, and/or water is injected or ingresses from a connected aquifer. The remaining volume of oil — measured at stock tank conditions — some time later is given by

$$N - N_p = \phi N_G S_o V / B_o. \tag{2.2}$$

In this equation N_p is the cumulative oil produced (measured at surface conditions, of course). Therefore the oil volume remaining in the reservoir is $N - N_p$, as indicated in Eq. (2.2).

The recovery factor, R_f, is defined as the ratio of the oil produced to the initial oil present:

$$R_f = N_p / N. \tag{2.3}$$

Dividing Eq. (2.2) by Eq. (2.1) and using Eq. (2.3) we find

$$1 - R_f = \frac{B_{oi} S_o}{B_o S_{oi}}, \tag{2.4}$$

from which we derive

$$R_f = 1 - \frac{B_{oi} S_o}{B_o S_{oi}}. \tag{2.5}$$

In any recovery process, we wish to make the recovery factor as high as possible, subject to economic and engineering constraints. How do we do this? From Eq. (2.5), we wish to make the oil saturation S_o as low as possible — we need to remove as much oil as possible from the pore space. In general, the oil saturation is a function of what and how other fluids are injected or ingress into the reservoir, and is largely independent of pressure. But what about pressure? From Eq. (2.5) it is also evident that we wish B_o to be as large as possible. If we had a gas field, then this simply means that we drop the pressure as low as is feasible and allow the gas to expand. However, for an oil field, the situation is different. As explained in the previous section, B_o is a maximum at the bubble point; regardless of the injection strategy therefore, it makes sense to drop the pressure to, but not below, the bubble-point pressure.

This was a first example to show how a simple accounting for volume, combined with conversion from reservoir to surface, can be used to obtain useful, quantitative insights into reservoir management. We will now apply more sophisticated versions of the material balance equation for both oil and gas fields.

2.1. Material Balance for Gas Reservoirs

It is unusual for a reservoir simulation study to be performed for gas fields — particularly smaller fields. Hence, material balance remains the principal method for engineering analysis in this case.

Assume that we have a dry gas — no oil produced, and no aquifer movement. If there is no aquifer movement, then we can assume that the gas saturation remains constant; we will relax this approximation later. We repeat Fig. 2.1 in Fig. 2.2, but simply change the nomenclature for gas (we replace the subscript o by g).

If you can immediately derive the material balance equation, similar functionally to Eq. (2.5), then fine. However, it is always instructive to go through this step by step until the derivation comes easily. If you know how to derive these equations, then you do not need to learn by rote what the equations are for every specific

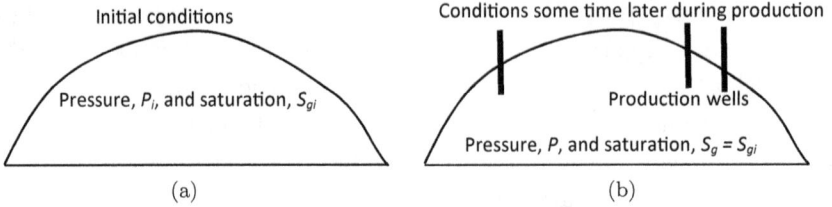

Figure 2.2. Schematic diagram showing a gas field at initial pressure and saturation (a), and at sometime later (b), during production, when the pressure has decreased. We assume no aquifer movement in this case and so the gas saturation remains the same.

case, and you have the ability to derive new equations for new circumstances (shale gas, gas condensates?).

The initial gas in place measured at surface conditions, G, is given by (see Eq. (2.1))

$$G = \phi N_G S_{gi} V / B_{gi}. \tag{2.6}$$

The remaining volume of gas — measured at stock tank conditions — some time later is given by

$$G - G_p = \phi N_G S_{gi} V / B_g, \tag{2.7}$$

where G_p is the cumulative gas produced (measured at surface conditions, of course). Therefore, the gas volume remaining in the reservoir is $G - G_p$, Eq. (2.7).

In this case the recovery factor, R_f, is defined as the ratio of the gas produced to the initial gas present: $R_f = G_p/G$. Then from Eqs. (2.6) and (2.7) we obtain

$$R_f = \frac{G_p}{G} = 1 - \frac{B_{gi}}{B_g}. \tag{2.8}$$

Traditionally, this equation is written in terms of the Z-factor. Using Eq. (1.7) we find

$$R_f = \frac{G_p}{G} = 1 - \frac{P Z_i}{P_i Z}. \tag{2.9}$$

For a producing field, we wish to know how much gas was originally in place (G) and how much gas will be produced at a certain pressure (G_p).

Table 2.1. Production data for $Z_i = 0.8$ and $P_i = 200$ atm.

G_p (10^8 scf)	P (atm)	Z	P/Z (atm)
0	200	0.80	250
1.53	180	0.85	212
2.56	160	0.86	186
3.78	140	0.90	156

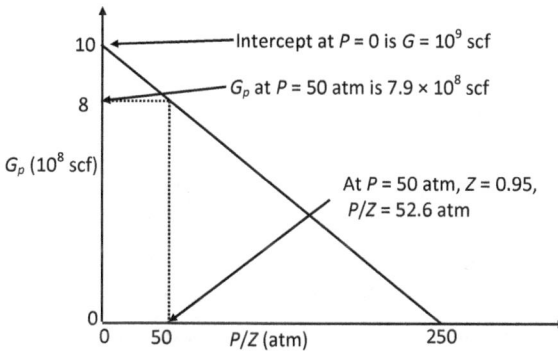

Figure 2.3. Classic P/Z plot for the analysis of a gas field using material balance, for the simple example described in the text.

The way in which this is done is through the "P/Z plot" — a universal analysis for gas fields: plot G_p vs. P/Z. From Eq. (2.9), the slope is GZ_i/P_i and intercept at $P = 0$ is G.

Let's consider a simple example: $Z_i = 0.8$ and $P_i = 200$ atm. The remaining production data is shown in Table 2.1.

In Fig. 2.3, we consider a problem where we find G, and G_p at the abandonment pressure of 50 atm when $Z = 0.95$.

The mechanics of this example are simple. This is a useful analysis, as it allows gas sales contracts to be made, based on a sound production-based estimate of reservoir size. The problem is that you do need some significant period of production before reliable estimates can be made.

Before continuing, it is also worth discussing the nature of the data.

Production data. G_p is the cumulative gas produced from one connected reservoir, not a single well. Often P/Z plots are studied on a well-by-well basis, assuming that — to a good approximation — each well drains a separate section of the field. However, this is not how the material balance equation was developed. We discuss later how material balance can be used to determine compartmentalisation of the field.

Pressure. Pressure can be measured down-hole during production. However, while the well is producing, the pressure is lowest at the well. Strictly, the pressure used in the material balance equation is the average for the entire field; this is the pressure at the well when there is no flow, at equilibrium. The average reservoir pressure is normally determined during periods of shut-in using standard procedures in well test analysis (i.e. using a flow model to extrapolate the pressure response to infinite time).

The abandonment pressure is not 1 atm. Why is this? Additional pressure is required for the gas to flow to the surface, through the surface equipment, and along pipelines. This pressure is also related to the desired (or contracted) production rate. Gas can be compressed and pumped, but this requires additional costs and energy, which have to be compared with the benefits of additional production. Even so, recovery factors can be very high — 79% in this case, and higher in many other fields — as the gas expansion can be considerable.

Z-factor. As mentioned previously, the Z-factor can be measured directly in the laboratory from gas samples taken from the well. More usually, it is estimated using correlations tuned to data and the measured composition of the gas. It needs to be known as a function of pressure at the reservoir temperature. If data is unavailable, it is possible to estimate Z (or B_g) from a sensible extrapolation from previous values.

Consistency with other data. As mentioned at the outset, the essence of good reservoir engineering is to assess information from different sources. If there is a seismic survey, then — in principle — G can be estimated from this survey, a determination (usually from logs) of the gas–water contact and log measurements of porosity, saturation and net-to-gross. This needs to be compared with the value obtained from material balance. If the two values are consistent within the bounds of uncertainty, then this gives confidence that the production wells are contacting the entire field. If the seismic value is considerably greater, this implies that the field could be compartmentalised — i.e. the production wells are not contacting all the producible gas — and that the field is divided into distinct compartments or reservoirs. If the seismically derived value is smaller, then — assuming that the seismic data is sound — this implies that there are other mechanisms contributing to production (such as aquifer influx) that need to be considered. This is addressed next.

2.1.1. *Connate Water and Pore Volume Compressibility*

Our analysis has ignored any contribution to recovery from the expansion of connate (initial) water in the pore space and the compaction of the rock itself. *Compressibility* measures the fractional change in volume per unit decrease in pressure. Technically, this is called "compressibility", even though in our case the fluids are expanding. An analogy would be the air in a bicycle tyre. If you release the pressure, the air expands and flows out of the valve. More compressible fluids expand more as the pressure is dropped, pushing out more oil or gas. In the reservoir, the fluid pressure drops and the fluids (oil, gas and water) expand, pushing oil and gas out through the rock pores and into the well. The rock also gets compressed — like squeezing a sponge to release water — and this adds to the production. Unlike sponges or air, the compressibility of the rock and fluids we consider is much lower, so the change in volume is relatively smaller.

Figure 2.4 shows a schematic of what is meant by the compressibility of fluids and rock, and how it contributes to production. As the pressure drops, the hydrocarbon and water expand. The rock grains — held apart by the high fluid pressure — begin to crush together, resulting in a decrease in porosity.

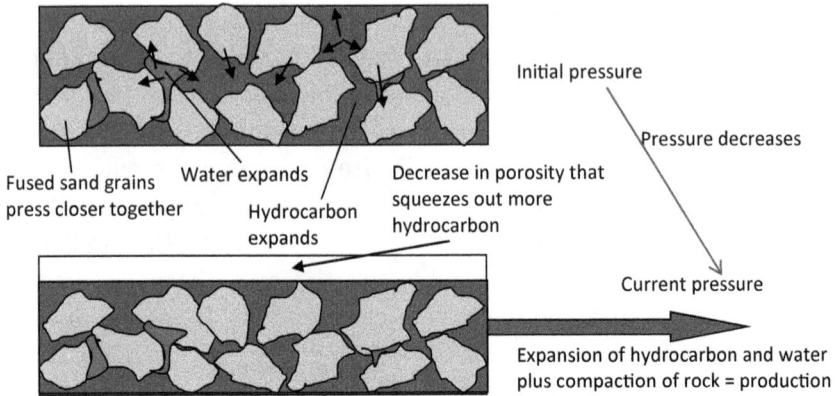

Figure 2.4. Schematic explanation of compressibility in a sandstone reservoir. As the oil or gas pressure drops, oil and water expand, while the rock compresses. The arrows indicate the expansion of the hydrocarbon and water and the collapse of the sand grains into the pore space (this effect has been exaggerated for clarity). All three phenomena contribute to production. The change in volume equals the production.

We have explicitly accounted for gas (and oil) compressibility through the use of formation volume factors, which convert reservoir volumes to surface volumes at different pressures, but we also need to account for the additional production from the expansion of water and the decrease in porosity.

The pore volume of gas, V_g, is given by

$$V_g = (1 - S_{wc})\phi V, \qquad (2.10)$$

where S_{wc} is the connate water saturation (strictly the average initial saturation in the reservoir and $S_g = 1 - S_{wc}$). V_g can be written as: $V_g = V_p - V_w$, where V_p is the pore volume ϕV and V_w is the water volume $S_{wc}\phi V$. The change in pore volume with pressure has two contributions from the compressibility of the formation and water expansion

$$\frac{\partial V_g}{\partial P} = \frac{\partial V_p}{\partial P} - \frac{\partial V_w}{\partial P} = V\frac{\partial \phi}{\partial P} - \frac{\partial V_w}{\partial P}. \qquad (2.11)$$

The first term on the right-handside of Eq. (2.11) represents the rock compressibility and the second the expansion of water.

In general, compressibility, c, is defined as the fractional change in volume, V, with pressure:

$$c = -\frac{1}{V}\frac{\partial V}{\partial P}. \qquad (2.12)$$

The minus sign indicates that in normal substances, volume decreases as pressure increases.

The water compressibility can be written as

$$c_w = -\frac{1}{V_w}\frac{\partial V_w}{\partial P}, \qquad (2.13)$$

with typical values around $5 \times 10^{-10}\,\mathrm{Pa^{-1}}$.

For the solid, we can define the rock compressibility, c_ϕ, as

$$c_\phi = \frac{1}{\phi}\frac{\partial \phi}{\partial P}. \qquad (2.14)$$

Note that there is no minus sign, as porosity decreases as the fluid pressure decreases, see Fig 2.4. More precisely, this is the pore volume compressibility induced by a decrease in pore pressure, with compaction only in the vertical direction. For further discussion of this point, I recommend *Compressibility of Sandstones* by R.W. Zimmerman, Elsevier Science Publishers, New York, NY, USA, ISBN 0444-88325-8, (1991).

Rock compressibilities vary from values similar to — or even lower than — water for highly consolidated sandstones, to values as high as $10^{-7}\,\mathrm{Pa^{-1}}$ for unconsolidated sands.

We now return to Eq. (2.11), which can be written in terms of the compressibilities of water and rock as

$$\frac{1}{V_g}\frac{\partial V_g}{\partial P} = \frac{c_\phi + S_{wc}c_w}{1 - S_{wc}}. \qquad (2.15)$$

The fractional change in pore volume as a result of the two effects of rock compaction and water expansion is usually small and largely outweighed by the gas expansion.

Consider, for instance, an example where $P_i = 300\,\mathrm{atm}$ and $S_{wc} = 0.2$, with a pressure drop of $100\,\mathrm{atm}$ ($\Delta P = 10^7\,\mathrm{Pa}$). If we take $c_w = 5 \times 10^{-10}\,\mathrm{Pa^{-1}}$ and $c_\phi = 10^{-9}\,\mathrm{Pa^{-1}}$ and assume that they are constant as a function of pressure, then from Eq. (2.15) the fractional change in volume can be written

$$\frac{\Delta V_g}{V_g} = \frac{c_\phi + S_{wc}c_w}{1 - S_{wc}}\Delta P \sim 1.4\%.$$ (2.16)

Rock compressibility contributes about 1% to the change in volume and initial water expansion 0.1%; these are relatively small effects.

Another way to see this is to consider the non-ideal gas, Eq. (1.15). From this and the definition of compressibility, Eq. (2.12), we can write

$$c = -\frac{1}{V}\frac{\partial V}{\partial P} = \frac{1}{P} - \frac{1}{Z}\frac{\partial Z}{\partial P}.$$ (2.17)

A useful guide is that — taking the first term in Eq. (2.17) — a gas compressibility is approximately given by the inverse of pressure. So, for a representative pressure of 200 atm (2×10^7 Pa) the compressibility is 5×10^{-8} Pa^{-1}, which is 50 times greater than rock and 100 times greater than water.

2.1.2. *Water Drive Gas Reservoirs*

We will now ignore the (usually small) effects of rock and connate water compressibility and consider the impact of aquifer influx on the behaviour.

Before the equations are presented, it is valuable to emphasise the approach that is taken in this chapter. Firstly, why are we now ignoring rock compressibility? Surely it is more "accurate" to include it? Reservoir engineers, often dazzled by elaborate software, often consider that a model with more variables and parameters is superior to one that is simpler. However, often the story of the field — the production mechanism and the controls on recovery — can be obscured behind a number of poorly understood inputs. You can ignore, for instance, rock compressibility if it is a small effect, or include it, but you should in both cases have a clear understanding of its impact. Secondly, we will include it in the full material balance equation presented later. In any event, it is preferable to understand, clearly, the concepts than simply and unthinkingly to apply complex equations or software.

In this section, we consider aquifer influx. We assume that connected to the gas field is a large body of water contained in porous rock. As the gas pressure decreases, water moves into the gas field in response. We will consider only one, simple model of this response; there are many other aquifer models that can be used, and I recommend other textbooks for details of them. Again, though, it is most important to understand the assumptions made and the conceptual picture behind their development.

Reservoir engineers spend most of their time in an exercise called *history matching*. This is when a model of the reservoir is reconciled with production data. Although the details can be very different, conceptually a mathematical representation of the key production process is tuned to match the data: the properties of the model are found that give the best match to the data. Once this is done, predictions of future performance can be made. In this section, we will specifically match a model of water influx to production data.

W_e is defined as the volume of water encroachment measured at *reservoir conditions*. If water is produced then it should be subtracted from W_e with an appropriate conversion ($1/B_w$ in this case) from surface to reservoir conditions.

Compared to the case without water influx, aquifer support decreases the reservoir volume of gas present in the field by W_e, and the surface volume by W_e/B_g. Hence, instead of Eq. (2.7) the remaining gas volume is

$$G - G_p = \frac{\phi N_G S_{gi} V}{B_g} - \frac{W_e}{B_g}. \tag{2.18}$$

Initial conditions   Conditions some time later during production

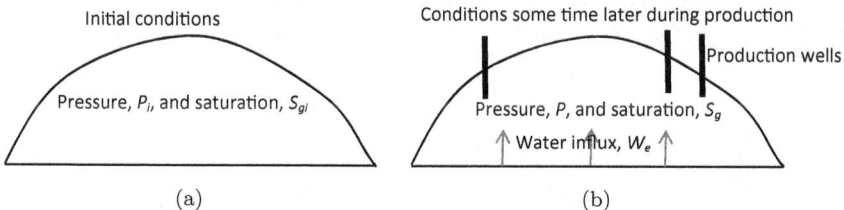

Production wells

Pressure, P_i, and saturation, S_{gi} Pressure, P, and saturation, S_g

Water influx, W_e

(a)   (b)

Figure 2.5. Schematic diagram showing a gas field at initial pressure and saturation (a), and at some time later (b), during production, when the pressure has decreased. In this case we allow for water influx indicated by the arrows.

Then from Eq. (2.6) for G we find

$$G_p = G\left(1 - \frac{B_{gi}}{B_g}\right) + \frac{W_e}{B_g}. \qquad (2.19)$$

Note that water influx increases production for a given pressure drop.

W_e/GB_{gi} is the fraction of the original hydrocarbon pore volume flooded by water. It is always less than 1.

2.1.3. *Simple Aquifer Model*

As mentioned above, there are several ways in which the aquifer response can be modelled. The simplest approach is the so-called pot aquifer model; here it is assumed that the gas field is in contact with an aquifer that responds instantly to any decline in average reservoir pressure. This is appropriate for relatively slow production, smaller aquifers and well-connected (high-permeability) aquifers.

We assume that the aquifer influx is simply due to the expansion of the rock and the water that fully saturates the pore space (S_w=1). Let the aquifer have a porosity ϕ and gross rock volume V_a; the water volume of the aquifer, W, (measured at reservoir conditions) is given by $W = \phi V_a$. By analogy with Eq. (2.11), the change in aquifer volume — which is the water influx, or the amount of water that *leaves* the aquifer — is given by

$$W_e = \Delta W + V_a \Delta \phi. \qquad (2.20)$$

The first term is due to water expansion alone (it is the change in water volume), while the second accounts for the decrease in porosity with pressure that squeezes water out of the aquifer adding to the water influx. Both terms have a positive sign, since they both contribute to the water influx as the pressure drops. When the water expands and the pore space compresses, the water has to go somewhere — it invades the gas field and contributes to the water influx. Then using the definitions Eqs. (2.13) and (2.14) for compressibility:

$$W_e = \left(c_w + c_\phi\right) W \Delta P. \qquad (2.21)$$

Note the signs — $\Delta P = (P_i - P)$ is a pressure drop (a positive number indicates a decrease in pressure), while, for porosity, a decrease in porosity reduces the pore volume available for water and again *adds* to the water influx.[1]

We can simplify the equation above with an overall compressibility:

$$W_e = W\left(c_w + c_\phi\right)\Delta P = Wc\Delta P, \tag{2.22}$$

where c is the total compressibility of the water and rock. With increasing aquifer size, or rock compressibility, we see larger production for given pressure drop.

2.1.4. *Aquifer Fitting*

With sufficient pressure production history you can determine the gas initially in place if you have an aquifer model. For our pot aquifer model, the material balance equation, from Eqs. (2.19) and (2.22), is

$$G_p = G\left(1 - \frac{B_{gi}}{B_g}\right) + \frac{Wc\Delta P}{B_g}. \tag{2.23}$$

Then, rearranging Eq. (2.23), we find

$$\frac{G_p}{\left(1 - \frac{B_{gi}}{B_g}\right)} = G + \frac{Wc\Delta P}{(B_g - B_{gi})}. \tag{2.24}$$

This is an equation of a straight line, where the intercept is G and the slope is Wc (it makes sense to consider this combined variable the unknown — we do not necessarily need to know W and c separately).

Given production data (G_p) as a function of pressure P, plus a determination of B_g, you can determine G and the strength of the aquifer (Wc) as follows. Plot $\dfrac{G_p}{\left(1 - \frac{B_{gi}}{B_g}\right)}$ on the y-axis as a function of

---

[1]The appearing and then disappearing minus signs can be confusing upon first reading. Rather than puzzle over this, for each equation consider carefully what makes physical sense. The mathematics has to make the same sense, so this determines which terms are negative and which are not.

Table 2.2. Illustration of an aquifer analysis as described in the text.

G_p (million scf)	P (MPa)	B_g(rb/scf)	$x = \dfrac{\Delta P}{\left(B_g - B_{gi}\right)}$	$y = \dfrac{G_p}{\left(1 - \dfrac{B_{gi}}{B_g}\right)}$
0	34	0.00234		
72	33	0.00298	1562.5	335.25
112	32	0.00345	1801.801802	348.1081081
129	31	0.00354	2500	380.55
155	30	0.00399	2424.242424	374.8181818

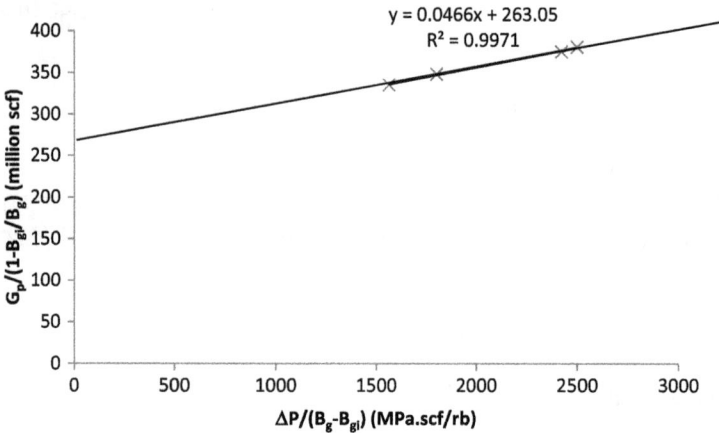

Figure 2.6. Graph plotting the data from Table 2.2. The slope gives the aquifer strength (Wc) while the intercept is the original gas in place, G. Be careful with units and do not quote answers with inappropriate accuracy.

$\dfrac{\Delta P}{(B_g - B_{gi})}$ on the x-axis. The intercept (value of y) when $x = 0$ is G, while the slope is Wc.

Let's give a simple example from Table 2.2. What is shown in bold is data — that is information you have to know or determine before you can perform the analysis. The other two columns are calculated from the data. The graph is plotted in Fig. 2.6.

It is straightforward to determine the unknown parameters from the graph — indeed Excel will give you a best-fit straight line automatically. However, there are two things to pay very careful attention to. The first is units: G has units (in this example) of

MMscf; Wc has the units of th y-axis (MMscf) divided by the units of the x-axis (MPa·scf/rb). In their own context both the MM and M stand for million, so cancel, leaving the units of Wc as rb/Pa. This does need to be thought through very carefully for every example.

The second common mistake is to assert a ludicrous precision and implied accuracy to your results. This is often encouraged by the confident quoting of many significant figures for the best-fit straight line. With real field data, a determination of G to within 10% by this method is reasonable — in no circumstances would quoting G to more than two significant figures be appropriate.

Hence, the final result, for this example, should be quoted as: $G = 260$ MMscf and $Wc = 0.047$ (or 0.05) rb/Pa. It is common for an engineer to first plot a P/Z plot and then determine if it is a straight line. A deviation from a straight line is said to indicate the presence of an active aquifer; at this stage plots with different aquifer models are attempted to find a match. However, this is not the approach that I recommend. Instead, perform this plot immediately, as deducing an aquifer from the P/Z plot is often not straightforward and an engineer is often tempted to assume that it is absent; it is easy to convince yourself that noisy data falls on a straight line. If there is no aquifer, the graph will be a horizontal straight line — any slope indicates an aquifer.

For these reasons the graph presented in Fig. 2.6 is a powerful tool in the analysis of a gas field, enabling the initial gas in place and the strength of the aquifer support to be determined.

2.1.5. *Impact of Residual Gas and Final Recovery*

The advantage of water influx is that for a given pressure decline there is more production — this leads to faster recovery, or a reduced need to compress or pump the gas. The disadvantage is that the water, when it displaces gas, traps residual gas as droplets in the pore space. This is discussed later. The residual gas leads to a lower final recovery — so the gas is produced faster, at a higher pressure, but the cumulative is generally lower.

The residual gas saturation is not known *a priori* — it is determined from core flood tests on samples of reservoir rock. Let S_{gr} be the *residual gas saturation*.

We will now calculate the water influx necessary to sweep the entire gas field to this residual — this represents the maximum theoretical recovery; in practice there will be significant water production before this point is reached. The gas saturation is initially $S_{gi} = 1 - S_{wc}$ and drops to S_{gr}. The change in saturation is $1 - S_{wc} - S_{gr}$. This is a change in reservoir gas volume of $V\phi(1 - S_{wc} - S_{gr})$. Hence, this represents a water influx:

$$W_e = Wc\Delta P = V\phi\left(1 - S_{wc} - S_{gr}\right). \tag{2.25}$$

We also know that $V\phi(1 - S_{wc}) = GB_{gi}$, since this is the reservoir-condition initial volume of gas. Hence,

$$Wc\Delta P = \frac{GB_{gi}\left(1 - S_{wc} - S_{gr}\right)}{(1 - S_{wc})}. \tag{2.26}$$

Equation (2.26) can be used to find the final pressure at which water influx will sweep the entire field. If $\Delta P = P_i - P_f$, where P_f is the final pressure, and recalling that we have determined G and Wc from the previous analysis:

$$P_f = P_i - \frac{GB_{gi}\left(1 - S_{wc} - S_{gr}\right)}{Wc\left(1 - S_{wc}\right)}. \tag{2.27}$$

Having found the final pressure, we can then determine the final recovery and recovery factor from Eq. (2.23) and using Eq. (2.26):

$$R_f = \frac{G_p}{G} = \left(1 - \frac{B_{gi}}{B_g}\right) + \frac{B_{gi}\left(1 - S_{wc} - S_{gr}\right)}{B_g\left(1 - S_{wc}\right)}$$

$$= 1 - \frac{B_{gi}S_{gr}}{B_g\left(1 - S_{wc}\right)}, \tag{2.28}$$

Note that this equation has the same form as Eq. (2.5) — our first application of material balance — if we replace the subscript o for oil with g for gas and note that the initial saturation is $1 - S_{wc}$ and the final saturation is S_{gr}. So, this equation can be derived simply and quickly, directly from a consideration of reservoir and surface volumes; you should be able to do this readily by now.

There is a subtlety, however; the value of B_g used here is the B_g at the final reservoir pressure, computed from Eq. (2.27), and not the B_g for the last production data point.

This concludes the discussion of gas fields. Material balance is a particularly effective tool for analysis in these cases, since there is a significant depletion in pressure accompanied by a non-linear expansion of gas, which can be distinguished from aquifer support. This analysis should always be performed, either as a complement to a reservoir simulation study, or on its own; there is never an excuse to ignore this straightforward calculation in favour of some fancy computer-based analysis.[2]

2.2. Material Balance for Oil Reservoirs

Here we will derive and apply the material balance equation, which was first derived and applied by Schilthuis in (1936) together with some simplifications. This is the general form of the material balance equation that accounts for oil, water and gas expansion, aquifer influx, and the compression of rock. We will use it to study oil fields that may or may not also have a gas cap present.

The conceptual picture is as follows and shown in Fig. 2.7. The reservoir has some pore volume which holds oil, water and gas. The reservoir is described as a tank with uniform properties that only vary with pressure. As the pressure drops, the fluids expand and the pore volume decreases as the rock compresses. How is this change in volume accommodated? The fluids have to go somewhere — they are produced. Thus the change in reservoir volume of rock and fluids is equal to the reservoir volume of the production. We construct the material balance equation by going through each contribution to production in turn; we find the increase in reservoir volume due to

---

[2]Even though, of course, there are software packages available to perform material balance analysis. This is fine, as long as the engineer retains a proper understanding of the results and approximations used.

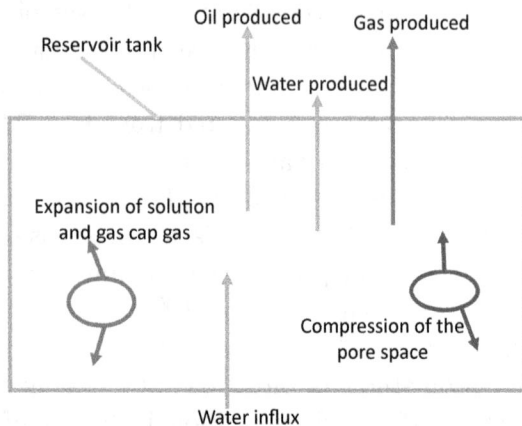

Figure 2.7. A diagrammatic representation of material balance. The reservoir is treated as a tank with uniform properties. The reservoir volume of the expansion of rock and fluids is equal to the reservoir volume of the fluids produced.

expansion. Don't worry about possible production — at the end we equate this to the reservoir volume of the fluids produced.

As previously, we define the initial volume of oil in place at standard conditions, N:

$$N = \frac{V\phi\left(1 - S_{wc}\right)}{B_{oi}},\qquad(2.29)$$

where V is the gross rock volume of the portion of the reservoir containing oil.

We define m as the initial hydrocarbon volume in the gas cap divided by the initial volume of oil. This is measured at *reservoir* conditions.

N_p is the cumulative oil production at standard conditions.

R_p is the cumulative gas/oil ratio. It is the cumulative gas production divided by the cumulative oil production. This is measured at standard conditions. Note that R_p is distinct from the solution gas/oil ratio R_s and the producing gas/oil ratio R (which measures the ratio of the rates of production, not the cumulative).

Initially, through these definitions, at reservoir conditions the oil volume is NB_{oi} and the gas cap volume is mNB_{oi}. The surface volume of the gas cap initially, defined as G in the previous section, is therefore

$$G = mNB_{oi}/B_{gi}. \tag{2.30}$$

We now consider the expansion — measured at reservoir conditions — of the fluids and rock at some time after a period of production.

Expansion of oil and solution gas, E_o. The expansion of liquid oil alone is the volume of oil now minus the initial volume, $N(B_o - B_{oi})$. This term can be negative — the oil volume may shrink because of the exsolution of gas. However, in this case there is another contribution, which is the expansion of gas liberated from solution. At initial conditions, if all the oil were brought to the surface, the volume of solution gas would be — by definition — NR_{si}. At some later time this volume would be NR_s, and would be a smaller volume if we are below the bubble point, because some gas has already come out of solution. The difference, $N(R_{si} - R_s)$, is the surface volume of solution gas that has exsolved. We convert this to a reservoir volume using the gas formation volume factor; hence the contribution of solution gas to expansion is $NB_g(R_{si} - R_s)$. The total expansion of oil and associated gas is therefore

$$NE_o = N \left(B_o - B_{oi} + B_g(R_{si} - R_s) \right). \tag{2.31}$$

This term is always positive — overall the combination of oil and solution gas expands as the pressure drops. Note that E_o alone is the fractional change in volume; for the change itself you need to multiply by N.

Gas cap expansion, E_g. The initial surface volume of the gas cap is G, given by Eq. (2.30). The initial reservoir volume of the gas cap is hence GB_{gi}. Sometime later, the surface volume of this gas is the same, G (remember we do not account for production here); this has a reservoir volume GB_g. The change in reservoir volume of gas — similar to that of oil — is then $G(B_g - B_{gi})$. Applying Eq. (2.30) we find

$$NmE_g = mNB_{oi}\left(\frac{B_g}{B_{gi}} - 1\right).\qquad(2.32)$$

Note that like E_o, E_g is the fractional change in volume — it is multiplied by mN to find the actual change in gas volume.

Initial water expansion and rock compressibility, E_r. These effects have already been considered in the context of gas reservoirs. Let V_h be the total hydrocarbon pore volume (oil and gas) and V_t be the gross rock volume of the reservoir — including both the oil column and the gas cap. We write for the hydrocarbon volume $V_h = V_t\phi(1 - S_{wc}), = V_t\phi - V_w$, where V_t is the gross rock volume of the oil field plus gas cap and V_w is the total water volume in the oil field plus gas cap. Then, similarly to Eq. (2.11) we can write

$$\frac{\partial V_h}{\partial P} = V_t\frac{\partial \phi}{\partial P} - \frac{\partial V_w}{\partial P}.\qquad(2.33)$$

As before, the first term on the right-handside of Eq. (2.33) represents the rock compressibility and the second the expansion of water. We apply the water and rock compressibilities given by Eqs. (2.13) and (2.14) respectively to find, similar to Eq. (2.15),

$$\frac{1}{V_h}\frac{\partial V_h}{\partial P} = \frac{c_\phi + S_{wc}c_w}{1 - S_{wc}}.\qquad(2.34)$$

Then for constant compressibilities and defining, as before, $\Delta P = P_i - P$ (and being very careful over signs — again use physical intuition to tell you what is correct and then the mathematics will look after itself) — we have

$$\Delta V_h = V_h\left(\frac{c_\phi + S_{wc}c_w}{1 - S_{wc}}\right)\Delta P$$

$$= (1 + m)NB_{oi}\left(\frac{c_\phi + S_{wc}c_w}{1 - S_{wc}}\right)\Delta P,\qquad(2.35)$$

using the earlier definitions to state $V_h = (1+m)NB_{oi}$. This change in reservoir pore volume of hydrocarbon also contributes to production and we define this expansion from Eq. (2.25) as

$$(1+m)\,NE_r = (1+m)\,NB_{oi}\left(\frac{c_\phi + S_{wc}c_w}{1 - S_{wc}}\right)\Delta P. \qquad (2.36)$$

Water influx, W_e. This has also already been discussed in the context of gas reservoirs. In general, this is simply left as the term W_e and later different putative aquifer models are fitted to the data. In this chapter we consider only the pot aquifer model with $W_e = Wc\Delta P$, Eq. (2.22).

Total production, F. At surface conditions we have produced N_p oil, $G_p = R_p N_p$ gas (defining R_p) and W_p water (again this defines W_p). We need to convert these volumes to reservoir conditions using the relevant formation volume factors. For the gas, we need to remove the amount that dissolves in oil — this is $R_s N_p$. Hence,

$$F = N_p\,(B_o + (R_p - R_s)B_g) + W_p B_w. \qquad (2.37)$$

Notice the introduction of the water formation volume factor B_w. This is generally close to 1, but may be larger thanks to the presence of dissolved gases in the brine (such as carbon dioxide) at reservoir conditions.

Final material balance equation. The material balance equation is now constructed by making the reservoir volume of produced fluids equal to the reservoir volume of the different expansion terms:

$$\begin{aligned}
F &= N(E_o + mE_g + (1+m)E_r) + W_e \\
&\quad \times N_p\,(B_o + (R_p - R_s)B_g) + W_p B_w \\
&= N\,(B_o - B_{oi} + B_g\,(R_{si} - R_s)) + mNB_{oi}\left(\frac{B_g}{B_{gi}} - 1\right) \\
&\quad + (1+m)NB_{oi}\left(\frac{c_\phi + S_{wc}c_w}{1 - S_{wc}}\right)\Delta P + Wc\Delta P. \qquad (2.38)
\end{aligned}$$

The advantage of the material balance analysis is that it is a simple, quick application to real reservoirs, when only the phase behaviour and production history as functions of pressure are known. It provides valuable insight into the size of the connected reservoir and the production process.

The major disadvantages of the method are the lack of time dependence, and that the properties are reservoir averaged. Reservoir simulation on large fields often does a much better job, particularly if fluids are injected to maintain the pressure. Material balance may also make poor or uncertain predictions if there is not much history.

We will now use the material balance equation to study recovery for different reservoir production mechanisms; material balance can be used to determine the relative contribution to recovery of different production processes. In theory, we can take the full material balance Eq. (2.38), and find the best match to the three unknowns: N, m and Wc. We can also determine the relative sizes of each of the expansion terms contributing to production. This is routinely performed by software; here, instead, we take a simpler approach and will consider examples when one mechanism dominates and we have only two unknown parameters to find, which is achieved by plotting the data as a straight line.

2.2.1. *Production Above the Bubble Point*

Here we assume that there is no gas cap initially present. In this case the production is caused by the expansion of oil and solution gas plus water influx.

Firstly, why might we suspect that there is no gas cap? Consider the initial reservoir pressure — particularly near the top of the formation. If a gas cap is present, then the oil pressure in contact with the gas must be the bubble point pressure; the gas pressure must be the dew point. Why is this? The oil and gas have remained in thermodynamic equilibrium for millions of years. If the oil were at a pressure above the bubble point, then gas would dissolve in the oil until the bubble point was reached. Any excess gas would migrate upwards to the gas cap. Similarly, the gas is at the dew point; if it were at a higher pressure, then oil would dissolve and any excess oil migrates downwards to the oil column. The only way for gas and oil to be in mutual equilibrium is as described. Hence, an oil whose initial pressure is well above the bubble point is unlikely to have a connected gas cap — any gas in the field is unlikely to be in contact with the oil.

What is meant by "well above"? First, consider the uncertainty in the measurement or prediction of the bubble point. Second, the datum depth may not be the top of the formation, and you need to correct for this (we discussed how pressure varies with depth previously). If, even accounting for this, it is highly unlikely that the oil pressure at the gas/oil contact can be at the bubble point, then it is also highly unlikely that there is a gas cap. The converse though is not necessarily the case — if the oil pressure is close to the bubble point pressure there is not necessarily a gas cap present; this could simply be coincidence.

The second inference for a gas cap comes from direct log measurements — is a gas/oil contact detected – and seismic measurements, which may be able to infer the presence of gas directly. A good engineer combines information and develops a model consistent with all the data. Trying to match production data with a gas cap because this is more "general" or "accurate" is senseless if there is other clear evidence that a gas cap is absent.

Determining the likely presence of an aquifer is more difficult, as most fields have an oil column in direct contact with water-saturated rock. Of course, if this is not the case, then there is not an aquifer and trying to assume one is again a pointless exercise. However, the presence of water in contact with oil does not prove an aquifer in an engineering sense — how large is the connected aquifer and is there sufficient permeability to make a significant difference to production? This is often difficult, if not impossible, to determine definitively from log and seismic information alone. Having a sophisticated simulation model does not help here either, since you have to input the aquifer in the first place — the behaviour is determined by the assumptions you make and does not magically emerge from the model. The sensible use of reservoir analogues may be some guide; well-test analysis may be able to determine the locations of reservoir boundaries to infer or preclude the presence of an aquifer. In the end though, it is often a material balance analysis that is used to determine if an aquifer is present and its strength — something that we will demonstrate below.

Above the bubble point, Eq. (2.38) simplifies to

$$F = E_o + E_r + W_e$$

$$N_p B_o + W_p B_w = N(B_o - B_{oi}) + N B_{oi} \left(\frac{c_\phi + S_{wc} c_w}{1 - S_{wc}} \right) \Delta P + W c \Delta P.$$

$$(2.39)$$

The first term — the oil expansion — can be written in terms of oil compressibility using Eq. (1.12). $N(B_o - B_{oi})$ is the change in the reservoir volume of oil. Assuming a constant compressibility we can write

$$\frac{B_o - B_{oi}}{B_{oi}} = c_o \Delta P. \qquad (2.40)$$

Then, Eq. (2.39) can be written

$$N_p B_o + W_p B_w = N B_{oi} \left(c_o + \frac{c_\phi + S_{wc} c_w}{1 - S_{wc}} + \frac{Wc}{N B_{oi}} \right) \Delta P. \qquad (2.41)$$

While this looks complicated, physically Eq. (2.41) states that the reservoir volume of produced fluids is proportional to the pressure drop. Hence, the production simply scales with the pressure decline. But what is the constant of proportionality? This is the weighted contribution of the compressibility of oil, water, rock and the aquifer, which is constant with pressure, to a good approximation. While rock, water and oil compressibility can be measured, the relative contribution of the aquifer compared to the oil cannot be determined from this analysis alone, since we do not know W and N independently. Hence, in this case material balance fails to distinguish between strong production from a large oil field (and relatively little aquifer support) or a small field and a large aquifer. Additional data — such as a seismically derived estimate of N — are necessary to understand the production mechanism in this case. It is only from the non-linear expansion of gas compared to the normally linear (constant compressibility) expansion of oil and water that the production process can be determined with any certainty.

Again, it is more important to understand this concept than to rely on software output that might find a "best match", but one with very little real accuracy or confidence.

If there is no aquifer support and we ignore water production, then Eq. (2.41) can be further simplified to

$$N_p B_o = N B_{oi} c_e \Delta P, \tag{2.42}$$

where c_e is the effective compressibility (and note that $1\text{-}S_{wc} = S_{oi}$):

$$c_e = c_o + \frac{c_\phi + S_{wc} c_w}{1 - S_{wc}}. \tag{2.43}$$

Using example values $c_w = 0.5 \times 10^{-9}\,\text{Pa}^{-1}$, $c_\phi = 10^{-9}\,\text{Pa}^{-1}$, $c_o = 1.5 \times 10^{-9}\,\text{Pa}^{-1}$, $S_{wc} = 0.2$ and $S_{oi} = 0.8$, we find an effective compressibility $c_e \approx 3 \times 10^{-9}\,\text{Pa}^{-1}$.

Then, for instance, we can calculate the recovery at the bubble point:

$$R_f = \frac{N_p}{N} = \frac{B_{oi}}{B_o} c_e \Delta P. \tag{2.44}$$

If we take a case where the initial reservoir pressure is $100\,\text{atm}$ ($10^7\,\text{Pa}$) above the bubble point, then the recovery from Eq. (2.44) is around 3%, assuming a relatively small change in B_o with pressure.

This analysis indicates that recovery factors from reservoirs without a gas cap or significant aquifer are only a few percent, if the pressure is maintained above the bubble point; to achieve significant recovery, water (or gas) must be injected to maintain pressure and displace the oil.

While it is usual to consider material balance as an analysis only to be performed once production data is available, it does also act as a simple aid to injection design even before development. One key uncertainty in reservoir development is often whether or not there is an active aquifer that could contribute significantly to production. Unfortunately, investment decisions are frequently made based on optimistic hunches or experience with "similar" fields, which often turn out to be incorrect. It is always valuable to perform a worst-case scenario — one with no aquifer — and allow flexibility in the design of production facilities to allow for water injection in this case. How long will it be before water injection is required to maintain pressure? The evasive answer is that you can't tell, because material balance does not include time dependence. But now — like any good reservoir engineer — additional information needs to be incorporated. Normally, facilities are designed with a target production rate — say Q (in stb/day). There will also be an estimate (or better still a range) of initial oil in place, N. So the time to reach the bubble point is simply NR_f/Q (in days), where R_f is given by Eq. (2.44). Normally you expect to produce around 2% ($Q/N = 0.02$) of the field (at least) per year, so the time is generally around 1–2 years. You need to ensure that, if the pressure does indeed decline as predicted from the no-aquifer case, it is possible to incorporate water injection facilities in time, otherwise you may be constrained on production rate, or fall below the bubble point and damage the reservoir by introducing a gas phase. The same approach can also be used — as shown below — if there is an aquifer model. With the target production rate, how rapidly do you predict the reservoir pressure to decline and do you have a plan in place to maintain pressure if needed?

2.2.2. *Solution Gas Drive*

Now consider that we drop the pressure below the bubble point. Here recovery is dominated by the expansion of gas liberated from solution — this is called a solution gas drive. As we have shown in the context of gas reservoirs, the gas compressibility is typically many times greater than that of water and rock and so — for simplicity — it is reasonable to ignore the rock and connate water terms in the material balance analysis. They can always be included if there is good data for rock compressibility, or if this compressibility is high (say a poorly consolidated formation).

If we assume no water influx (let us take a scenario where the observed rapid decline in pressure to the bubble point has precluded the existence of an active aquifer), then Eq. (2.38) reduces to

$$F = NE_o$$

$$N_p\left(B_o + (R_p - R_s)B_g\right) = N\left(B_o - B_{oi} + B_g\left(R_{si} - R_s\right)\right). \quad (2.45)$$

The way to identify a solution gas drive is that the ratio F/E_o is a constant with pressure — its value is the initial oil in place, N; later we show a better way to analyse this including the possible presence of a gas cap. The recovery factor is given by

$$R_f = \frac{N_p}{N} = \frac{B_o - B_{oi} + B_g\left(R_{si} - R_s\right)}{B_o + (R_p - R_s)B_g}. \quad (2.46)$$

The expansion of gas contributes significantly to recovery until the gas saturation reaches some critical value, S_{gc}, at which point the gas is connected through the pore space and is produced preferentially to oil. Then excessive quantities of gas are produced and the field essentially becomes a gas field, leaving the vast majority of the more valuable oil behind. In principle S_{gc} can be measured in the laboratory, although defining a representative value is challenging; often, mistakenly, in simulation models a high value of S_{gc} is assumed, below which the gas relative permeability is set to zero. This, by construction, allows the pressure to drop below the bubble point for a time without any gas production; the reality is that there is some gas flow even at very low saturation, but this is significant only when there is good connectivity of the gas. An accurate model of the gas relative permeability is required and this is rarely achieved through the simplistic assignment of an optimistic critical gas saturation (it is often better to set it to zero in simulation models); relative permeability is discussed further later.

The average gas saturation in the field can be estimated from material balance once the value of N has been determined. The initial volume of oil in the reservoir is NB_{oi}(measured at reservoir conditions). The oil volume some time later, during production is $(N-N_p)B_o$ (the oil that has not been produced converted to reservoir conditions at the current pressure). If we ignore changes in the pore volume from rock compression and connate water expansion (this is a relatively small effect in the presence of gas), then the change in volume is accommodated by gas. We convert this to a gas saturation by dividing by the pore volume $NB_{oi}/(1 - S_{wc})$. Hence, we find[3]

$$S_g = \frac{NB_{oi} - (N - N_p)B_o}{NB_{oi}/(1 - S_{wc})} = \left(1 - \left(1 - \frac{N_p}{N}\right)\frac{B_o}{B_{oi}}\right)(1 - S_{wc}).$$

(2.47)

In terms of saturation, we can rearrange Eq. (2.47) for the recovery factor:

$$R_f = \frac{N_p}{N} = 1 - \frac{B_{oi}}{B_o}\frac{1 - S_{wc} - S_g}{1 - S_{wc}}.$$

(2.48)

This is a complex way to derive the expression that can be obtained directly from material balance — see Eq. (2.5)

Let's take a typical example with $S_g = S_{gc} = 0.3$, $S_{wc} = 0.2$ and $B_o/B_{oi} = 0.88$. This gives $R_f = 29\%$. Recoveries from a solution gas drive are generally in the 20%–30% range before the field has to be abandoned due to excessive gas production. Furthermore, it is then difficult to recover more oil from the field; water injection below the bubble point requires an increase in pressure to boost recovery, so the water simply compresses gas and is initially very inefficient in recovering additional oil.

In modern reservoir engineering, a solution gas drive is only a good option as a tertiary recovery mechanism. A field is produced initially under primary production to the bubble point. Then it is waterflooded and pressure is maintained. Then, and only then, when the field is mainly residual oil, is the pressure dropped. This is achieved simply and cheaply by producing oil and not injecting water. The field is then essentially managed as a gas field to liberate the solution gas. This is a good option for light oils where there is a network to sell the produced gas. This concept was pioneered by Shell for the Brent field in the North Sea and is now frequently considered for mature fields in provinces with a good gas pipeline infrastructure.

---

[3]Note that the standard textbook in this area, Dake (1991) on p. 86 gives the wrong expression for gas saturation.

2.2.3.   *Gas Cap Drive*

Here we consider expansion of a gas cap, but we ignore water influx and the (relatively small) contribution of rock compressibility and connate water expansion. As mentioned previously, the initial reservoir pressure is the bubble point and, as production proceeds, the pressure drops below this. As a result there is recovery due both to the expansion of gas in the gas cap itself, and solution gas. The material balance equation, Eq. (2.38), becomes

$$F = N(E_o + mE_g)N_p\,(B_o + (R_p - R_s)B_g)$$
$$= N\left(B_o - B_{oi} + B_g\,(R_{si} - R_s) + mB_{oi}\left(\frac{B_g}{B_{gi}} - 1\right)\right).$$
$$(2.49)$$

The approach here is to reduce this to an equation of a straight line:

$$F/E_o = N + NmE_g/E_o,\qquad(2.50)$$

so we take the production data and fluid properties as a function of pressure. We compute F, E_o and E_g for each pressure value. We then plot F/E_o on the y-axis and E_g/E_o on the x-axis.[4] The intercept when $y = 0$ is N, the original oil in place. A slope to the data plotted this way indicates the presence of a gas cap (if there is no slope we have a pure solution gas drive and there is no active gas cap). The slope itself has a value Nm, from which the relative gas cap size, m, can be easily found. Here the units should be straightforward — even if field units are used, no conversions are needed, as we are dealing in ratios of quantities. N is measured in stb while m is a dimensionless ratio.

For clarity, a simple example is shown in Table 2.3. As for the gas reservoir case, the data — information you need to perform the analysis — are shown in bold, while the other quantities are calculated from the data.

The graph is plotted in Fig. 2.8. The best fit to the data is $N = 250\,$MMstb and the slope of the line gives $mN = 7.6\,$MMstb, or

---

[4]There is no point for the initial condition, as the ratios are 0/0.

Table 2.3. An example material balance analysis for an oilfield with no aquifer but the possible presence of a gas cap.

N_p (MMstb)	G_p (MMscf)	P (MPa)	R_s (scf/stb)	B_o	B_g (rb/scf)	F	E_o	E_g	$y = F/E_o$	$x = E_g/E_o$
0	0	32	400	1.356	0.000187					
1.41	480	30	400	1.361	0.000199	1.902294	0.005	0.087016	380.4588	17.40321
1.98	1568	28	370	1.355	0.000213	2.86084	0.00539	0.188535	530.7681	34.97862
3.41	3016	26	345	1.349	0.000251	5.061817	0.006805	0.464086	743.8379	68.19773
5.78	6890	24	295	1.335	0.000302	9.28214	0.01071	0.833904	866.6797	77.86216

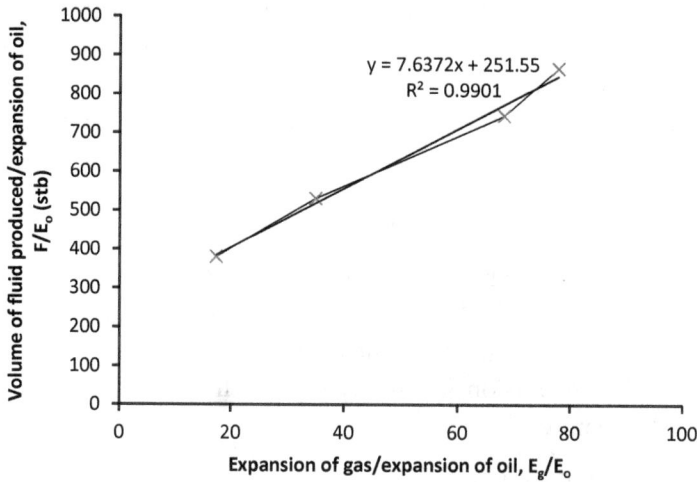

Figure 2.8. Graph plotting the data from Table 2.3. The intercept gives N, the value of the initial oil in place, while the slope is mN, where m is the relative size of the gas cap.

$m = 0.03$. In this case we have a relatively small gas cap (only 3% of the total oil volume), yet the expansion of gas is considerably larger than the expansion of oil. There is a large expansion of the relatively small volume of gas in the gas cap. Note that the overall recovery is only around 2%, so the actual amount of gas expansion is small.

However, the best way to understand this is to perform the exercise yourself; at the end of this work there are many exam questions that can be used for practice. Once the mechanics of the analysis are known, you can use software and apply the methods to real field cases with some confidence. My recommendation is to use this plot, even if you are not sure about the presence of a gas cap — if there is none, then the points simply lie on a horizontal line.

The presence of a gas cap can allow higher recoveries than solution gas alone; the gas is at the top of the reservoir and when it expands, it pushes the oil downwards. With wise well completions near the base of the oil column (or horizontal wells low in the oil column), the problem of excessive gas production can be mitigated, although withdrawing at a high rate encourages coning and an increase in the gas/oil ratio. Typical recoveries are 25%–35%, but values as high as 70% are possible for very slow production, allowing the gas to displace oil down to very low final saturations. The mechanism for this — oil layer drainage — is discussed later. In our example however, with a very small gas cap, the recovery is very modest (only around 2%) for a significant drop in pressure; here we have principally the production of solution gas from the reservoir.

2.2.4. *Natural Water Drive*

Natural water drive, or aquifer drive, has already been discussed in the context of gas fields. Here, for simplicity, we assume that there is no gas cap and we assume that the effects of rock compressibility and connate water expansion are relatively small. The material balance Eq. (2.38) becomes

$$F = NE_o + Wc\Delta P,$$

$$N_p\left(B_o + (R_p - R_s)B_g\right) + W_p B_w$$
$$= N\left(B_o - B_{oi} + B_g\left(R_{si} - R_s\right)\right) + Wc\Delta P. \qquad (2.51)$$

Here we have assumed a simple pot aquifer model, although others can be used to match the data more accurately, where needed.

Again, the approach is to plot the data as an equation of a straight line:

$$F/E_o = N + Wc\Delta P/E_o. \qquad (2.52)$$

As for a gas cap drive, we plot F/E_o on the y-axis. On the x-axis we plot $\Delta P/E_o$. The intercept where $y = 0$ is, again, N, the initial oil in place. The slope is Wc. Here, as with the analysis of a gas reservoir, care needs to be taken with the assignment of units. In field units E_o is given in rb/stb. If ΔP is reported in psi, then Wc has units of rb/psi. Once again, the best way to understand this is through working through some of the example exam questions — in this case I will not give a worked example.

2.2.5. *Compaction Drive*

The final primary recovery mechanism is compaction drive, where the rock compressibility is significant. An example is the Bachaquero field in Venezuela where $c_\phi = 15 \times 10^{-9}\,\text{Pa}^{-1}$ and compaction accounts for 50% of oil recovery; rock compression is also significant, for instance for the poorly consolidated and unconsolidated sandstones in the Gulf of Mexico.

If rock compression dominates over aquifer support (and connate water expansion) with no gas cap, then for Eq. (2.38) we can write

$$F = N(E_o + E_r),$$

$$N_p\left(B_o + (R_p - R_s)B_g\right) + W_p B_w$$
$$= N\left(B_o - B_{oi} + B_g\left(R_{si} - R_s\right)\right) + N B_{oi} c_\phi \Delta P / S_{oi}. \quad (2.53)$$

By this stage, you should see the approach to use: once more we plot F/E_o on the $y-$axis. On the x-axis, we plot $B_{oi}\Delta P/E_o$.

$$\frac{F}{E_o} = N + N c_\phi B_{oi}\Delta P / S_{oi} E_o. \quad (2.54)$$

The intercept is N, as before, while the slope is $N c_\phi / S_{oi}$ from which the rock compressibility can be found. Of course, it is possible to measure rock compressibility from rock samples taken from the field; however, this is not necessarily representative of the field as a whole. The advantage of this method is that production data is used to deduce a flow-averaged compressibility — this should be compared with the measured values for consistency, but it is the value from material balance that is the more robust, as it represents the average behaviour of the whole field under production.

This now concludes the analysis of different primary recovery production mechanisms using the material balance analysis. As previously mentioned, the data can usually be analysed using commercial software; this is fine and allows the consideration of more than one production mechanism simultaneously, but is no excuse for not having a clear understanding of the production process yourself.

2.2.6. *Rate Dependence*

The material balance analysis does not take rate — fluid flow — directly into account. It therefore assumes that the recovery behaviour is insensitive to the rate at which the field is being produced. This is a reasonable approximation for solution gas drives (dominated by gas expansion and evolution) and a strong water drive with a well-connected aquifer.

However, reservoirs above the bubble point with a weak aquifer drive are sensitive to the production rate. If the pressure is dropped very rapidly, then the aquifer cannot expand sufficiently quickly in response, and so the recovery is lower (for a given pressure drop) than that of a slower production that allows the aquifer time to move into the oil reservoir.

Other production processes that are rate-sensitive include any situation where gravity segregation is important (for instance where gas evolves and rises to the top of the formation). Rapid production does not allow sufficient time for the fluids to segregate, resulting in poorer production (in general) than cases where the fluids do separate. This is also the case where coning is significant — higher production rates lead to more coning and increased production of water and/or gas.

This needs to be borne in mind when designing target production rates; in the end, though, a reservoir simulation approach is needed to assess the sensitivity of recovery to rate.

2.2.7. *Recap of Material Balance*

While a material balance analysis should be performed for any field with a substantial pressure drop, it is the main method of analysis for primary production data for small fields, including most gas fields. Production history is used to predict oil and gas reserves and ultimate recovery. The more production there has been, the more accurate the predictions.

The approach is conceptually similar to that seen in other areas of petroleum engineering, such as well test analysis. The first — and often most difficult — step is to decide on the reservoir drive process (model identification). For this the engineer should incorporate all the data that are available to make a sensible assessment of the likely production mechanism.

In these notes, I have shown how to plot the data so that they lie on a straight line, with the slope and intercept giving you the required parameters. Once again, to understand this more fully, I recommend attempting some of the homework problems given at the end.

Chapter 3

Decline Curve Analysis

Decline curve analysis is an empirical way to forecast production decline from measurements of production rate; it acts as a simple complement to material balance analysis, where rate dependence is not included.

The method is important for project economics. The rate of decline will depend on the wellbore condition, well spacing, surface facilities, porosity, permeability, reservoir thickness, fractures, relative permeability, formation damage, drive mechanism, compressibility and gas production. Ideally — and indeed as shown in later volumes in this series — the production data should be matched to a physically based flow model (either analytical or simulation). Here briefly, we simply introduce the nomenclature and some example types of decline; however, this is not a replacement for a more rigorous approach based on flow modelling.

Most of the equations presented here were first proposed by Arps (1956) before the advent of reservoir simulation and a modern approach to understanding flow processes. Hence, while it provides a simple analysis of production, its predictions are highly unreliable unless associated with a proper model of flow. Decline curve analysis is currently used to interpret and predict the behaviour of shale oil and gas reservoirs — this is often a reflection that the flow processes are not fully understood and such analyses are of dubious accuracy at best and need to be rooted in an understanding of flow processes.

The analysis assumes that you have stable operation at capacity. Below capacity, no decline may be seen at all (plateau).

The nominal decline rate is defined as follows:

$$b = -\frac{1}{Q}\frac{dQ}{dt},\tag{3.1}$$

where Q is the oil production rate (stb/day) from either the entire field, or a single well — the analysis can be performed on either.

3.1. Exponential Decline

This is the simplest type of decline where b is assumed to be constant. In this case it should be obvious that the production rate is given by

$$Q(t) = Q_0 e^{-bt},\tag{3.2}$$

where Q_0 is the initial production rate at some nominal $t = 0$.

How do we check if we have exponential decline? Either plot the computed value of b from Eq. (3.1) against time, or better (since it removes the need to compute numerical derivatives from the data) check if $\ln Q$ varies linearly with time, t. If so, then the slope is $-b$.

The cumulative production is given by

$$N_p = \int_0^t Q dt = \int_0^t \frac{dQ}{b} = \frac{Q_0 - Q}{b}.\tag{3.3}$$

The cumulative production varies linearly with flow rate; this is another check for exponential decline.

It is possible to relate analytical solutions for flow in a homogeneous medium to different types of decline, which gives some more confidence in the predictions. This type of decline is seen in reservoirs above the bubble point with no strong aquifer drive; in solution gas drives; and in the latter stages of recovery from fractured reservoirs and shale gas.

3.2. Hyperbolic Decline

Here the decline rate is proportional to a power of rate:

$$b = -\frac{1}{Q}\frac{dQ}{dt} = cQ^{1/a},\tag{3.4}$$

for some constant a. Here the analysis is rather more cumbersome. It is common simply to compute a decline rate and assume it constant (exponential decline as the default in the absence of a flow simulation) or presume a value of a based on analytical solutions to the relevant flow equations.

In general, though, we can write

$$ct = -\int Q^{-\frac{1}{a}-1} dQ, \qquad (3.5)$$

and hence

$$ct = aQ^{-1/a} + \text{constant}. \qquad (3.6)$$

The constant is found by setting $Q = Q_0$ at $t = 0$:

$$ct = a\left(Q^{-1/a} - Q_0^{-1/a}\right) = aQ_0^{-1/a}\left(\left(\frac{Q}{Q_0}\right)^{-1/a} - 1\right), \qquad (3.7)$$

$$\left(\frac{Q}{Q_0}\right)^{-1/a} = \frac{ct}{a}Q_0^{1/a} + 1, \qquad (3.8)$$

$$Q = \frac{Q_0}{\left(1 + \frac{ctQ_0^{1/a}}{a}\right)^a} = \frac{Q_0}{\left(1 + \frac{b_0 t}{a}\right)^a}, \qquad (3.9)$$

where b_0 is the initial decline rate (at $t = 0$).

The cumulative production is

$$N_p = \int_0^t Q\,dt = \int_0^t \frac{Q_0}{\left(1 + \frac{b_0 t}{a}\right)^a}\,dt. \qquad (3.10)$$

This can be evaluated as

$$N_p = \frac{a}{a-1}\frac{Q_0}{b}\left(1 - \left(1 + \frac{b_0 t}{a}\right)^{1-a}\right)$$

$$= \frac{a}{(a-1)b}\left(Q_0 - Q\left(1 + \frac{b_0 t}{a}\right)\right). \qquad (3.11)$$

Strictly hyperbolic decline is $a = 2$ and occurs for gravity drainage and in gas reservoirs. $a = 1/2$ is seen in the early production behaviour of shale gas and fractured reservoirs. You should be able to demonstrate this later using the methods developed later in these notes.

Harmonic decline is a special case when $a = 1$. This is observed for high-viscosity oil and a water drive. It may also be seen for high water/oil ratio (WOR) and constant fluid rate, such as in thermal projects and steam soak.

In this case for Eq. (3.9) we can write

$$Q = \frac{Q_0}{1 + b_0 t},\tag{3.12}$$

and

$$N_p = \frac{Q_0}{b_0} \ln(1 + b_0 t).\tag{3.13}$$

The treatment here of decline curve analysis has been deliberately very brief; it is an empirical approach that should, where possible, be substantiated with numerical simulation or analytical results. However, for this we need to understand fluid flow, which is the subject of the subsequent sections.

Chapter 4

Multiple Phases in Equilibrium

We now divert our attention away from oil fields and petroleum recovery to the details of the science of how fluids are arranged in the pore space of rocks at the micron scale. This is necessary to have a good understanding of how multiple fluids — oil, water and gas — are configured in the pore space of the rock and how they flow.

We start with a presentation of the fundamental equations that govern contacts between fluids and solids and the meniscus that separates fluid phases.

4.1. Young–Laplace Equation

If we have multiple phases present in a porous medium, then there is a pressure difference across the interfaces between these phases. The pressure difference is given by the Young–Laplace equation:

$$P_c = \sigma \left(\frac{1}{r_1} + \frac{1}{r_2} \right), \tag{4.1}$$

where r_1 and r_2 are the principal radii of curvature of the fluid interface (the radii of curvature measured perpendicular to each other and σ is the interfacial tension between the phases — it has the units of a force per unit length (N/m) or an energy per unit area (J/m^2).[1]

---

[1]Sometimes σ can be called the surface tension. Strictly this is only valid for a liquid in equilibrium with its vapour. This is a situation that we will rarely encounter in real situations, so it is always preferable to refer to the interfacial

The non-wetting phase has the higher pressure.

It is possible to derive this equation using principles of force or energy balance (see Dullien, 1992; or de Gennes *et al.*, 2002), but is somewhat cumbersome, as it involves some obscure mathematical results concerning the 3D geometry of curved surfaces. It is, however, relatively straightforward to derive specific cases from first principles. This we will do later when we consider the pressure difference between two phases in a circular cylindrical tube. For now, though, it is helpful to assert that the Young–Laplace equation is valid.

The second major concept we need to introduce is that of contact angle. While the Young–Laplace equation considers the pressure across an interface, it does not address how that interface interacts with a solid surface.

4.2. Equilibrium at a Line of Contact

Consider a wetting fluid (say water) resting on a solid surface surrounded by a non-wetting phase (such as oil or gas) shown in Fig. 4.1.

A horizontal force balance gives

$$\sigma_{so} = \sigma_{sw} + \sigma_{ow} \cos \theta. \tag{4.2}$$

This is the Young equation. The contact angle is given by

$$\cos \theta = \frac{\sigma_{so} - \sigma_{sw}}{\sigma_{ow}}. \tag{4.3}$$

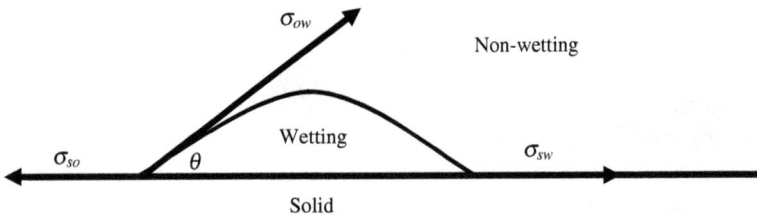

Figure 4.1. Contact angle, θ, between two phases, measured through the denser phase.

tension — the energy per unit area of a boundary between two phases, regardless of composition.

Figure 4.2. Spot the difference. One is an insightful English physicist; the other a brilliant French mathematician, clinging to a post-revolutionary title.

What is the vertical force balance? Intermolecular forces in the solid counteract the vertical tension, as the solid is very slightly perturbed.

Thomas Young, Fig. 4.2, was an early-19th-century English physicist and all-round genius who worked on everything from deciphering hieroglyphics to the wave theory of light. Young's modulus (in elasticity) and the Young double slit experiment recognise just two of his contributions to science.

Pierre-Simon (the so-called Marquis de) Laplace was a brilliant French mathematician and physicist of the same era. His contributions include astronomy, statistics and the Laplace transform. The Laplace equation is generally considered to be $\nabla^2 \phi = 0$; Eq. (4.1) is sometimes called the "Laplace equation" which is confusing; and Eq. (4.2) the "Young–Laplace equation," which possibly understates the relative contribution of Young to this subject.

4.3. Spreading Coefficient

Define a spreading coefficient as

$$C_s = \sigma_{so} - \sigma_{sw} - \sigma_{ow}. \tag{4.4}$$

If $\sigma_{so} > \sigma_{sw} + \sigma_{ow}$ or $C_s > 0$, then there is no solution for θ in Eq. (4.3); the water spreads over the solid surface. This is complete wetting and $\theta = 0$.

4.4.   Two Fluids in A Capillary Tube

Now consider two fluids in a capillary tube of radius r, as shown in Fig. 4.3.

The radius of curvature is given by $R = r/\cos\theta$ (we can derive this using simple trigonometry).

Thus we find

$$P_c = P_{nw} - P_w = \frac{2\sigma}{r}\cos\theta. \tag{4.5}$$

Imagine now water rising up in a capillary tube of radius r, as shown in Fig. 4.4.

$$\rho g h = \frac{2\sigma}{r}\cos\theta. \tag{4.6}$$

What would happen for mercury/air? The interface would be lower than the free (flat) surface, as in this case mercury is the non-wetting phase.

Equation (4.6) can be derived without the Young–Laplace equation, using an energy balance. Water rises up the tube, since this is energetically favourable; there is a lower energy for water to coat the solid (glass) surface than air. This though is balanced by potential energy, as the water rises against gravity. So, the water continues

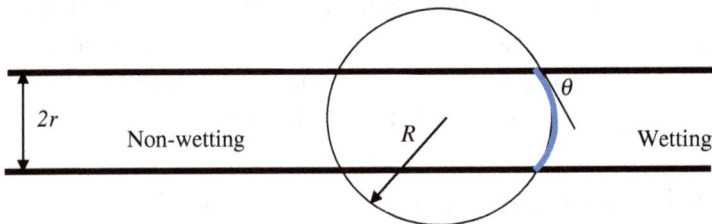

Figure 4.3. Two fluids in equilibrium in a cylindrical tube. The blue arc is the interface between the wetting and non-wetting phases.

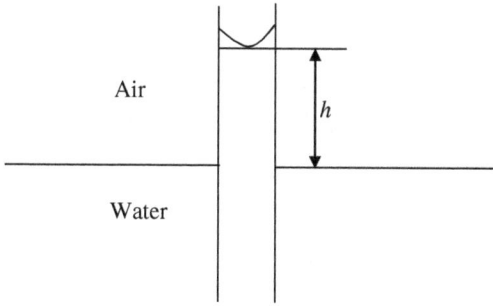

Figure 4.4. Capillary rise in a tube; here water is the wetting phase.

to rise until the change in potential energy is just balanced by the change in interfacial energy.

Imagine that the height of the interface changes by a small amount; if this lowers the energy, then the interface will move until it reaches a position of equilibrium. The equilibrium height is found when changes in potential and interfacial energy are equal for small fluctuations in this height — this is the stable configuration.

The gravitational potential energy of a mass m at a height h is mgh. From a height h to $h + dh$, the water in the tube has a mass $A\rho dh$, where A is the area (πr^2). Then if we change the height from h to $h + dh$, the change in potential energy is

$$\Delta E_p = \pi r^2 \rho g h dh. \tag{4.7}$$

This must be equal to the energy gained when the water moves up the tube. Now consider that σ_{sa} is the energy per unit area (interfacial tension) of an interface between the solid and air, while σ_{sw} is the energy per unit area of interface between solid and water. Then the energy change for the water to move from h to $h + dh$ is

$$\Delta E_s = 2\pi r (\sigma_{sa} - \sigma_{sw}) dh = 2\pi r \sigma \cos \theta dh, \tag{4.8}$$

since $2\pi r h$ is the area of wetted solid (glass) surface, and we have used the Young Equation (4.2). Conservation of energy asserts that the energies in Eqs. (4.7) and (4.8) are the same, leading to Eq. (4.6) directly:

$$\Delta E_s = \Delta E_p,$$
$$2\pi r \sigma \cos \theta dh = \pi r^2 \rho g h dh,$$
$$2\sigma \cos \theta = r \rho g h, \tag{4.9}$$
$$\frac{2\sigma \cos \theta}{r} = \rho g h.$$

4.5. Wettability

The contact angle is traditionally measured through the denser phase (water for oil/water and gas/water and oil for gas/oil).

If $\theta = 0$, then we have complete wetting.

If $\theta < 90°$, then the fluid is wetting (water-wet).

If $\theta \approx 90°$, then the fluid is of intermediate or neutral wettability.

If $\theta > 90°$, then the fluid is non-wetting (oil-wet).

4.5.1. *Wettability Alteration*

Clean rocks are generally water-wet, since the polar surface of the solid — say quartz or calcite — interacts strongly with the water, making its interfacial tension lower than the tension (energy per unit area) with oil or gas, which has less interaction with the surface.

Why then are most reservoir rocks not completely water-wet? In contact with a solid surface, surface active components of the oil — high molecular weight molecules called asphaltenes — adhere to the solid surface rendering it less water-wet. Regions of the solid surface that are not directly contacted by oil remain water-wet. Thus, many oil reservoirs are what is known as mixed-wet or fractionally wet — different regions of the pore space have different wettabilities. This important concept is developed later; the wettability (or contact angle) is not a constant throughout the rock and depends precisely on the mineralogy of the surface, surface roughness, the oil and brine compositions, and the temperature and pressure of the reservoir.

In other settings — aquifers and soils — it is the presence of organic material, particularly surfactants or other compounds that adhere to the solid surface — that alter the wettability. In general

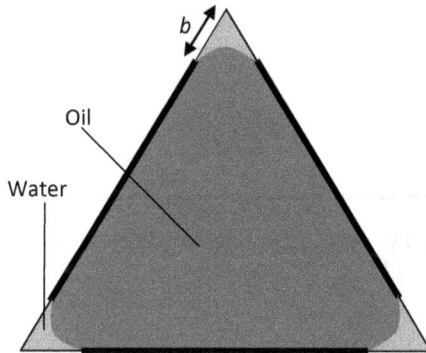

Figure 4.5. Oil and water in a triangular pore after primary drainage (oil migration into the reservoir). The areas directly contacted by oil (shown by the bold line) have an altered wettability, while the corners that are water-filled remain water-wet. *b* is the length of the water-wet surface.

the contact angle is governed by a subtle balance of surface forces between the fluids and the rock.

The wettability change typically takes around 1,000 hours to complete, and since oil has been in a reservoir over geological times, there has been plenty of time for the wettability alteration to occur. The wettability alteration also has sufficient time to take place in most polluted soils.

Figure 4.5 is a schematic of what happens in a single pore with an idealised triangular cross-section; regions of the surface directly contacted by oil have an altered wettability (this is not necessarily oil-wet, but is not strongly water-wet either) while the corners remain water-wet.

4.5.2. *Contact Angle Hysteresis*

The contact angle is also usually different depending on the flow direction, Fig. 4.6. The static contact angle (no movement of the solid/wetting/non-wetting contact), the advancing contact angle (denser phase advancing) and the receding contact angle may all be different.

The three reasons for this are wettability alteration (described above), chemical inhomogeneities on the surface, and small-scale surface roughness.

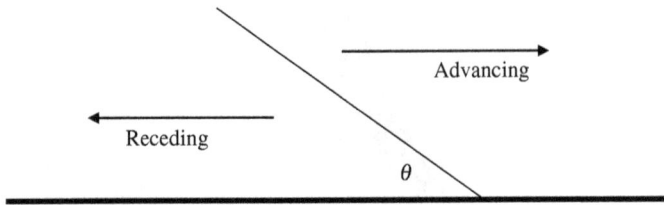

Figure 4.6. Advancing and receding contact angles.

Displacement — the movement of one phase across another — is always impeded by the most difficult step, or part, of the motion. This is an important concept that we will meet many times in these notes. So, if we imagine a surface that is chemically heterogeneous with water-wet and oil-wet patches, then water will invade the water-wet regions quickly and get impeded by oil-wet portions. The pressure — the highest pressure — in the water necessary to move across these patches will be the same as if the system were entirely oil-wet — hence the system looks to water as if it is oil-wet. Now consider the reverse: oil invasion. In this case, movement across the oil patches is easy (occurs at a low oil pressure) but is impeded by the water-wet regions. Hence, for oil invasion, the surface appears water-wet. In terms of contact angle, the (water) advancing angle is greater than 90°, while the receding angle is less than 90°; we see significant contact angle hysteresis.

The main effect in porous media, however, is usually small-scale surface roughness. Figure 4.7 shows the measured relationship between advancing and receding contact angle on a rough surface while the effect is shown schematically in Fig. 4.8. While the molecular-scale contact angle may indeed be the intrinsic value, the apparent angle viewed on the larger scale — and the angle that will control the capillary pressure (that is the interfacial curvature) for displacement — is different. As mentioned previously, displacement is impeded by the highest pressure in the displacing (advancing) phase necessary for the contact to move; hence the apparent angles

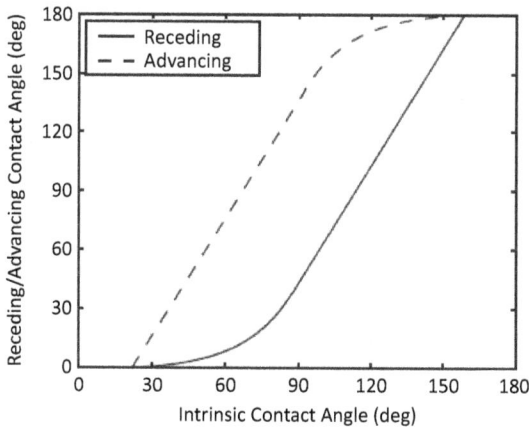

Figure 4.7. Relationship between intrinsic contact angle (the angle at rest on a smooth surface) and the advancing and receding contact angles, based on the measurements by Morrow (1975). The principal reason for this contact angle hysteresis is surface roughness — in a porous medium, the solid surfaces are not smooth.

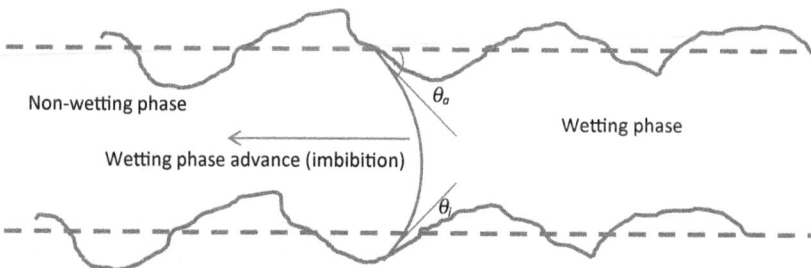

Figure 4.8. A schematic of surface roughness (the solid surface is indicated by the irregular line) and its impact on contact angle. θ_i is the intrinsic contact angle — the angle at the surface (regardless of orientation); in this example it is close to zero, indicating a strongly water-wet system. However, the effective dynamic contact angle for imbibition, θ_a, is larger — this is the angle that gives the correct curvature in the Young–Laplace equation and is observed at a larger scale with an apparently smoother solid surface (shown by the dashed line). The roughness impedes imbibition; a higher wetting-phase pressure is required for displacement than on a smooth surface. For drainage the displacement is limited by roughness oriented in the other direction and the apparent contact angle is smaller than θ_I, impeding the advance of the non-wetting phase.

are different in imbibition (wetting phase advancing) than drainage (non-wetting phase advancing). The interface is impeded at the points where the pressure in the advancing phase is largest — we need to capture this to calculate the correct capillary pressures for displacement.

Chapter 5

Porous Media

We now properly introduce multiple phases in a porous medium. Before proceeding with the description of macroscopic, averaged properties, we will first use some illustrative examples in an attempt to provide some insight into displacement and fluid configurations at the pore scale. The emphasis in this section will be on understanding and appreciating the complexity of the pore space at the micron (pore) scale.

5.1. X-ray Imaging

In recent years there has been a revolution in our ability to see inside the pore space of rocks using X-rays. The development of modern imaging methods relies on the acquisition of three-dimensional (3D) reconstructions from a series of two-dimensional (2D) projections taken at different angles; the sample is rotated and the absorption of the X-rays in different directions is recorded and used to produce a 3D representation of the rock and fluids. In the 1980s, these methods were first applied in laboratory-based systems to measure two- and three-phase fluid saturations for soil science and petroleum applications with a resolution of around 1 mm–3 mm. The first micro-CT (micron or pore-scale) images of rocks were obtained by Flannery and coworkers at Exxon Research using both laboratory and synchrotron sources (Flannery *et al.*, 1987). In a synchrotron, a bright monochromatic beam of X-rays was shone through a small rock sample. Several rocks were studied with resolutions down

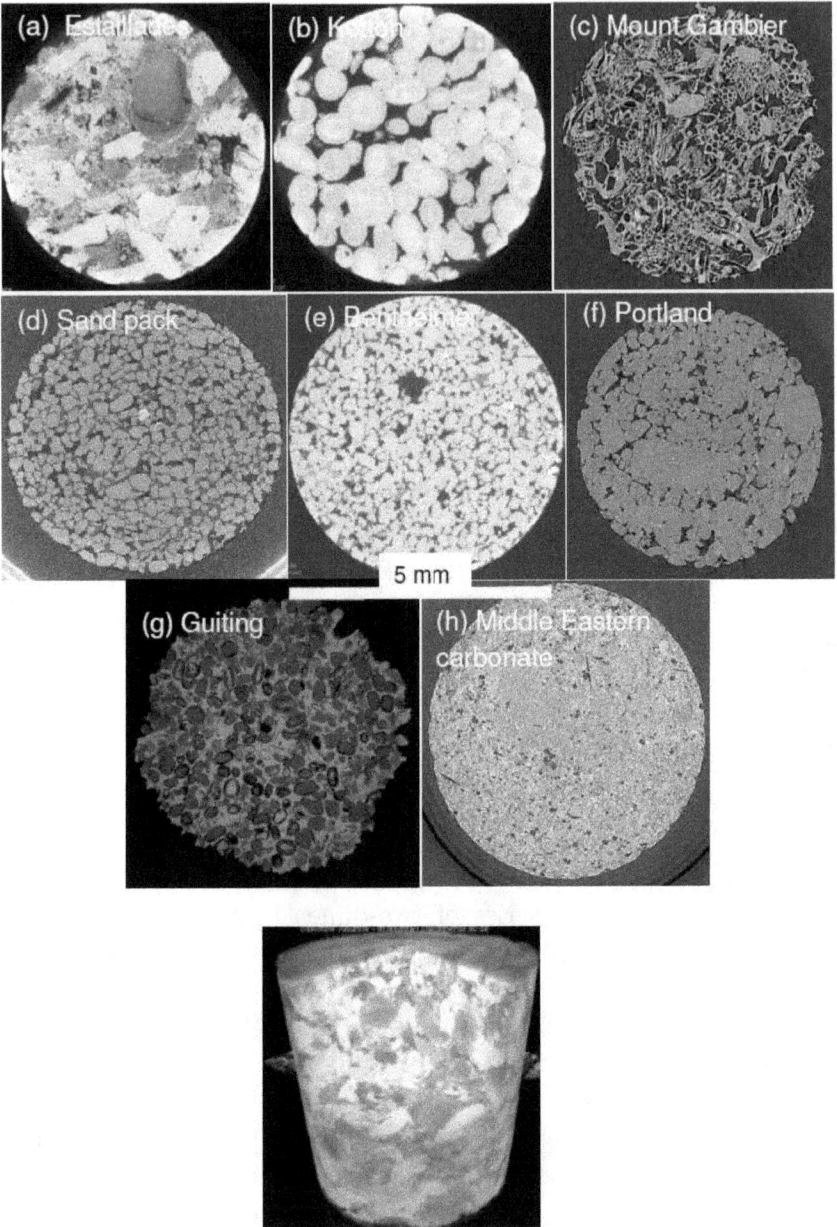

(a) Estaillades (b) Kentish (c) Mount Gambier (d) Sand pack (e) Bentheimer (f) Portland (g) Guiting (h) Middle Eastern carbonate

5 mm

to around $3\,\mu$m. Dunsmuir *et al.* (1991) extended this work to characterise pore space topology and transport in sandstones.

One of the pioneers of the continued development of this technology has been the team at the Australian National University in collaboration with colleagues at the University of New South Wales (see, for instance, Arns *et al.*, 2001, 2005). They have built a bespoke laboratory facility to image a wide variety of rock samples and then predict flow properties; this work is also now available as a commercial service. The base image is a 3D map of X-ray adsorption; this is thresholded to elucidate different mineralogies and clays and, principally, to distinguish grain from pore space.

The now-standard approach to imaging the pore-space of rocks is to use a laboratory instrument, a micro-CT scanner, which houses its own source of X-rays. A picture of the inside of our instrument at Imperial College is shown on the front cover of these notes together with the core holder into which a small cylinder of rock is fitted.

In a micro-CT scanner the X-rays are polychromatic and the beam is not collimated — the image resolution is determined primarily by the proximity of the rock sample to the source. These machines offer the advantage that access to central synchrotron facilities or a custom-designed laboratory is not required, and there is

Figure 5.1. 2D cross-sections of 3D micro-CT images of different samples. These are grey-scale images where the pore space is shown dark. (a) Estaillades carbonate. The pore space is highly irregular with likely micro-porosity that cannot be resolved. (b) Ketton limestone, an oolitic quarry limestone of Jurassic age. The grains are smooth spheres with large pore spaces. The grains themselves contain micro-pores that are not resolved. (c) Mount Gambier limestone is of Oligocene age from Australia. This is a high-porosity, high-permeability sample with a well-connected pore space. (d) A sand pack of angular grains. (e) Bentheimer sandstone, a quarry stone used in buildings, including the pedestal of the Statue of Liberty in New York. (f) Portland limestone. This is another oolitic limestone of Jurassic age that is well-cemented with some shell fragments. Portland is another building material used, for instance, in the Royal of School of Mines at Imperial College. (g) Guiting carbonate is another Jurassic limestone, but the pore space contains many more shell fragments and evidence of dissolution and precipitation. (h) Carbonate from a deep highly-saline Middle Eastern aquifer. The final figure (bottom) is a 3D view of the Estaillades limestone.

no constraint on the time taken to acquire the image, allowing signal-to-noise to be improved. The disadvantage is that the intensity of the X-rays is poor compared to synchrotrons, while the spreading of the beam and the range of wavelengths introduces imaging artefacts.

Figure 5.1 shows 2D cross-sections of 3D images for eight representative rock samples: several carbonates, including a reservoir sample, a sandstone and a sand pack, together with a 3D picture of one of the samples. The images were acquired either with a synchrotron beamline (SYRMEP beamline at the ELETTRA synchrotron in Trieste, Italy) or from a micro-CT instrument (Xradia Versa).

For the quarry carbonates shown in Fig. 5.1 (Estaillades, Ketton, Portland, Guiting and Mount Gambier), a connected pore space is resolved, although the details of the structure are complex and at least two of the samples — Ketton and Guiting — are likely to contain significant micro-porosity that is not captured with the resolution of the image. Also included is a carbonate from a Middle Eastern aquifer. In this case, while some pores are shown with a voxel size of almost $8\,\mu$m, it is likely that there is significant connectivity provided by pores that is below the resolution of the image.

Figure 5.2 shows example 3D images of three carbonates where only the pore space is shown. Ketton is a classic oolitic limestone composed of almost spherical grains with large, well-connected pores between them. Estaillades has a much more complex structure with some very fine features that may not be fully captured by the image. Mount Gambier has a very irregular pore space, but it is well connected and the porosity and permeability are very high. Overall, while a resolution of a few microns can resolve the pore space for some permeable sandstones and carbonates, many carbonates and unconventional sources, such as shales, contain voids that have typical sizes of much less than a micron.

Typical X-ray energies are in the range 30 keV–160 keV for micro-CT machines — with corresponding wavelengths 0.04 nm–0.01 nm — while synchrotrons have beams of different energies for which those with energies less than around 30 keV are ideal for imaging. Resolution is determined by the sample size, beam quality and the detector specifications; for cone-beam set-ups (in laboratory-based instruments) resolution is also controlled by the

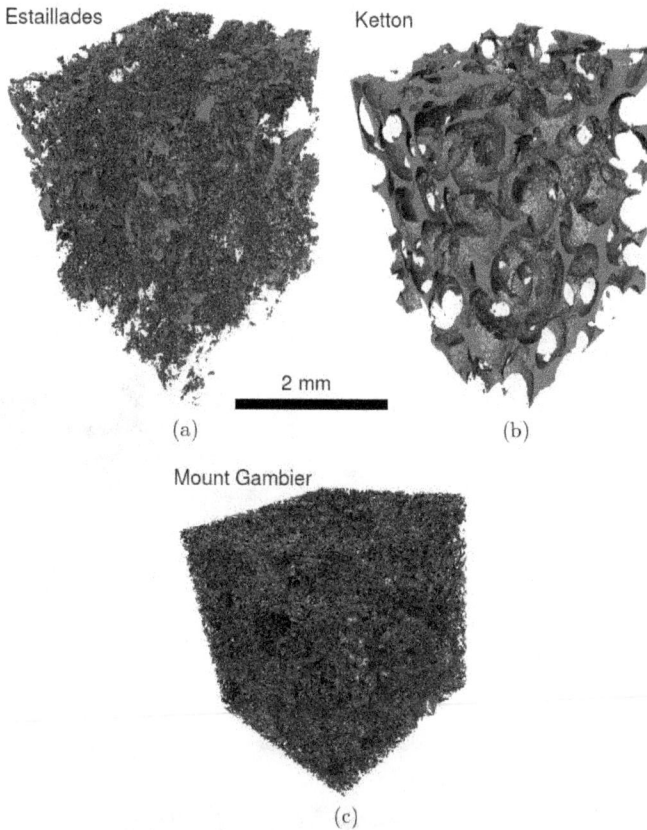

Figure 5.2. Pore-space images of three quarry carbonates: (a) Estaillades; (b) Ketton; (c) Mount Gambier. The images shown in cross-section in Figs. 5.1(a)–5.1(c) have been binarised into pore and grain. A central 1000^3 (Estaillades and Ketton) or 350^3 (Mount Gambier) section has been extracted. The images show only the pore space.

proximity of the sample to the beam, as mentioned previously, while detecting absorption at a sufficiently fine resolution. Current micro-CT scanners will produce images of around 1000^3 voxels–2000^3 voxels. To generate a representative image, the cores are normally a few mm across, constraining resolution to a few microns; sub-micron resolution is possible using specially designed instruments and smaller samples. Developments in synchrotron imaging may allow much larger images to be acquired, but at present most images have an approximately 1000-fold range from resolution to sample size.

5.2.   Electron Microscopy to Image Micro-porosity

Micro-porosity — small pores typically within larger grains with a size of around a micron or smaller — can be imaged using electron microscopy techniques. Figure 5.3 shows images of Ketton and

(a)

(b)

Figure 5.3. Scanning electron microscope images for Ketton limestone (a) and Indiana limestone (b) at 2000× and 4000× magnification respectively, showing micro-porosity: small pores within larger grains. The thick black lines below both images are 10 microns long.

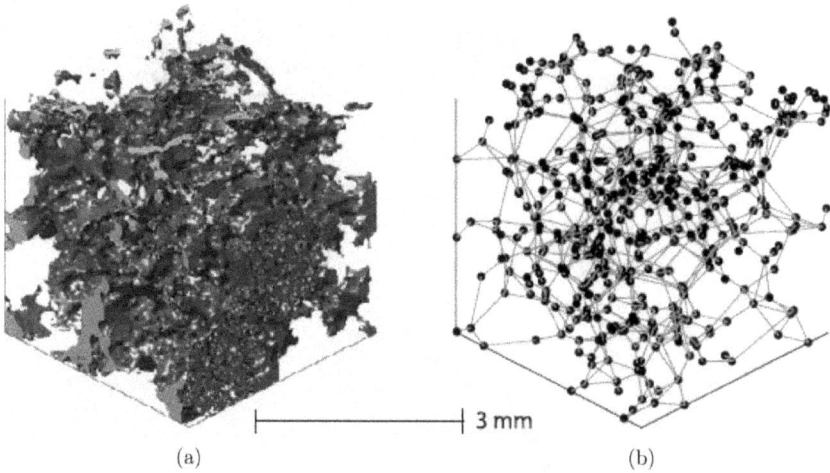

Figure 5.4. (a) The void space of a simple sandstone (Berea) and (b) the associated topologically representative network of pores and throats.

Indiana showing small pore spaces, smaller than 1 micron, that are generally below the resolution of micro-CT scanners, but which may contribute significantly to the connectivity and porosity of the rock.

5.3. Topologically Representative Networks

The final conceptual step is to describe the pore space of the rock in terms of a network. This is a topologically representative description of the pore space, where the larger voids between grains are called pores and these pores are connected together through narrower connections, called throats. Each pore and throat in reality has a complex shape in cross-section, but we will describe these — for simplicity — as triangles. This allows the wetting phase to reside in the corners of the pore space while the non-wetting phase occupies the centres. This way of viewing the rock allows us to understand multiphase flow and, in some cases, make quantitative predictions of flow and transport properties. I will not go through the details of how a network is extracted — there are several different methods to do this (see, for instance, Dong and Blunt, 2009); I will simply show some illustrative examples.

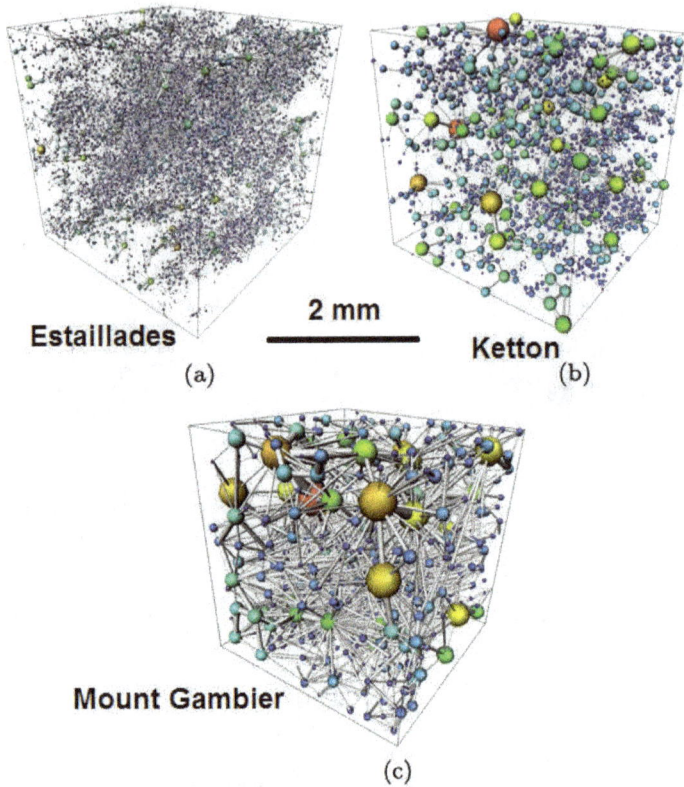

Figure 5.5. Pore networks extracted from the images shown in Fig. 5.2: Estaillades, Ketton, and Mount Gambier. For illustrative purposes, only a section of the Mount Gambier network is shown. The pore space is represented as a lattice of wide pores (shown as spheres) connected by narrower throats (shown as cylinders). The size of the pore or throat indicates the inscribed radius. The pores and throats have angular cross-sections — normally a scalene triangle — with a ratio of area to perimeter squared derived from the pore-space image.

Figure 5.4 shows the pore space and network for Berea sandstone, a benchmark used for many experiments and modelling studies.

Figure 5.5 shows the more complex networks for carbonates — based on the images shown in Fig. 5.2.

Chapter 6

Primary Drainage

Now consider a porous medium that is initially fully saturated with water, and is water-wet. Then a non-wetting phase (oil) enters the porous medium. Imagine that this is done sufficiently slowly that the pressure drop across the oil (from Darcy's law, described later) is small in comparison with the capillary pressure. This process is called *primary drainage* and is the process by which oil migrates from source rock to fill a reservoir. It is also the process by which injected carbon dioxide displaces brine in a storage aquifer.

If we return to the Young–Laplace equation, Eq. (4.1), then the non-wetting phase will preferentially fill the larger pore spaces, where the radius of curvature for the meniscus is larger, resulting in a lower capillary pressure. A lower capillary pressure means that — for a given wetting-phase pressure — a lower non-wetting phase pressure for invasion is needed. As the non-wetting phase pressure is increased, smaller regions of the pore space (lower radii of curvature) can be accessed. As a consequence, primary drainage proceeds as a sequence of filling events, accessing progressively smaller pores. In a network representation, filling pores is easy, since they are larger than the connecting throats. Hence, the invasion of the non-wetting phase is limited by the throat radius. The non-wetting phase will next fill the largest-radius throat that is connected to a pore already filled with non-wetting phase. This largest throat and the adjoining pore fill, and then again the largest-radius throat is filled. This is technically known as an *invasion percolation process* (Wilkinson and Willemsen, 1983): the pore network is filled in order of size, with the constraint that the invading (non-wetting) phase must be connected to the inlet. There are subtleties associated with trapping the wetting phase (how does the water escape?), but this is a good model of primary drainage and motivates why a network representation of the pore space is useful for the understanding of fluid displacement.

At a macroscopic (core) scale, if we average the behaviour over millions of individual pores and throats, we can plot the capillary pressure (the pressure difference between the phases) as a function of water saturation, see Fig. 6.1. Similar to the previous discussion of contact angle, the process is always limited by the most difficult step, or the invasion with the highest capillary pressure. In theory, for a slow displacement at a fixed flow rate, the capillary pressure can decrease when larger pores, or the large throats connected to them, are filled. However, in general, in experiments the capillary pressure

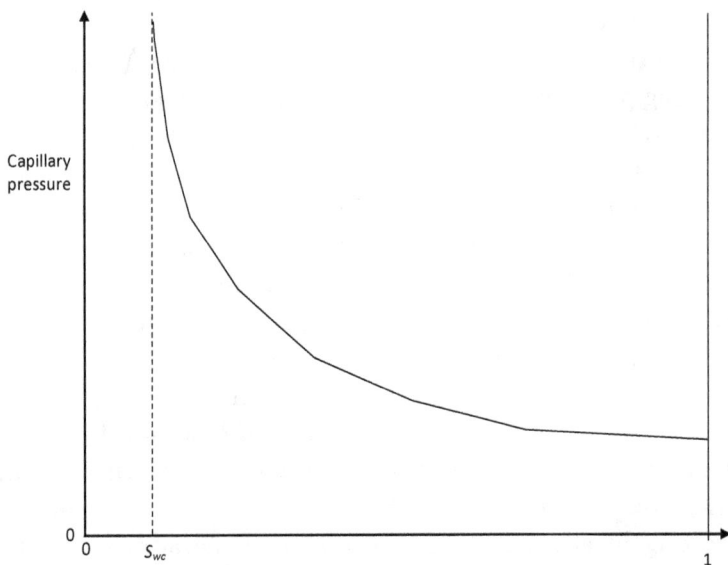

Figure 6.1. Schematic of the primary drainage capillary pressure. Experimental examples will be shown later. S_{wc} is the connate or irreducible water saturation.

simply increased in increments, and the amount of the pore space accessed at that pressure was recorded.

The oil invades progressively smaller regions of the pore space. We find a connate or irreducible water saturation, S_{wc}, where further increases in capillary pressure result in little or no decrease in water saturation. At this point the water may either be trapped in wetting rings around rock grains, or (more likely) contained in roughness, grooves and corners of the pore space. While, in theory, this water could be displaced, it would require a huge amount of time and a very high capillary pressure to do so.

During and after primary drainage, regions of the pore space that come in direct contact with oil may alter their wettability as described before (see Sec. 4.5).

6.1. Typical Values of the Capillary Pressure

In most reservoir sandstones, pore radii, R, are in the range $1\,\mu$m–$100\,\mu$m — see Fig. 5.3. $P_c \approx 2\sigma/R$ (see Sec. 4 — assuming a contact angle close to zero). $\sigma_{ow} \approx 50\,$mN/m for say alkane/water. P_c is around $0.1/R$ or in the range $10^3\,$Pa–$10^5\,$Pa. These values are typically ten times higher for mercury/air, since σ is higher (the interfacial tension is around $480\,$mN/m).

Figure 6.2 shows the effect of heterogeneity and permeability on the capillary pressure: lower permeability normally reflects smaller pore (throat) sizes and a higher capillary pressure, while a more heterogeneous structure leads to a wider range of capillary pressures as more of the pore space is invaded.

6.2. How is Capillary Pressure Measured?

The standard method to measure primary drainage capillary pressure is through mercury injection. Here a small, dry rock sample — usually around 5 mm across — is placed under vacuum and then mercury is injected as the non-wetting phase. The volume of mercury that enters the rock sample is recorded as a function of the imposed pressure. Here we do not see any irreducible saturation, so the mercury can, in theory, invade the entire (connected) pore space.

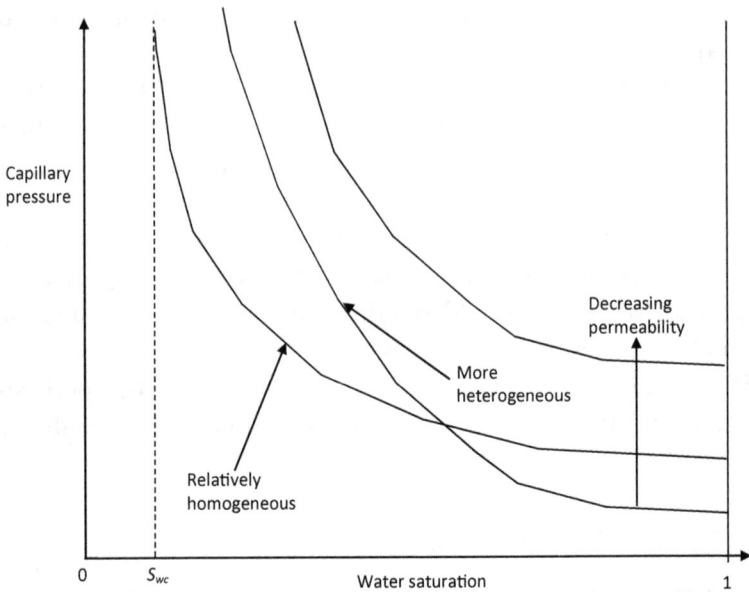

Figure 6.2. Schematic of the primary drainage capillary pressure showing the effect of permeability and heterogeneity in the pore size distribution.

It is also possible to measure capillary pressure on experiments with two fluids — such as water and oil — although this is more difficult. Later I will show some mercury injection capillary pressure curves measured on different rock samples. I will also present capillary pressures measured when a core is initially fully saturated with brine and then oil (or carbon dioxide) is injected at some known pressure and the volume that is injected is recorded. There is a porous plate at one end — a porous ceramic disc with a very high capillary pressure — that prevents any injected non-wetting phase from leaving the system.

To help illustrate the results, Fig. 6.4 shows pictures of the rock samples studied taken with X-rays at a resolution of around $7\,\mu$m, where the pore space is clearly visible. We have seen some of these samples previously. Then the measured capillary pressures for primary drainage are shown in Fig. 6.5; the apparatus used to make the measurements is shown in Fig. 6.3. Note how there is a region

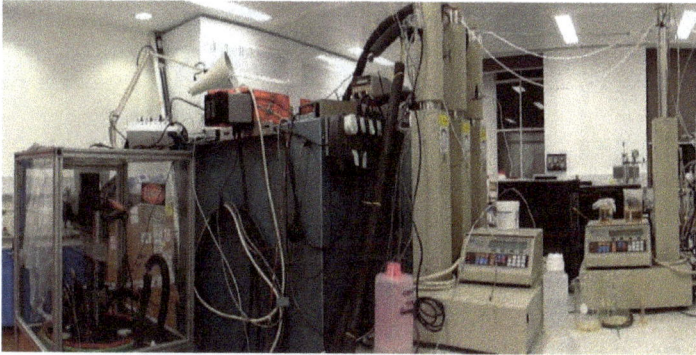

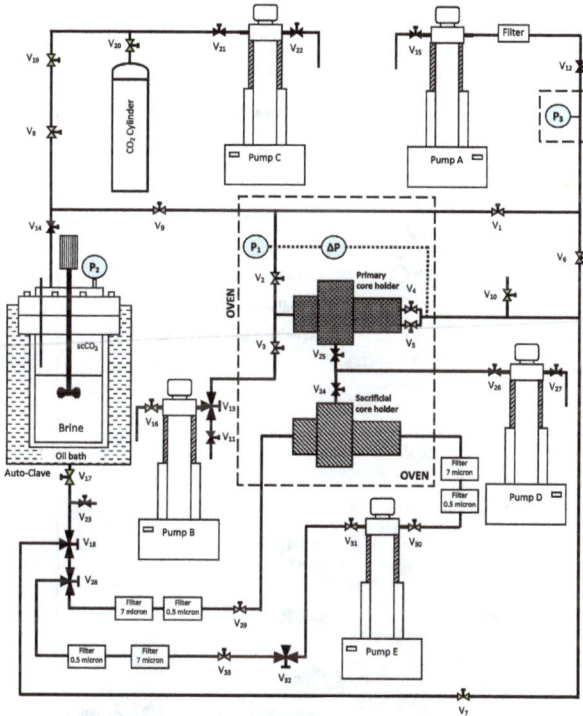

Figure 6.3. An apparatus to measure displacement and capillary pressure in a rock sample. This apparatus is designed specifically to study displacements involving carbon dioxide. The upper picture shows a photo of the apparatus at Imperial College, while the lower diagram shows the apparatus and flow loops. Note the complexity of the experiment; flow is controlled through pumps and each displacement cycle takes many days to complete, reproducing the slow flow conditions usually seen in reservoirs (from Rehab El-Maghraby's PhD thesis, 2013).

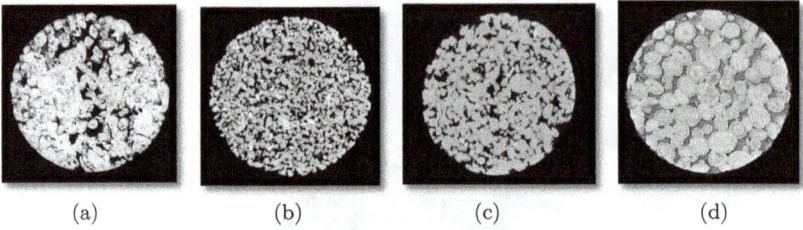

Figure 6.4. Micro-CT images of four rock samples (from (a) to (d): Indiana, Berea, Doddington, Ketton) on which capillary pressure was measured. Note how the trend in permeability (in Table 6.1) is evident from the grain sizes: in all cases the diameter of the sample is 5 mm.

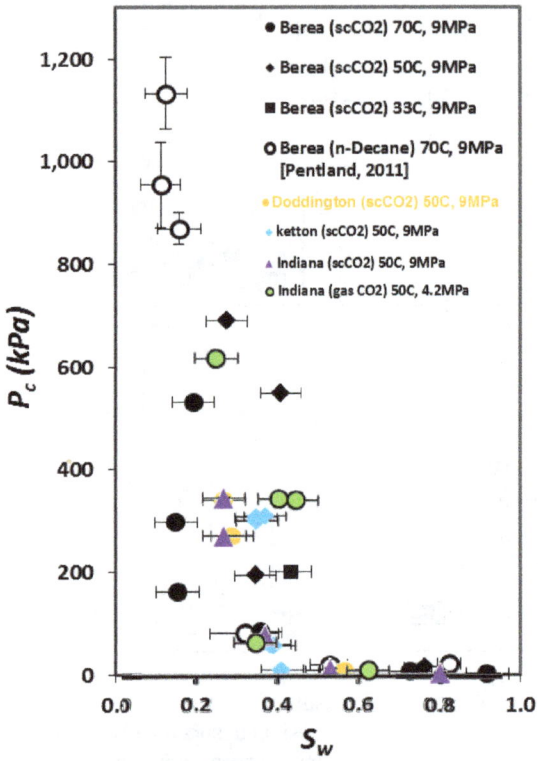

Figure 6.5. Measured primary drainage capillary pressure for Indiana limestone. Note how the magnitude of the pressure relates to the average pore size, evident in the micro-CT images, Fig. 6.4.

Table 6.1. Properties of the rock samples in Fig. 6.4.

	Porosity	K_{brine} (m^2)	K_{brine} (mD)
Berea	0.2188	4.6×10^{-13}	460
Doddington	0.214	1.565×10^{-12}	1565
Ketton	0.2337	2.81×10^{-12}	2809
Indiana	0.1966	2.4×10^{-13}	244

of relatively low pressure when the non-wetting phase invades the larger pores followed by a steep rise where smaller and smaller pores and throats are invaded. The magnitude and shape of the curve is an indicator of pore size and structure.

Chapter 7

Imbibition

Imbibition is the opposite process to drainage, where wetting fluid invades a porous medium containing non-wetting fluid. We normally only consider secondary imbibition, which is the invasion of wetting fluid into non-wetting fluid after primary drainage; i.e. there is some wetting fluid initially present in the porous medium. This is the process that occurs when water is injected to displace oil in a reservoir, when the aquifer encroaches into a gas field during production, or when brine displaces stored carbon dioxide, as the carbon dioxide rises in the storage aquifer.

The capillary pressure for imbibition is always lower than for primary drainage. There are three reasons for this:

1. trapping of non-wetting fluid;
2. contact angle hysteresis;
3. different displacement mechanisms at the pore scale.

In drainage, the non-wetting fluid advances through the porous medium by a connected piston-like advance; i.e. regions can only be filled with non-wetting fluid if they are adjacent to a region that also contains non-wetting fluid. Both wetting and non-wetting phases remain connected. In imbibition, the displacement process is different, as described in the next section.

7.1. Pore-scale Displacement, Trapping of the Non-wetting Phase and Snap-off

In imbibition, it is possible for the wetting phase to trap the non-wetting phase. This is through *bypassing* and *snap-off*. Bypassing is when invading fluid surrounds and strands a ganglion of non-wetting phase; this occurs due to local inhomogeneities in the pore structure (or local capillary pressure). In snap-off, the more important process, water flows through wetting layers and fills narrow regions of the pore space in advance of the main wetting front. This is the principal mechanism by which non-wetting phase is surrounded and trapped.

For reference Figs. 7.1 and 7.2 show some schematic pictures of pore-scale displacement. As the pressure in the wetting phase increases, wetting layers in the pore space thicken. There comes a point at which the meniscus between the wetting and non-wetting phases loses contact with the solid — it is no longer possible to place the interface in the pore space. At this point, the throat fills rapidly with wetting phase.

In imbibition, piston-like advance is favoured in the narrow throats, but impeded by the wide pores (the water wants to be in the narrow regions of the pore space). However, there is a subtlety: it is easier (i.e. the radius of curvature is smaller) for the wetting phase to fill a pore if more of the surrounding throats are also full of water. This is shown in Fig. 7.1. In a series of classic papers Roland

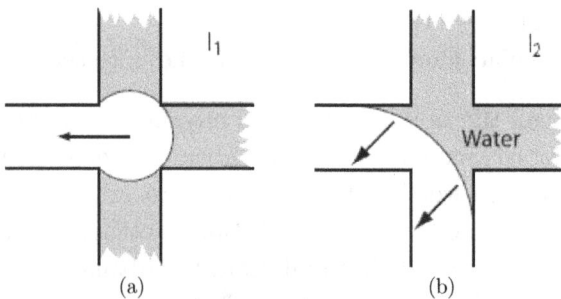

Figure 7.1. Pore-filling processes in water (wetting phase) injection. (a) I_1, when all but one of the connected throats is full of water. This is the most favourable displacement. (b) I_2, where two throats are initially filled with oil. This is less favoured as the threshold capillary pressure is lower — the radius of curvature of the interface as it invades is larger.

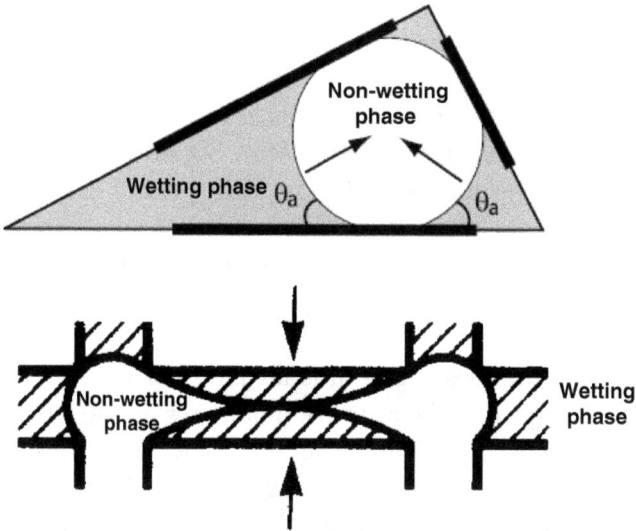

Figure 7.2. The snap-off process. Here water flows along the corners of the pore space indicated in the top figure, showing a throat in cross-section. An instability occurs when the wetting layer in the corner loses contact with the surface — any further increase in water pressure leads to the rapid filling of the centre of the throat. The lower figure illustrates how the non-wetting phase is pinched-off when viewed along the length of the throat (from Lenormand and Zarcone, 1984).

Lenormand (see, for instance, Lenormand *et al.*, 1983) described these imbibition processes as I_n, where the n refers to the number of connecting throats filled with non-wetting phase; I_1 is more favoured than I_2, which in turn is more favoured than I_3. The result of this is that — at the pore scale — the wetting front tends to be flat, filling in any local channels filled with non-wetting phase. This tends to suppress trapping. Hence, without snap-off — described next in more detail — most of the non-wetting phase is recovered from the porous medium (which is good for oil and gas recovery, but bad for carbon dioxide storage).

In the snap-off process, as the wetting-phase pressure increases, the wetting layers in the corners of the pore space swell, as mentioned previously. It is possible that these layers swell to a sufficient thickness that they lose all contact with the solid. This always occurs in the narrowest portions of the pore space — the smallest throats.

This creates an unstable configuration, and the wetting phase then rapidly fills the throat. This process only occurs for slow flow — there has to be sufficient time for the wetting phase to flow along wetting layers, and for low contact angles and sharp corners (which allow the wetting layer to swell in the first place). So, all the narrow regions of the pore space — anywhere in the rock — are filled. If we fill all the throats around a pore, then the non-wetting phase in the pore is trapped — it cannot escape. This is the origin of residual saturation, a very important concept in multiphase flow that will be discussed in further detail below.

As mentioned above, snap-off is favoured for low contact angles, while connected piston-like advance dominates as the contract angle becomes larger (but is still less than 90°, as we only consider water-wet systems here). Equation (4.5) gave the capillary entry pressure for piston-like advance for invasion of a cylindrical throat of radius r. We now consider which process is favoured — snap-off or piston-like advance. For snap-off, we do have to have an angular pore. If the pore is square in cross-section, then it is easy to calculate the critical radius of curvature at which the meniscus first loses contact with the solid.[1] This gives a critical capillary pressure,

$$P_c = \frac{\sigma}{r} \cos \theta (1 - \tan \theta), \tag{7.1}$$

where r is now the inscribed radius of the pore (or throat). The ratio of the capillary pressure for snap-off to piston-like advance for the same throat is

$$\frac{P_{c,\text{snap-off}}}{P_{c,\text{piston}}} = \frac{1}{2}(1 - \tan \theta). \tag{7.2}$$

This is always less than one; physically this is because in piston-like advance we have curvature in two directions — the hemispherical meniscus — whereas in snap-off we only have curvature normal to the length of the throat.

---

[1]This is a geometric calculation which can be generalized for any half-angle α of the throat. In this case the $\tan \theta$ term in Eq. (7.1) is multiplied by $\tan \alpha$.

The process with the highest capillary pressure is favoured in imbibition and hence — if possible — piston-like advance will occur rather than snap-off. However, this requires that an adjoining pore is also full of wetting phase, which may not be the case. So, snap-off does occur, but only when pore filling is suppressed because of the large pore size and different cooperative pore-filling mechanisms. This means that porous media with a large difference between pore and throat sizes will see a lot of snap-off and trapping, while porous media with similar sized pores and throats will see less trapping and a more connected wetting phase advance. Furthermore, Eq. (7.2) indicates that snap-off becomes less favourable as the contact angle increases. This is true even if we include different pore shapes and the effects of cooperative pore filling: as the contact angle increases there is less snap-off and a lower residual saturation.

During imbibition the water pressure *increases*, meaning that the capillary pressure *decreases*. Imbibition ends at a water saturation $1 - S_{or}$, where S_{or} is the residual oil saturation. This residual oil is very important, as it determines how much oil can be recovered from a reservoir. It is also significant in carbon dioxide storage, since this residual is trapped, cannot move and therefore cannot escape back to the surface.

7.2.   Pore-scale Images of Trapped Phases

We can image residual saturations directly using micro-CT scanning. Figure 7.3 shows these trapped clusters (the non-wetting phase is dense carbon dioxide at high pressures and temperatures) in Doddington sandstone. As mentioned above, the principal process governing the amount of trapping is snap-off.

Figures 7.5 and 7.6 illustrate the saturation distributions of carbon dioxide after primary drainage and brine injection in a carbonate — Ketton, whose pore space has been shown previously. Last, Fig. 7.7 shows images from different rock samples. In all cases approximately two-thirds of the saturation initially present in the pore space is trapped.

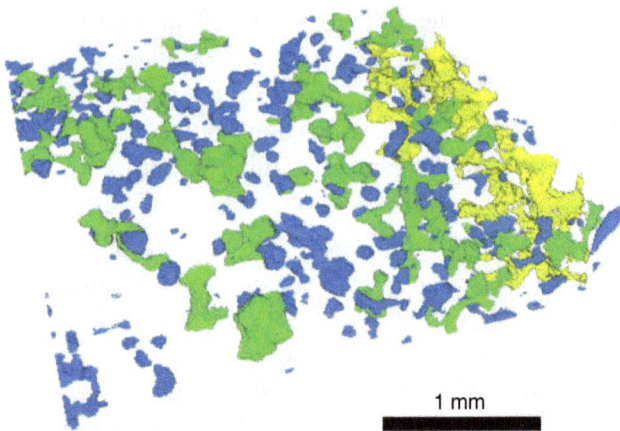

Figure 7.3. A micro-CT image of Doddington sandstone showing the residual CO_2. The colours indicate the size of trapped cluster. The image has a resolution of approximately $10\,\mu m$ and water and rock are not shown. The overall residual saturation is 25% (from Iglauer *et al.*, 2011).

These figures show that ganglia of many sizes are trapped, from clusters filling a single pore to large clusters that almost span the system. Indeed we see an approximately power-law distribution of cluster sizes consistent with percolation theory, which indicates that, as expected, the pore space is filled in order of size with the smaller regions of the pore space filled first by water, trapping the non-wetting phase in the bigger pores (Andrew *et al.*, 2013, 2014).

Figure 7.4 shows how the micro-CT images are processed to identify trapped non-wetting phase. The analysis consists of four steps: filtering of the raw image (top figures), cropping the image to the desired size, watershed seed generation to identify different phases — the seeds are placed on voxels that are clearly one phase or another (middle row) — and the application of the watershed algorithm (bottom).

7.3. Typical Capillary Pressure Curves and Secondary Drainage

The final sequence of saturation change is secondary drainage, where non-wetting fluid re-invades the porous medium after imbibition.

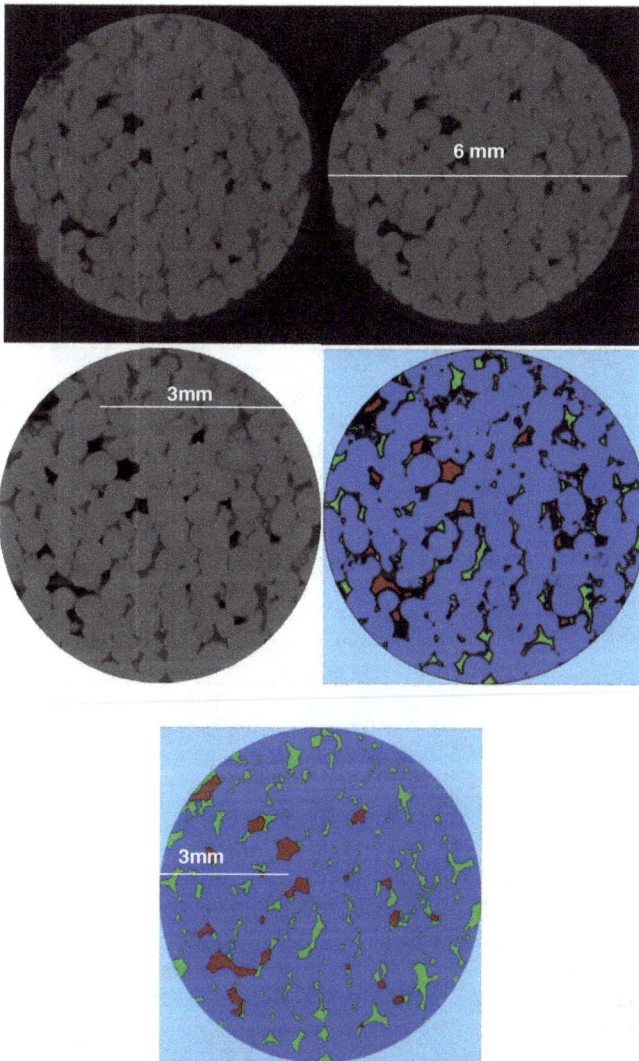

Figure 7.4. Processing micro-CT images to identify non-wetting phase. The raw image (top left) shows super-critical carbon dioxide, $scCO_2$ (the darkest phase), brine (the intermediate phase) and the rock grains (the lightest phase). The rock grains are around $700\,\mu m$ across. The processed image shows rock as dark blue, brine in green and carbon dioxide — the non-wetting phase — in red (from Andrew *et al.*, 2013). Here 2D cross-sections of 3D images are shown.

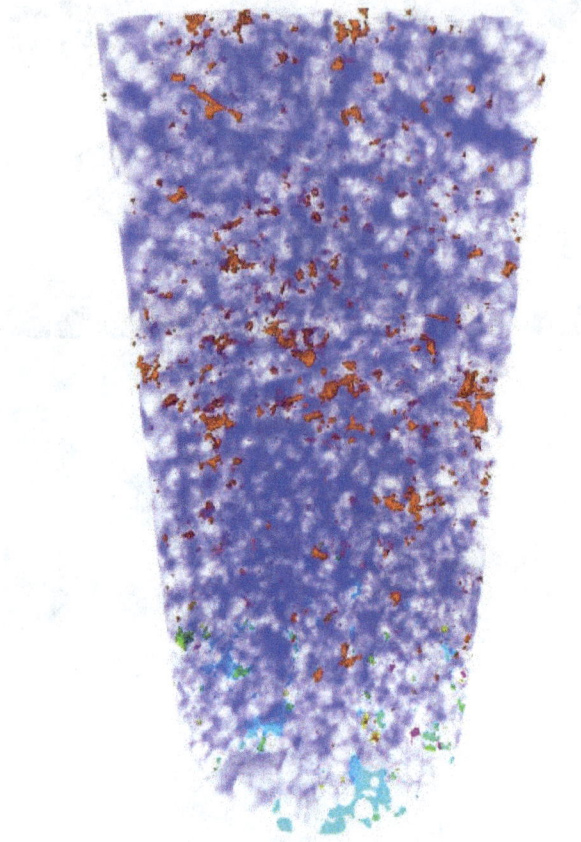

Figure 7.5. Visualisation of the fluids in the pore space of Ketton limestone at the end of primary drainage. The pale blue represents a connected cluster of non-wetting phase (carbon dioxide, CO_2, as a dense, supercritical phase, at high pressure and temperatures, typical of conditions in a deep storage aquifer). The other colours represent smaller disconnected clusters of the carbon dioxide (Andrew *et al.*, 2013).

Figure 7.8 shows typical capillary pressure curves for the flooding sequence — primary drainage, waterflooding (imbibition in this case) and secondary drainage — for a water-wet rock.

Why is the capillary pressure for secondary drainage lower than that for primary drainage? This has to do with the trapped non-wetting phase that becomes reconnected during secondary drainage. Rather than think of the curve as being lower, consider the secondary

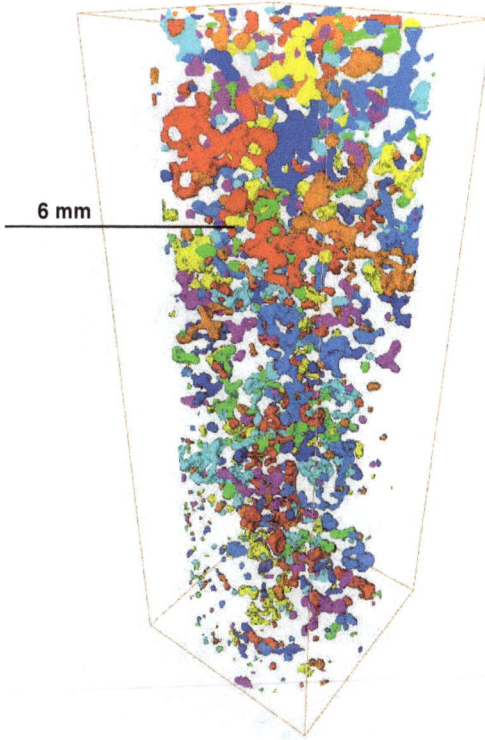

Figure 7.6. 3D rendering of CO_2 after brine injection. Each unique CO_2 ganglion is displayed as a different colour. Each ganglion is isolated, and so is trapped (Andrew *et al.*, 2013).

drainage shifted along the saturation axis, to represent the trapped saturation.

The key features to note in the capillary pressure are the irreducible wetting phase saturation, the residual non-wetting phase saturation, the shapes of the curves and their relative magnitude.

7.4. Different Displacement Paths and Trapping Curves

Figure 7.9 shows a schematic of different saturation paths, where the non-wetting phase is injected to an initial saturation and then wetting phase is injected. The amount of trapping depends on the

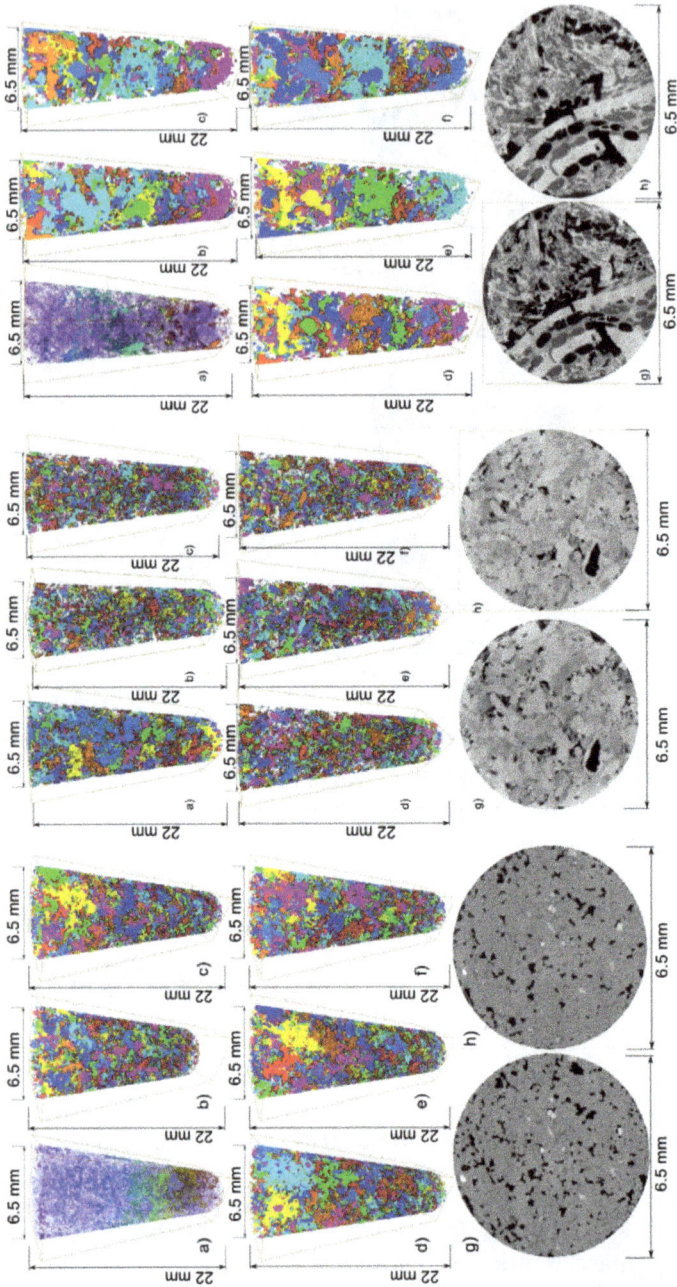

Figure 7.7. 3D rendering of CO$_2$ after brine injection. Each unique CO$_2$ ganglion is displayed as a different colour. Each ganglion is isolated, and so is trapped. (Left) Bentheimer sandstone; (middle) Estaillades limestone; (right) Mount Gambier limestone. The results from five experiments from each rock type are shown: the top left image shows the fluid distribution after primary drainage, while the other five are shown after waterflooding. The bottom row shows 2D slices of the raw images. From Andrew *et al.* (2014).

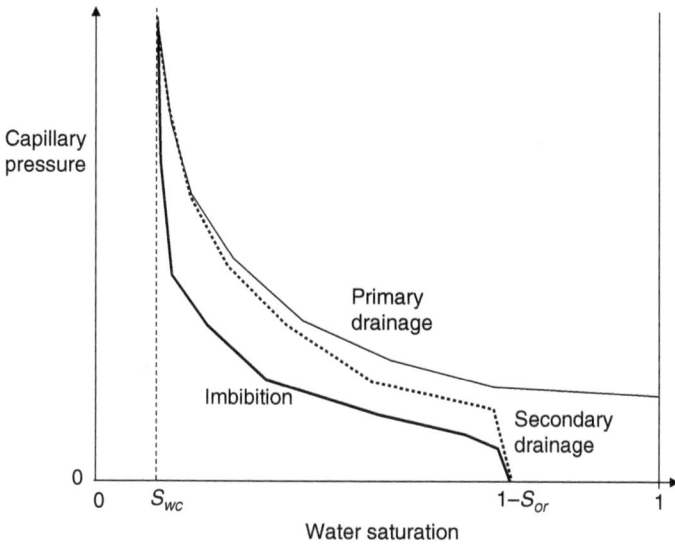

Figure 7.8. Illustrative capillary pressure curve showing primary drainage, imbibition and secondary drainage.

initial saturation; as the non-wetting phase invades progressively more of the pore space, there are more places where it can be trapped. This phenomenon is observed in the transition zone of oil fields (discussed later) and during CO_2 injection, where it is unlikely that the injected CO_2 will completely fill the pore space everywhere, as illustrated in Sec. 7.3.

The physical picture is as follows. During primary drainage the non-wetting phase fills progressively smaller portions of the pore space. If primary drainage stops at some intermediate saturation, then only the larger pores have been filled. During waterflooding (imbibition), trapping occurs preferentially in the larger pore spaces. Hence, the more of these pores that have been filled initially with non-wetting phase, the more that can be trapped. However, notice the characteristic curvature of the trapping curve (the relationship between initial and residual saturation shown in Fig. 7.10); at low initial saturations, only the very largest pores are invaded and these are trapped, so the curve has a slope of almost one (which represents everything being trapped), but the slope decreases with

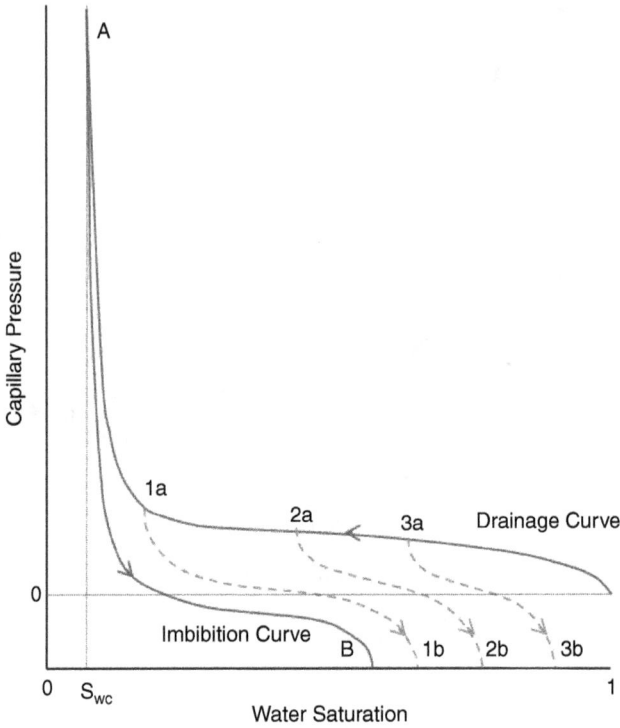

Figure 7.9. A schematic of primary drainage and imbibition capillary pressure curves where primary drainage ends at different initial saturations. This in turn determines the amount trapped during subsequent imbibition (waterflooding). From Pentland *et al.* (2010).

initial saturation. When the initial saturation is high, only small pores are being filled and these contribute very little to the overall amount of trapping.

Figure 7.11 shows an experimentally measured trapping curve for Berea sandstone and Indiana limestone from the Imperial College PhD thesis of Rehab El-Maghraby (see also El-Maghraby and Blunt, 2013). The upper set of points are for Berea where we observe lots of trapping. A different degree of trapping is observed when the non-wetting phase is $scCO_2$; here it is hypothesised that the contact angles change, larger values represent a weakly water-wet system that has, consequently, less snap-off.

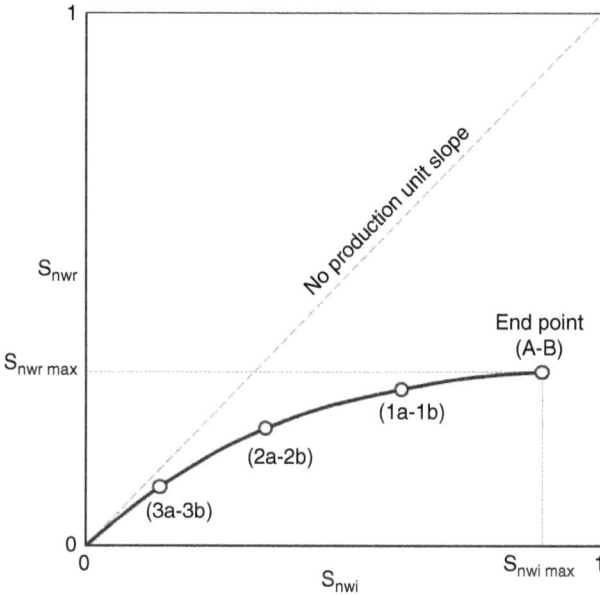

Figure 7.10. The trapping curve — the relationship between initial and residual non-wetting phase saturation — based on the capillary pressure curves shown previously. From Pentland *et al.* (2010).

Trapping curves are normally fit by empirical curves; these have no physical significance, but are a convenient way to make sense of the data and provide convenient input into numerical reservoir simulators. The most used model is due to Land (1968) and was originally developed for the trapping of gas. The residual saturation is written as

$$S_r^* = \frac{S_i^*}{1 + CS_i^*},$$

(7.3)

where C is a constant fit to the data and S^* is a normalized saturation, defined by

$$S^* = \frac{S}{1 - S_{wc}},$$

(7.4)

where S_{wc} is the connate or irreducible water saturation.

Another model by Spiteri *et al.* (2008) — used in the figures above to match the data — is to assume a quadratic (parabolic)

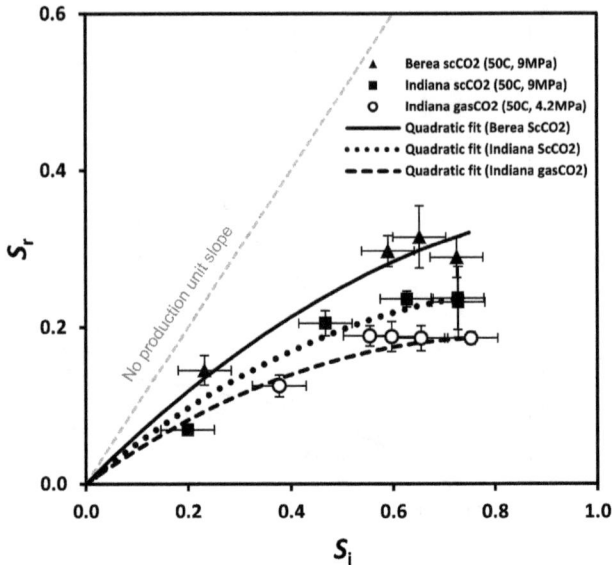

Figure 7.11. An experimentally measured trapping curve for Indiana limestone, compared to Berea, with supercritical (high-pressure) CO_2 as the non-wetting phase. Here — in contrast to Berea — more non-wetting phase is trapped for the CO_2 system (El-Maghraby and Blunt, 2013).

match as follows:

$$S_r = \alpha S_i - \beta S_i^2, \qquad (7.5)$$

where α and β are parameters chosen to reproduce the data.

As discussed above, in general, less trapping implies less snap-off and hence a less strongly water-wet system (larger contact angles). When the system becomes oil-wet or mixed-wet, we see a different behaviour which is discussed later.

Chapter 8

Leverett J-function

The Leverett J-function is a way of expressing the capillary pressure in dimensionless form, which takes account for different average pore size and interfacial tensions. This is a very useful scaling for dealing with laboratory measurements that may be performed with fluid pairs and at conditions — in terms of average porosity and permeability — different from in the field.

The capillary pressure is written as follows:

$$P_c(S_w) = \sqrt{\frac{\phi}{K}} \sigma \cos \theta J(S_w), \tag{8.1}$$

where J is the dimensionless J-function, which is a function of saturation (as is capillary pressure). The motivation behind this expression is the capillary pressure for a single tube, Eq. (4.5), where a typical pore radius is written as $\sqrt{(K/\phi)}$. This last expression can be derived from considering a bundle of capillary tubes of radius R a distance d apart.

The J-function only includes information about the geometry of the porous medium.

Sometimes the $\cos \theta$ term is ignored, and it is assumed that the system is strongly water-wet ($\cos \theta = 1$) for primary drainage. For imbibition, the $\cos \theta$ scaling is no longer appropriate, since other displacement mechanisms (snap-off and cooperative pore filling) control the behaviour and the contact angle term is neglected.

Figure 8.1 shows a measured mercury injection capillary pressure rescaled as the J-function for Berea sandstone. Mercury is always the

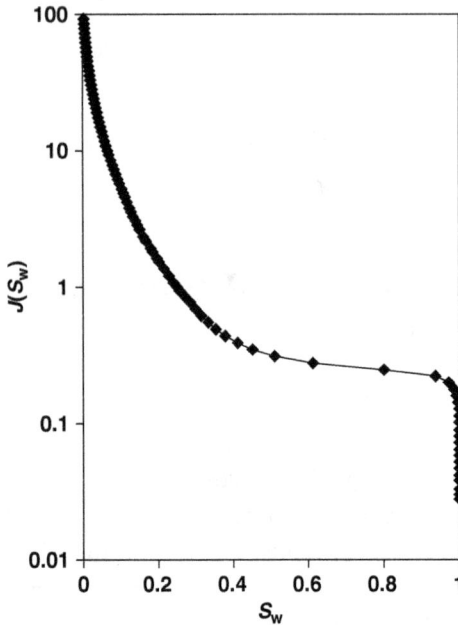

Figure 8.1. An example Leverett J-function measured by mercury injection during primary drainage on Berea sandstone. Note that the majority of the displacement occurs for $J < 1$. Also, since mercury displaces a vacuum there is no irreducible or connate wetting phase saturation in this experiment. This and the other figures in this section are taken from the Imperial College PhD thesis of Rehab El-Maghraby.

non-wetting phase, and so this represents a drainage displacement. Note that the minimum value of the J-function — the dimensionless entry pressure — is typically less than 1 (around 0.2–0.3) in the example above. By a J-function value of 1, most of the pore space has been accessed by the non-wetting phase. Much larger values are possible, as the non-wetting phase forces its way into narrow corners and cracks of the pore space as well as the smallest pores themselves. However, often these high values are simply experimental artefacts; a huge capillary pressure is imposed and insufficient time is given to achieve capillary equilibrium. In general, be cautious about using J-function values much above 1 in quantitative calculations, as they may not correspond to a true state of equilibrium.

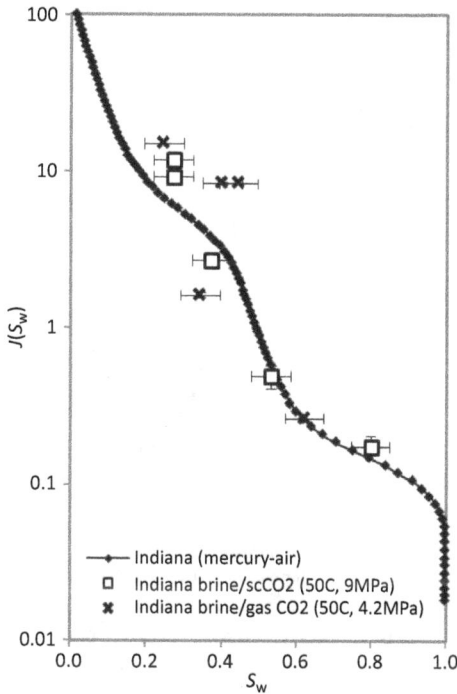

Figure 8.2. The Leverett J-function measured by mercury injection during primary drainage, and by displacement of brine by CO_2 on Indiana limestone. Notice that when different measurements with different fluids are represented as a J-function they lie on the same curve to within experimental error. Also note that the curve shows regions where the saturation changes rapidly with pressure, indicting at low pressure the intergranular porosity and, at high pressure, the micro-porosity within grains.

In carbonates with micro-porosity, however, there may be significant displacement for $J > 1$, as the non-wetting phase enters these small pores, as shown in Figs. 8.2 and 8.3.

8.1. Capillary Pressure and Pore Size Distribution

It is possible to relate the capillary pressure to the pore size distribution; this is routinely performed on the results of mercury injection tests. First, Eq. (4.5) is used to convert the capillary pressure into a throat radius, assuming piston-like displacement into

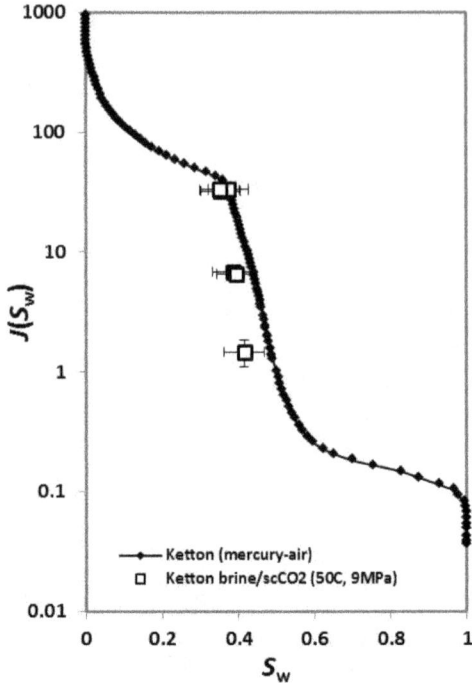

Figure 8.3. The Leverett J-function measured by mercury injection during primary drainage, and by displacement of brine by CO_2 on Ketton limestone. As in Indiana, there is a clear signature of intergranular (macro) porosity and intragranular (micro) porosity.

a circular tube. Hence, instead of P_c as a function of saturation (the wetting phase saturation, even though for mercury injection this is a vacuum), we define an effective radius as a function of saturation:

$$r(S) = \frac{2\sigma \cos \theta}{P_c(S)}. \tag{8.2}$$

Then the radius distribution is computed. Usually this is done by defining

$$G(r) = \frac{dS}{d(\ln r)} = r\frac{dS}{dP_c}\frac{dP_c}{dr} = -P_c\frac{dS}{dP_c} = -\frac{dS}{d\ln P_c}. \tag{8.3}$$

$G(r)$ is an indication of the number of throats of radius r. Logarithmic axes are used since there is typically a wide variation in

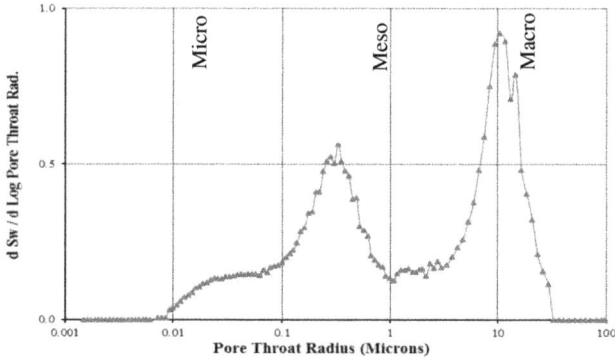

(a)

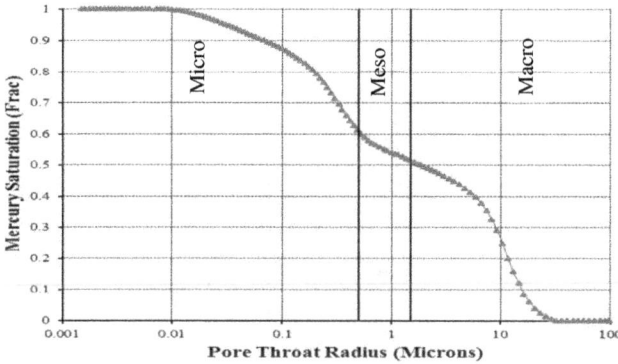

(b)

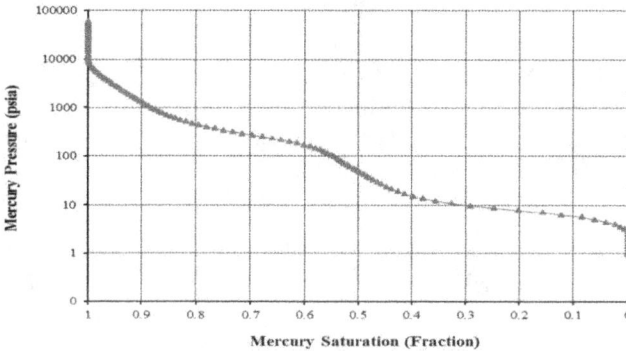

(c)

Figure 8.4. The mercury injection capillary pressure for Indiana limestone (c). This is converted into saturation as a fraction of effective throat radius (b) from which a throat size distribution is computed (a) using Eqs. (8.2) and (8.3). Here we see a very wide range of throat size with a clear indication of macro- and micro-porosity.

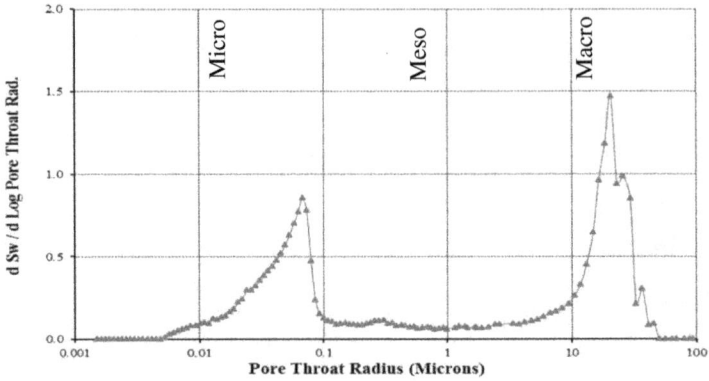

(a)

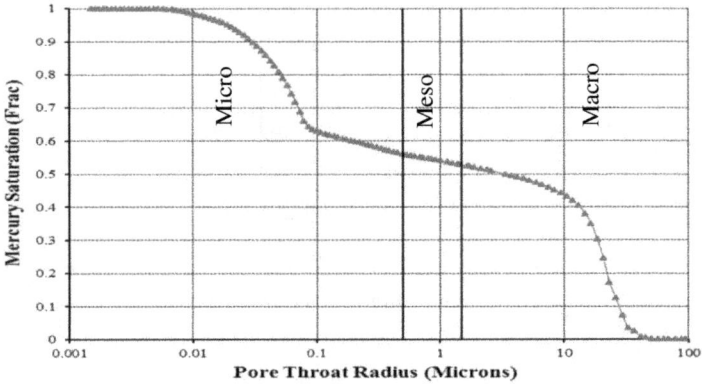

(b)

Figure 8.5. The relationship between apparent radius and saturation, together with the inferred throat size distribution for Ketton limestone. Here there is a clear distinction between the large, intergranular, pores and the much smaller intragranular micro-porosity.

capillary pressure and hence effective pore size. It is not strictly the throat size distribution, since the displacement process during primary drainage — technically similar to invasion percolation — allows the filling of regions with wide pores and throats, which are only accessed through a smaller throat at a high pressure.

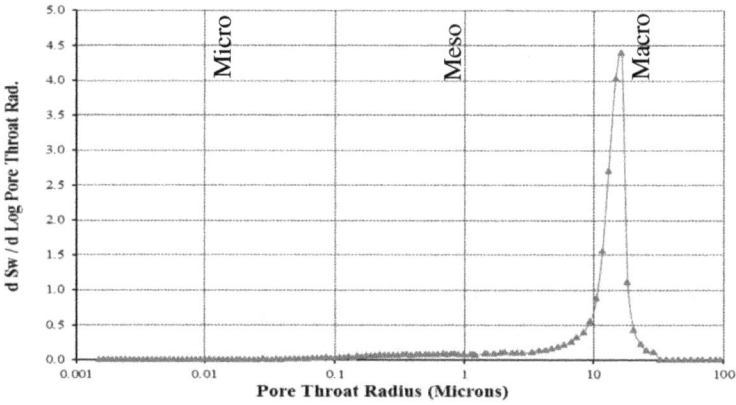

(a)

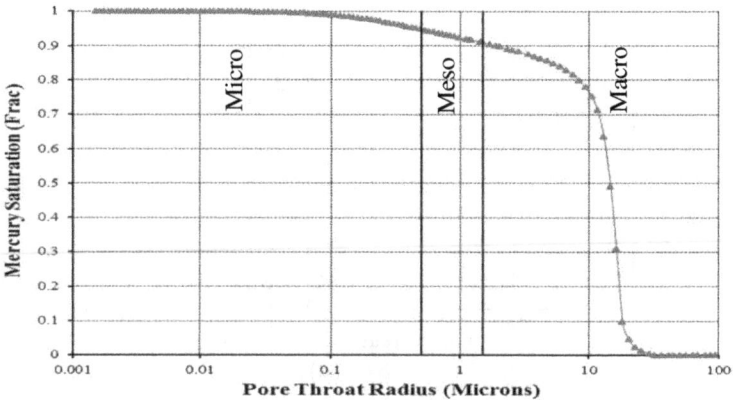

(b)

Figure 8.6. The throat size distribution for Doddington sandstone. Here there is a relatively narrow distribution of large pore spaces — hence there is no micro-porosity.

However, it does give some indication of the range of pore sizes in the material.

Figures 8.4–8.7 show some example distributions on our example rock types; note that for the carbonates we see a bimodal distribution showing macro-pores and micro-porosity.

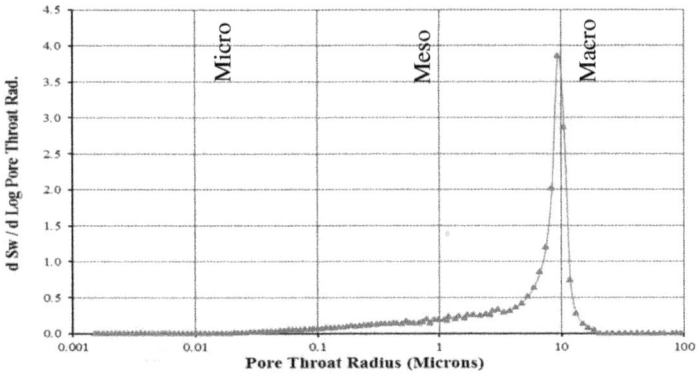

(a)

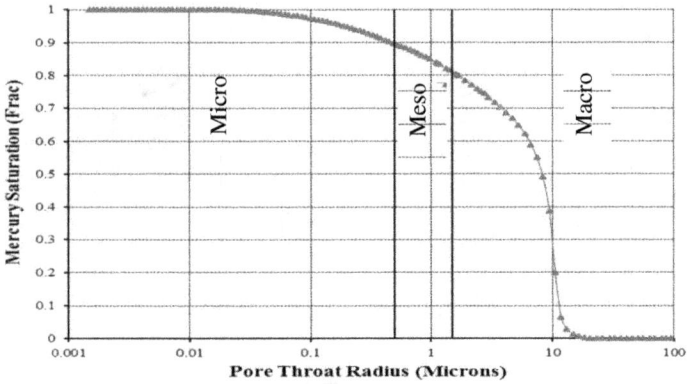

(b)

Figure 8.7. The throat size distribution for Berea sandstone. Here again there is a relatively narrow distribution of large pore spaces, but they are smaller on average, and have a larger range of size than for Doddington.

Chapter 9

Displacement Processes
in Mixed-wet Media

As we discussed before, most reservoir rocks contain both oil-wet and water-wet regions. This means that during water invasion a negative capillary pressure (a water pressure higher than the oil pressure) needs to be applied to force oil out of the oil-wet regions.

Figure 9.1 shows some classic experiments (Killins *et al.*, 1953) illustrating the effect of wettability. The curves are good for illustrative purposes but do not show the correct residual for the oil-wet case, since this is difficult to measure with any accuracy experimentally. I will provide a physical explanation in this section. The emphasis is how to relate the macroscopic properties — in this case capillary pressure — to the pore-scale physics and the configuration of fluids at the micron scale.

9.1. Oil Layers

The degree of trapping is dependent on the presence and connectivity of oil layers sandwiched between water in the corners and water in the centres of oil-wet pores. This is illustrated in Fig. 9.2. If the surface is oil-wet, then water becomes the non-wetting phase. This means that it preferentially fills the centres of the largest pores. However, water is still retained in the corners after primary drainage. Hence, between water in the corners and water in the centre, there is an oil layer.

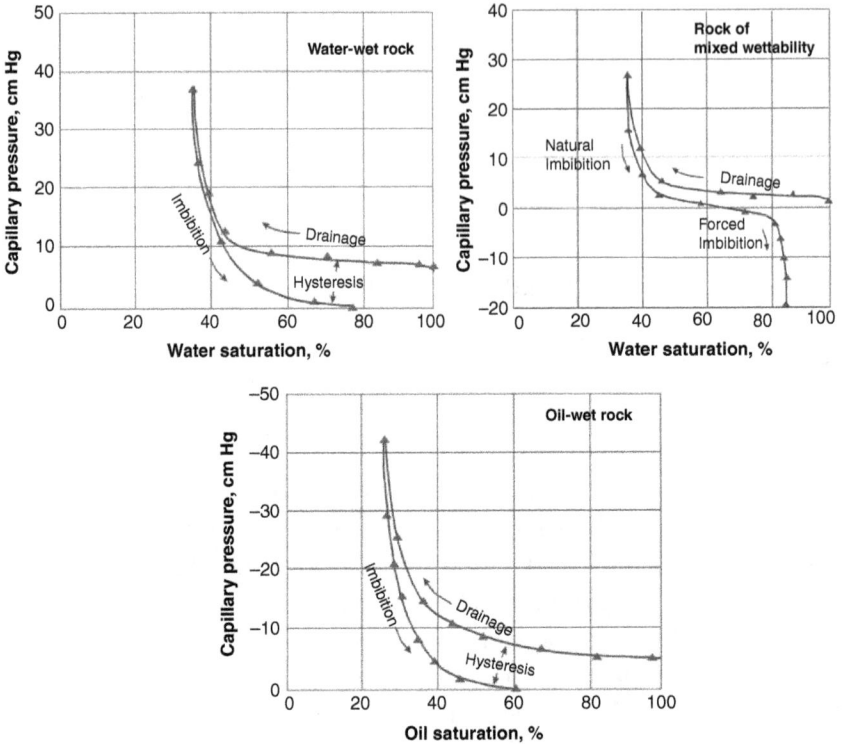

Figure 9.1. Capillary pressure (P_c) curves, for water-wet, mixed-wet and oil-wet rock. Note that the capillary pressure becomes negative if the sample is not water-wet (Killins *et al.*, 1953). The term "imbibition" is here incorrectly used to describe forced displacement at a negative capillary pressure — we confine imbibition to refer only to a spontaneous process occurring at a positive capillary pressure.

These oil layers maintain the connectivity of the oil phase down to low saturation. As the water pressure increases, the oil layers become increasingly thin and will eventually become unstable, allowing trapping. However, the flow rate through these layers is very low and so — experimentally — you have to wait a very long time to see the oil drain down to its true residual saturation. This situation is similar to the irreducible water saturation in primary drainage; again we have layer flow and waiting longer can allow lower water saturations to be achieved.

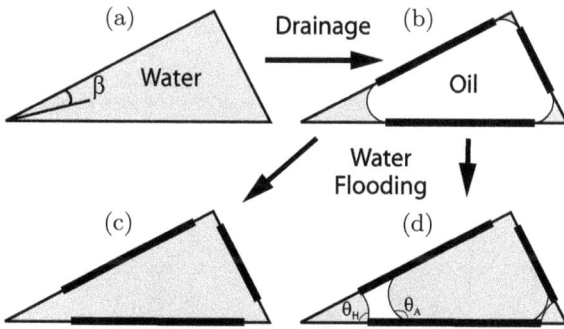

Figure 9.2. Possible configurations of oil and water in the pore space of a single pore or throat. Initially the porous medium is completely saturated with water. After drainage, the non-wetting phase (oil) resides in the centres of the pore space, with water confined to the corner. The regions of the pore space directly contacted by oil may change their wettability. When water is injected, the water can fill the entire pore space, or — if the altered wettability surface is oil-wet — a layer of oil can form sandwiched between water in the corner and water in the centre. These oil layers allow the oil to remain connected and drain to very low saturation, albeit very slowly. From Valvatne and Blunt (2004).

9.2. Effect of Wettability on Capillary Pressure

9.2.1. *Weakly Water-wet Media*

Figure 9.3 shows a weakly water-wet system. Here some water is displaced during forced water injection; however, the medium does not spontaneously imbibe any oil, indicating that there are no connected oil-wet pathways through the system.

The region of the water invasion curve where the capillary pressure is negative is called forced water injection. Where the capillary pressure is positive we have, as before, spontaneous imbibition or spontaneous water injection. To avoid confusion, imbibition and drainage are used only when the corresponding capillary pressure is positive.

The capillary pressure becomes negative, even though no part of the pore space has an intrinsic contact angle greater than 90°; instead, thanks to contact angle hysteresis, or the exact meniscus configuration during pore filling, some of the displacement pressures are negative, meaning that a higher pressure in the water than the non-wetting phase (oil) is required for invasion.

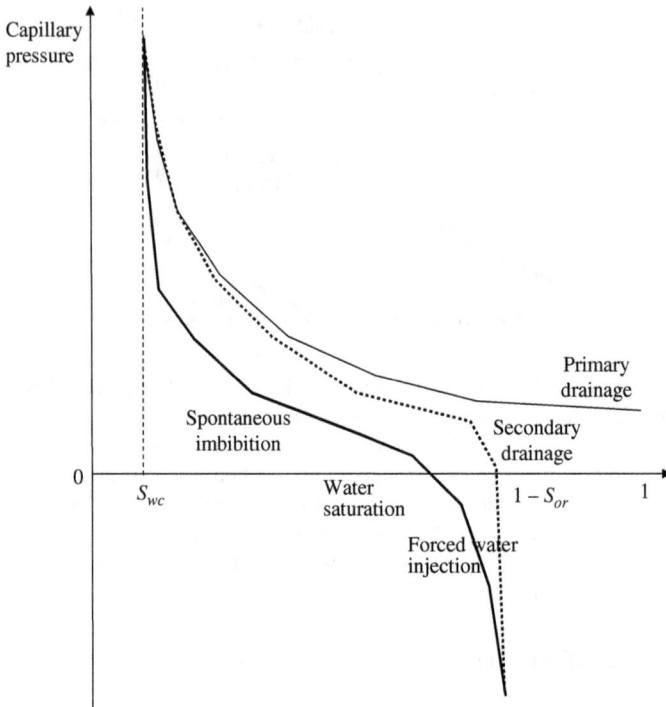

Figure 9.3. Typical capillary pressures for a weakly water-wet system. Note that there are some parts of the pore space that require forced water injection to access; it is unusual in natural samples to have a capillary pressure curve during waterflooding that is entirely positive. Only the waterflooding and secondary drainage curves are shown; generally primary drainage shows only water-wet characteristics (either mercury injection is used, or the displacement occurs before oil has altered the wettability of the system).

9.2.2.   *Capillary Pressures for Mixed-wet Media*

A mixed-wet porous medium (Fig. 9.4) spontaneously displaces both oil and water — there are continuous pathways of both water-wet and oil-wet patches in the pore space. The oil imbibes in the same way as water in a water-wet system during secondary invasion of oil, with snap-off of water accommodated through oil layer flow. This can lead to considerable trapping of water as a non-wetting phase. In contrast, there is less trapping of oil, again due to the connectivity of oil layers.

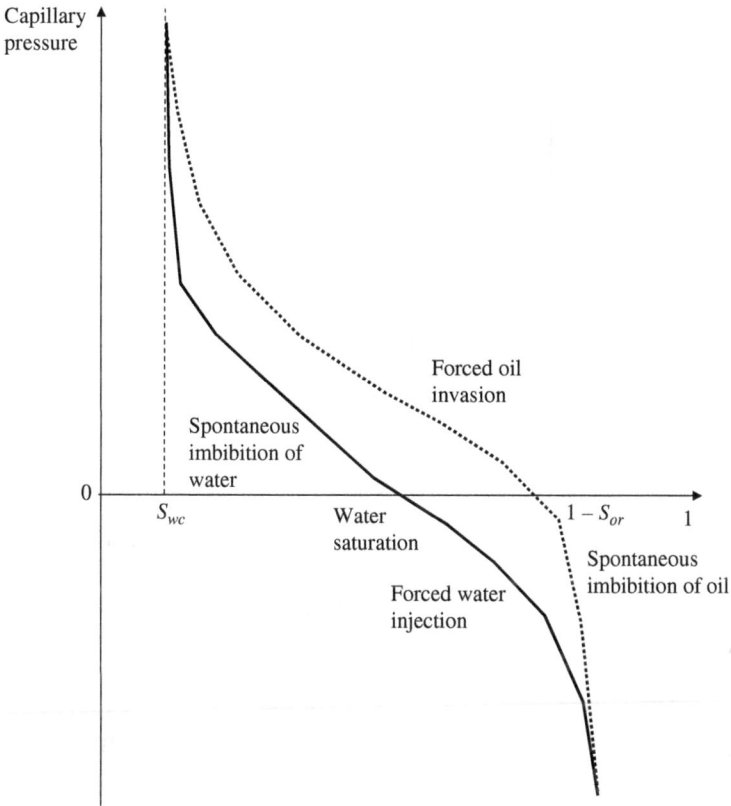

Figure 9.4. Typical capillary pressures for a mixed-wet system. Here there are connected regions of both water-wet and oil-wet pores and we see imbibition (spontaneous uptake) of both oil and water.

When the capillary pressure is negative during waterflooding, the largest oil-wet pores are filled preferentially in what is technically a drainage process, followed by progressively smaller regions. The capillary pressure curve tends to be relatively flat as it crosses zero (Fig. 9.5), since we transition from the filling of large water-wet pores (for spontaneous imbibition) to large oil-wet pores (for forced displacement); there is a large change in saturation associated with the filling of these pores, with relatively little change in capillary pressure.

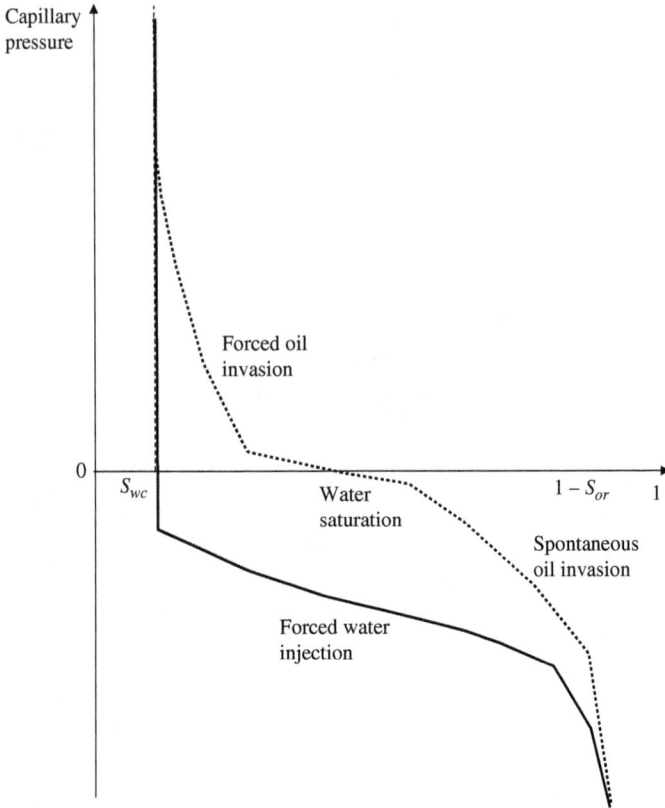

Figure 9.5. Typical capillary pressures for an oil-wet system. We see no water imbibition, but a significant amount of spontaneous displacement (imbibition) of oil.

9.2.3. *Oil-wet Systems*

Figure 9.5 shows a strongly oil-wet system that has no spontaneous imbibition of water. The residual oil saturation is lower than that of the water-wet medium; this is because the oil maintains connectivity down to low saturation thanks to the presence of oil layers. There is significant imbibition of oil during secondary oil invasion.

9.3. Trapping Curves in Mixed-wet Systems

In a mixed-wet system the relationship between initial and residual saturation becomes more complex than we showed before for water-wet media. First, the remaining oil saturation is controlled by the extent to which oil is allowed to drain from the system through layers. Hence, as more water is injected, more oil is produced and the remaining oil saturation decreases. As mentioned before, it is very difficult to obtain a true residual, or minimum, saturation experimentally, as the oil continues to flow, very very slowly, until low saturations are reached.

Second, the relationship between initial and remaining oil saturation is non-monotonic. For low initial saturations, the remaining saturation increases with initial saturation for the obvious reason that the more oil initially present, the more oil that can be trapped. However, at higher initial saturations, the residual decreases. This surprising phenomenon is due to the presence and stability of oil layers. At higher initial oil saturation (and hence imposed capillary pressure in primary drainage), the water is pushed farther into the corners of the pore space. This makes oil layers, formed during subsequent waterflooding, thicker and more stable, meaning a more negative capillary pressure is required (higher water pressure) to collapse them. This extends the layer drainage regime, allowing more oil to be recovered. At the very highest initial saturations, the remaining oil saturation may increase again, as there is more oil present, and there is a subtle competition between displacement and connectivity in the oil-wet and water-wet regions of the pore space, and the layers do, eventually, lose connectivity.

This curious behaviour has been seen experimentally, and can be predicted using pore-scale modelling. Some experimental data is shown in Fig. 9.6.

9.4. Transition Zones

During primary oil migration (primary drainage) into a hydrocarbon reservoir, both the saturation and capillary pressure vary with height above the free water level — defined as the depth when the oil and water pressures are the same — so the capillary pressure is zero. Above the free water level, the capillary pressure increases with depth, as analysed earlier for a meniscus in a capillary tube, Eq. (4.9). For a reservoir with variable permeability and porosity,

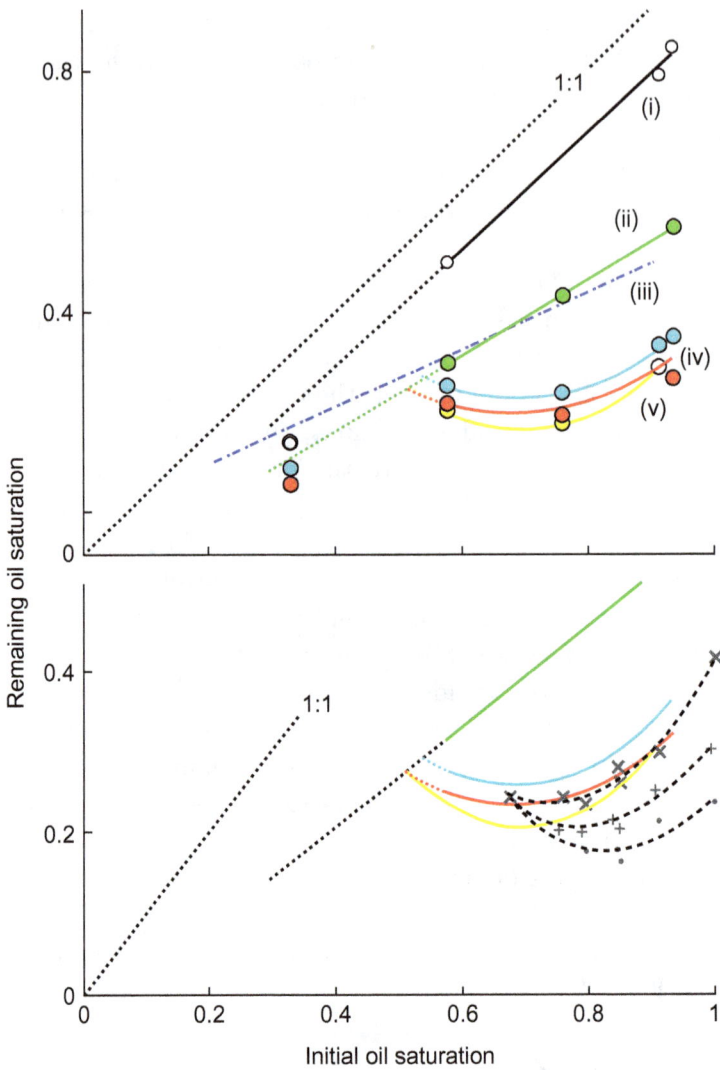

Figure 9.6. Experimental data showing the variation of remaining oil saturation as a function of initial oil saturation in mixed-wet Indiana carbonate. The top graph (a) is data from Imperial College (Tanino and Blunt, 2013) while the bottom (b) is a replotting of data from Salathiel (1973). The different curves represent different amounts of waterflooding from 1 pore volume to over 200 pore volumes injected.

we invoke J-function scaling to determine the water saturation at a given height h above the free water level. It is the saturation such that

$$\Delta \rho g h = P_c(S_w) = \sqrt{\frac{\phi}{K}} \sigma \cos \theta J(S_w), \qquad (9.1)$$

where $\Delta \rho$ is the density difference between oil and water. This can be rewritten

$$S_w(h) = J^{-1} \left(\frac{\Delta \rho g h}{\sigma \cos \theta} \sqrt{\frac{K}{\phi}} \right). \qquad (9.2)$$

This initial distribution of saturation — normally established from a primary drainage capillary pressure curve — is shown in Fig. 9.7 for a homogeneous medium. The initial saturation, in turn, affects the wettability for subsequent waterflooding once the oil has aged the surface; higher capillary pressure (low initial saturation) forces the oil into contact with more of the solid, favouring oil-wet conditions. Intermediate saturations may lead to a mixed-wet rock, while a high initial water saturation means that there is little initial

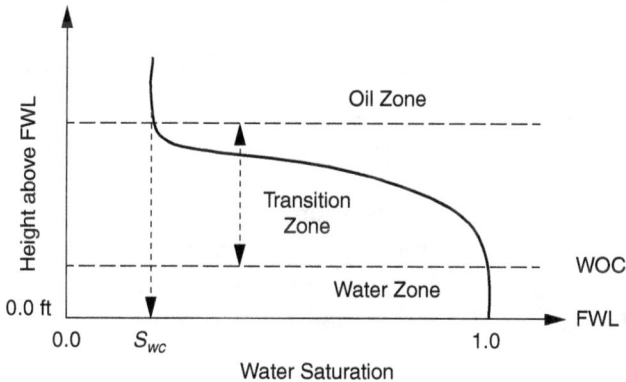

Figure 9.7. A schematic of the transition zone, where the saturation varies as a function of height above the free water level (FWL), defined as where the capillary pressure is zero. The water–oil contact (WOC, or the oil–water contact, OWC) is found from down-hole log measurements of resistivity where there is first a noticeable presence of water (low resistivity) in the formation. This distribution of saturation also affects the wettability during waterflooding.

contact between the oil and solid and the medium remains largely water-wet. Later we will explore the implications for oil recovery by waterflooding, once we have discussed how multiple fluids flow in the pore space. For now it is sufficient to note that many oil reservoirs experience a transition from water-wet to oil-wet conditions with height above the oil-water contact; in low permeability rocks this transition zone may extend several tens of metres as evident when representative numbers are put into Eq. (9.1).

Imagine, for instance, a permeability of around $10\,\mathrm{mD}$ $(10^{-14}\,\mathrm{m}^2)$, a porosity of 0.16, a density difference of $200\,\mathrm{kg\cdot m^{-3}}$ and $\sigma \cos \theta = 0.025\,\mathrm{N\cdot m^{-1}}$. Then, the water saturation will decline to close to its irreducible (connate) value for $J = 1$. This then gives a height of around $50\,\mathrm{m}$ — essentially there is a variation of saturation with height across the entire reservoir and it is wrong to assume that the initial saturation is the irreducible value.

9.5.   Amott Wettability Indices

The *Amott wettability index* is a quantitative and useful measure of wettability measured on core samples and first described by Amott (for a more modern description, see Anderson, 1986, 1987). Remember that it is difficult to measure contact angles *in situ* (although see Andrew *et al.*, 2014b), and in any event, this does not directly relate to capillary pressure, so a macroscopic description of wettability is valuable. Start with a core at waterflood residual oil.

1. Perform spontaneous oil invasion ($P_c < 0$).
2. Then perform forced injection of oil ($P_c > 0$).
3. Then spontaneous imbibition of water ($P_c > 0$).
4. Then last forced water injection ($P_c < 0$).

The wettability indices for oil and water are defined by:

A_o = increase in oil saturation in step 1/total increase from steps 1 and 2.
A_w = increase in water saturation in step 3/total increase from steps 3 and 4.

Clean sand and rock will have $A_o = 0$ and $A_w = 1$. We rarely see $A_o = 1$. Sometimes we have rocks where one value is low (around 0.1) and the other is 0 — little or no spontaneous displacement, implying contact angles everywhere close to 90°. However, the most common situation, for a reservoir sample, is mixed-wettability where both oil and water spontaneously imbibe and neither index is zero.

Many researchers, rather depressingly, are unable to cope with two numbers to represent wettability and so use instead the (completely useless!) Amott–Harvey index, $A_w - A_o$. Please never do this — the concept was introduced when it was considered that there was one uniform contact angle in the rock, and so, by definition, one of the Amott indices had to be zero. However, this is extremely unhelpful for mixed-wet systems; there is a significant difference between a mixed-wet rock with an Amott–Harvey index of 0 (meaning that around half the pores are oil-wet and the other half water-wet) and an intermediate-wet rock with uniform contact angles close to 90° that also has an Amott–Harvey index of 0. Be sure to distinguish between these two cases.

9.6. Example Exercises

1. Return to our discussion of the oil/water transition region in oil reservoirs. In many reservoirs — such as Prudhoe Bay off the North Slope of Alaska — the reservoir is weakly water-wet near the oil/water contact and becomes more oil-wet with height away from the contact. Explain in your own words why we see this wettability trend.

2. Consider equilibrium at a line of contact for the following three situations: oil/water, gas/oil and gas/water. Derive a relationship between the oil/water, gas/oil and gas/water contact angles and interfacial tensions. This will be derived later when we consider three-phase flow.

3. Consider a dome-shaped gas reservoir. The gas is trapped by a layer of shale that has permeability 0.1 mD and porosity 0.2. The gas is in pressure communication with the surrounding aquifer. Estimate the height of gas in the reservoir. The gas

density is $200\,\mathrm{kg/m^3}$ and the water density is $1000\,\mathrm{kg/m^3}$. The gas/water interfacial tension is $60\,\mathrm{mN/m}$. (Hint: Draw a cartoon of the reservoir. Gas can accumulate until there is a sufficient capillary pressure for gas to enter the shale layer. Use the Leverett J-function and a dimensionless entry pressure of 0.3. The capillary pressure is the pressure difference due to density differences.)

4. Explain why a water-wet porous medium has a higher residual oil saturation than an oil-wet medium. Explain why a porous medium with oil/water contact angles close to $90°$ gives a lower residual oil saturation than a porous medium with a contact angle close to 0. Two porous media have an Amott–Harvey index of 0, indicating that they are neither oil-wet nor water-wet. One medium has $A_w = A_o = 0$, while the other has $A_w = A_o = 0.4$. Which system do you expect to have the lower residual oil saturation? Explain your answer carefully.

Chapter 10

Fluid Flow and Darcy's Law

We now introduce flow into our analysis — in particular Darcy's law for flow in a porous medium — and provide a pore-scale basis for the relationship between flow rate and pressure gradient. We start with a consideration of single-phase flow (one fluid — by default water — saturates the pore space completely, and we consider the water movement and that of solutes dissolved in the water) before extending the analysis to multiphase flow (where oil, water and gas may all be present).

10.1. Stokes Flow

There are two fundamental concepts that we will use to describe fluid flow. These are not specifical to flow in a porous medium, but rather apply to the movement of any fluid. The first concept is conservation of mass, written as

$$\nabla \cdot \rho \boldsymbol{v} = 0. \tag{10.1}$$

I will not derive this equation here — which is simple to do considering flow into an arbitrary volume and applying Green's theorem — but I will return to similar derivations for multiphase flow later.

The second concept is the conservation of momentum for a fluid. This is the Navier–Stokes equation that is used to describe the flow of everything from volcanic magma to air and oceans. It can be written as

$$\rho\left(\frac{\partial \boldsymbol{v}}{\partial t} + \boldsymbol{v} \cdot \nabla \boldsymbol{v}\right) = -\nabla P + \mu \nabla^2 \boldsymbol{v}, \qquad (10.2)$$

where ρ is density, $\boldsymbol{v} = (u, v, w)$ is the vector of velocity, P is pressure, and μ is the viscosity.

In our case we can make a number of simplifications. If we consider relatively incompressible fluids, such as oil or water, or, for gas, flow in a small domain (the pore scale) where changes in pressure — and hence density — are small compared to the overall pressure, then the density is constant, or is almost constant to a very good approximation. Instead of Eq. (10.1) we can write

$$\nabla \cdot \boldsymbol{v} = 0, \qquad (10.3)$$

which is an expression of conservation of volume and can be derived directly (assuming incompressible fluid), in much the same way as the expression for conservation of mass.

We also consider flows where the flow field changes slowly over time, and so we can neglect any explicit time dependence in the Navier–Stokes equation. Furthermore, flow is very slow, compared to, say, air flows, and in this limit the term with the velocity multiplied by itself can be ignored (physically it means that viscous forces dominate over inertial forces — this is discussed further below — and means that we can dismiss some of the more complex effects, namely turbulence, which occur in other, more unconfined flows). Then we are left with the steady-state Stokes equation:

$$\mu \nabla^2 \boldsymbol{v} = \nabla P. \qquad (10.4)$$

With modern linear solvers and fast computers, it is possible to solve these last two equations numerically on the pore-space images we have introduced previously. We are now able to use standard desktop computers to solve billion-cell problems.[1]

---

[1]We use the OpenFoam library to solve the Navier–Stokes equation; there are a number of excellent public domain solvers, readily downloaded from the internet, that can be employed.

10.2. Reynolds Number and Flow Fields

The concept of slow flow can be quantified through the introduction of the Reynolds number, R_e, which is the ratio of inertial to viscous forces for fluid flow:

$$R_e = \frac{\rho v L}{\mu}, \qquad (10.5)$$

where ρ (as before) is the fluid density, μ its viscosity, v a characteristic flow speed and L a characteristic length. In a porous medium the fluid moves slowly through the labyrinthine interstices between rock grains, or through a complex fracture network. The characteristic length is that of a pore throat, or a narrow restriction between larger pore spaces, that impedes the flow. L is typically around $10\,\mu\text{m}$–$100\,\mu\text{m}$ ($10^{-5}\,\text{m}$–$10^{-4}\,\text{m}$) for consolidated rock and of order $10^{-3}\,\text{m}$ for sand and gravel. The flow speed typically $10\,\text{m/day}$ (roughly $10^{-4}\,\text{m·s}^{-1}$) or less for groundwater movement (natural flow speeds in arid areas can be 1 m per year or smaller) and is one or two orders of magnitude lower in oil reservoirs. For water $\rho = 10^3\,\text{kg·m}^{-3}$ and $\mu = 10^{-3}\,\text{kg·m}^{-1}\text{s}^{-1}$, giving $R_e \approx 10^{-1}$–10^{-3}. Viscous forces dominate and we have laminar flow in porous media (turbulent flow generally occurs for $R_e \approx 1000$).

It is possible to average the Navier–Stokes equation, which describes flow of a single fluid, for slow, laminar flow past many obstacles. The flow is averaged over a representative volume element containing several rock or sand grains. Since the Reynolds number is low, viscous forces predominate. A pressure gradient, or force, is required for flow at a constant velocity. It is possible to derive a linear relation between volumetric flow rate and pressure gradient, known as Darcy's law (Bear, 1972), which we will consider later.

First though, I will show some illustrative examples of the pore space, pressure distribution and flow fields (Figs. 10.1–10.5). The flow is relatively uniform in the more homogeneous systems, such as a bead pack, but is confined to a few tortuous channels in the more heterogeneous media, as seen in many of the carbonates we study.

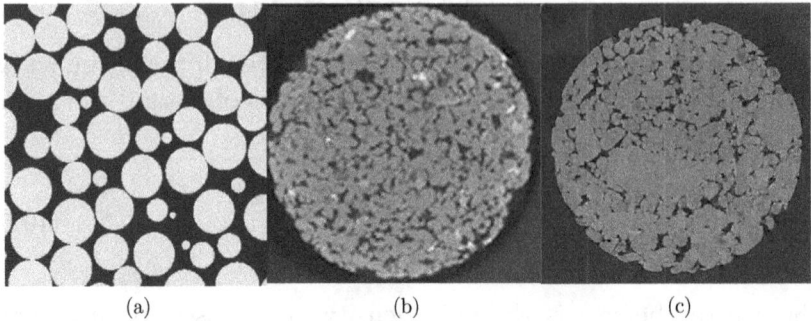

(a)   (b)   (c)

Figure 10.1. 2D cross-sections of 3D images of (a) a bead pack, (b) Bentheimer sandstone, and (c) Portland limestone. The diameter of the samples is around 5 mm and the images have a voxel size of between 5 μm and 9 μm.

10.3. Averaged Behaviour and Darcy's Law

The local velocity in the pore space is highly variable; indeed, we typically see eight-orders-of-magnitude variation in flow speed in the samples shown above. However, there is a way to simplify this if we are only interested in the average flow.

We find, empirically, that there is a linear relationship between flow rate and pressure gradient; this can be derived rigorously mathematically as well. This relationship is Darcy's law:

$$q = -\frac{K}{\mu} \left(\nabla P - \rho \boldsymbol{g} \right). \tag{10.6}$$

q is not a local flow velocity, even though it has the units of a speed; it is the volume of fluid flowing per unit area (and this area includes both solid and pore) per unit time. It is, in essence, the sum of all the highly variable local speeds in the overall direction of flow multiplied by the porosity. The minus sign indicates the physically obvious fact that flow goes from high to low pressure — i.e. along a negative gradient in pressure. Notice that in Eq. (10.6) I have also included the effect of gravity: g is the vector of gravitational acceleration.[2]

---

[2]Darcy's law is named after Henry Darcy, a French Civil Engineer, who, in 1856, published a now-famous book "Les fontaines publiques de la ville de Dijon." The book is available in a 2004 English translation. The main body of the work

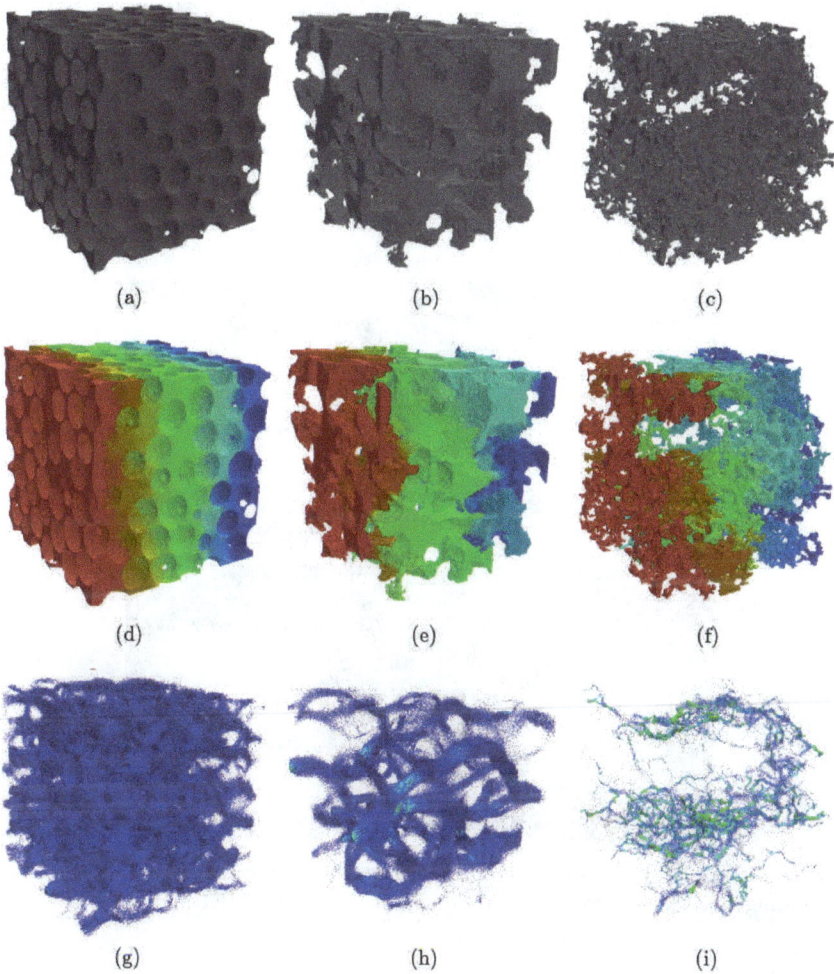

Figure 10.2. The pore space of the three porous media shown in Fig. 10.1: (a) bead pack; (b) Bentheimer; (c) Portland. Then the pressure field for flow from left to right is shown, with red representing high values and blue low values; flow goes from high to low pressure. The final row illustrates the flow field, with the regions of highest flow indicated. While flow is relatively uniform through the pore space of a bead pack, in the carbonate it is confined to a few tortuous channels (from Bijeljic *et al.*, 2011).

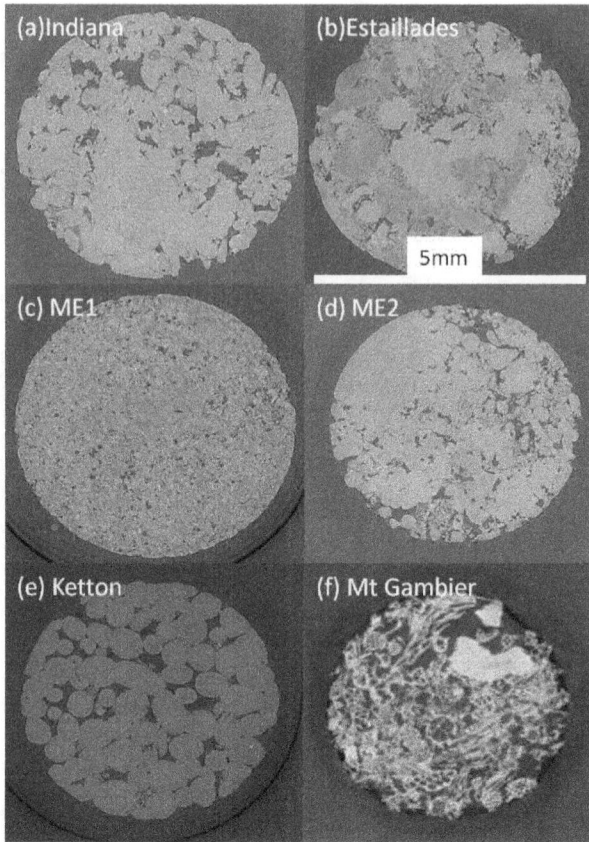

Figure 10.3. 2D cross-sections of 3D images of various carbonate samples, including two cases from a deep Middle Eastern aquifer ME1 and ME2.

concerns the design of a network of pipes to bring spring water into the city. As an apparent aside, and written in an appendix, is the first statement of this famous law, based on a series of flow experiments in sand filters. Flowing water through sand is an extremely effective way to remove bacteria (and even viruses) from the water; the larger organisms are simply trapped in the pore space as they (or clumped groups of them) cannot pass through the narrower pores, or are absorbed on the huge surface area a porous medium presents. This explains how, for instance, mountain streams run clear drinkable water, while the nearby fields are full of cow pats and sheep droppings. And last, why is he Henry and not Henri if he was French? Well, he had an English wife, so anglicised his first name!

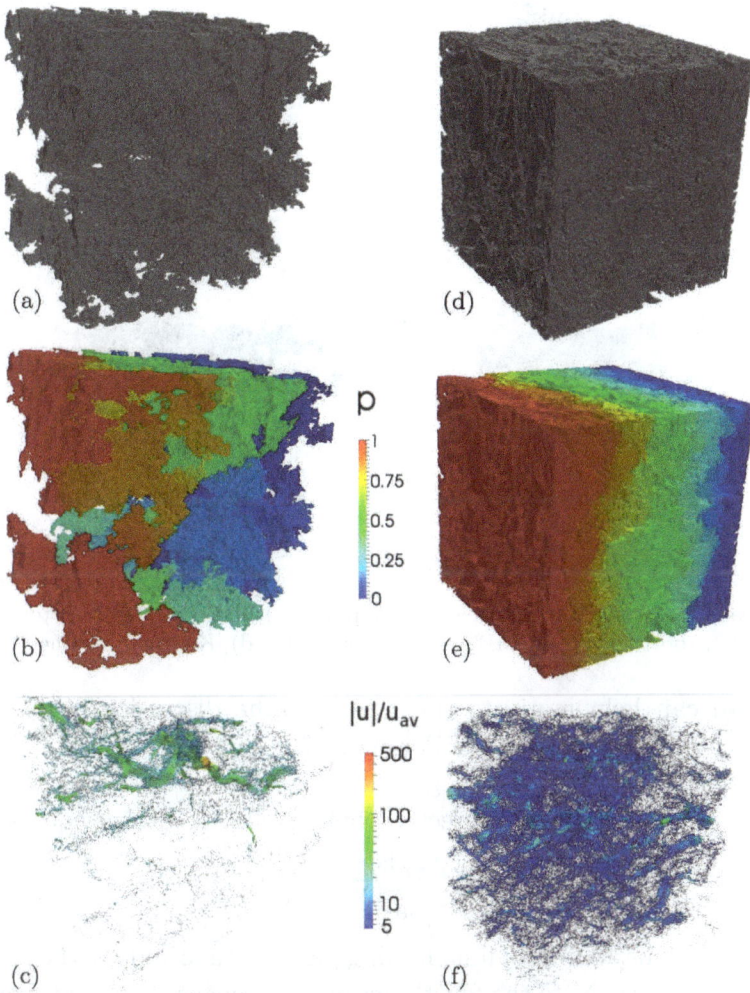

Figure 10.4. The pore space of Estaillades (left column, (a)–(c)) and Mount Gambier (right, (d)–(f)), with the corresponding computed pressure and flow fields. Estaillades has very tortuous flow paths indicating a poorly-connected pore space, while Mount Gambier, despite its geologically complex structure, is well-connected and supports a relatively uniform flow (Bijeljic *et al.*, 2013b).

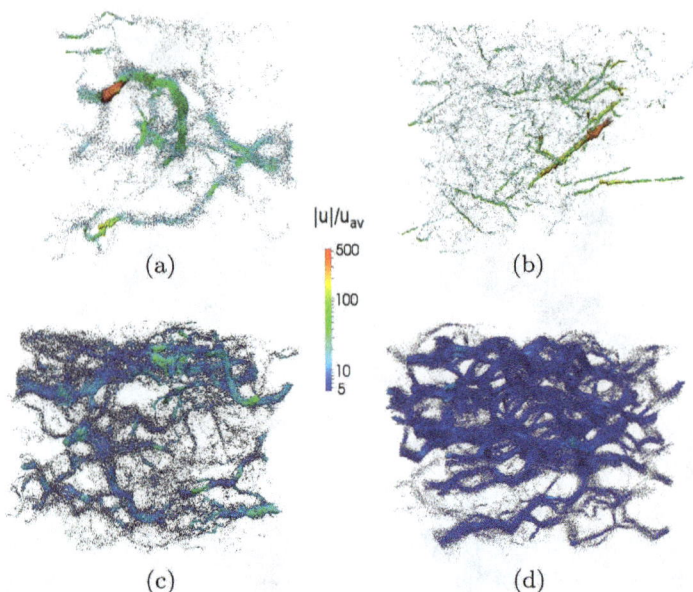

Figure 10.5. Normalised flow fields for: top left (a), Indiana limestone; top right (b), ME1; bottom left (c), ME2; and bottom right (d), Ketton limestone.

We can help motivate this, and aid the discussion of relative permeability, by quoting the Poiseuille law; this is an expression that relates flow rate to pressure gradient in a single circular cylindrical tube and can be derived directly from the Navier–Stokes equation:

$$Q = -\frac{\pi r^4}{8\mu}(\nabla P - \rho \boldsymbol{g}), \qquad (10.7)$$

where Q is the volume of fluid flowing per unit time and r is the radius of the capillary. Note the fourth power; this means that conductance is very sensitive to the size of the channel through which the fluid flows.

This fourth power of radius — or area squared — is unlike that encountered for, say, electrical current. The current in a wire — with a fixed potential (voltage) drop — is simply proportional to the area of the wire. Why? There is a flux of electrons — double the area to flow and the current doubles too. So, why a different relationship when it is fluid flow, not electrical current? The key distinction here is that the electrons move at a speed determined by the potential gradient and the metal in the wire — it is independent of the cross-sectional area of the wire. For fluid flow, however, the situation is different, since we have a no-flow $v = 0$ boundary condition on the solid surface. This means that flow speed increases away from the solid — governed by the viscosity — and so the flow is faster in large pores than in small pores. The total flux, Q, is related to the average speed multiplied by the area to flow. In this case, both terms increase with pore size.

We will meet this behaviour later for multiphase flow; how well each phase flows is exceptionally sensitive to the size of the pores that it moves through.

Now consider that we have an array of parallel tubes a distance d apart. Then the porosity is $\pi r^2/d^2$ and the Darcy velocity q is Q/d^2. Then we can write Eq. (10.7) as

$$q = -\frac{\phi r^2}{8\mu}(\nabla P - \rho g), \tag{10.8}$$

or, in equivalence to Darcy's law, Eq. (10.6):

$$K = \frac{\phi r^2}{8}. \tag{10.9}$$

Note that the factor $\phi/8$ is typically much less than one; if we account for tortuous flow through a less well connected pore space, the permeability is typically 1,000 times lower in magnitude than the area of a typical pore (radius squared). We will use this concept later when we discuss the effects of flow rate on displacement behaviour.

We can use Eq. (10.9) to estimate a typical pore size from the more easily measured macroscopic parameters K and ϕ. We find

$$r \sim \sqrt{\frac{K}{\phi}}. \tag{10.10}$$

This relationship was used for the derivation of the Leverett J-function (see Sec. 8) which explains the relationship between capillary pressure and its dimensionless form.

On a lighter note, below are pictures of some famous names in this subject (Fig. 10.6). I have been unable to find a picture of the most

Figure 10.6. George Stokes, Claude-Louis Navier and Henry Darcy. Spot the odd one out — the scientist who was not born in Dijon.

famous person specifically for this work: M.C. Leverett (of J-function fame). It is interesting to note that both Darcy and Navier were born in Dijon.

10.4.   Other Ways to Write Darcy's Law and Hydraulic Conductivity

We can write Darcy's law for flow in one direction, x, as follows:

$$q = -\frac{K}{\mu}\left(\frac{\partial P}{\partial x} - \rho g_x\right). \tag{10.11}$$

Similar equations can be written for flow in the y- and z-directions.

Often you need to calculate the total flow Q (with dimensions volume per unit time) through a system of cross-sectional area A. Since $q = Q/A$, Darcy's law may be written

$$Q = -\frac{KA}{\mu}\left(\frac{\partial P}{\partial x} - \rho g_x\right), \tag{10.12}$$

and for many linear flows the term $\partial P/\partial x$ can be substituted by $\Delta P/L$, where ΔP is a pressure drop over a distance L.

The hydrology literature is principally concerned with the flow of water and Darcy's law is often seen written in terms of a hydraulic conductivity, K_H, defined as

$$K_H = \frac{K\rho g}{\mu},$$ (10.13)

where g is the acceleration due to gravity and the density and viscosity are those of water. K_H has the dimensions length/time. K_H is $1\,\mathrm{m\cdot s^{-1}}$ if the volumetric flow rate of water moving vertically under gravity is $1\,\mathrm{m\cdot s^{-1}}$. Darcy's law can then be written

$$q = -K_H \frac{\partial p}{\partial x},$$ (10.14)

where

$$p = \frac{P}{\rho g} + z$$ (10.15)

is the sum of the pressure head $P/\rho g$ and the elevation head z, where z is the vertical coordinate (upwards). Since we will be concerned with the flow of air and oil, as well as water, we will use Darcy's equation in the form shown in Eq. (10.11).

10.5. Units of Permeability and the Definition of the Darcy

The permeability K is a property of the geometry of the porous medium. Except for gas flows at very high speeds (if we have Stokes flow at the pore scale, there is a linear relationship between pressure gradient and flow speed), the permeability is not a function of flow rate or the properties of the fluid, such as viscosity and density. K has the dimensions of length squared. Conventionally, permeability is measured in units of a Darcy (D); if there is a flow of $1\,\mathrm{cm^3\cdot s^{-1}}$ of a fluid of viscosity $10^{-3}\,\mathrm{kg\cdot m^{-1}s^{-1}}$ or $10^{-3}\,\mathrm{Pa\cdot s}$ (water) through a cube of rock $1\,\mathrm{cm}$ in all directions, the permeability is 1 Darcy if there is a pressure drop of 1 atmosphere across it ($1\,\mathrm{atm} \approx 10^5\,\mathrm{Pa}$). $1\,\mathrm{D} \approx 10^{-12}\,\mathrm{m^2}$. Although the Darcy is not an SI unit, it is a convenient measure of permeability. For consolidated rock, the mD (milliDarcy) unit is often used: $1{,}000\,\mathrm{mD} = 1\,\mathrm{D}$.

10.6.　Definition of Flow Speed and Porosity

Although q has the units of velocity (and is often called the Darcy velocity), it is strictly speaking not a real flow speed. The actual flow velocity in a system with porosity ϕ is q/ϕ. Remember that q is defined as the volume of fluid passing through the soil or rock per unit area per unit time. Imagine that we have a slab of rock $1\,\mathrm{cm}^2$ in cross-section with a porosity of 0.5. If $q = 1\,\mathrm{cm}\cdot\mathrm{s}^{-1}$, then each second $1\,\mathrm{cm}^3$ of fluid enters the rock. It fills the void space. If $\phi = 0.5$, then this $1\,\mathrm{cm}^3$ of fluid fills $2\,\mathrm{cm}^3$ of rock. Hence, each second the fluid encroaches a further $2\,\mathrm{cm}$ into the slab. This corresponds to a flow speed of q/ϕ or $2\,\mathrm{cm\cdot s}^{-1}$. For unconsolidated sand or gravel ϕ is approximately 0.3–0.35. For consolidated rock, deep underground, typical values are 0.1–0.2. For fractured rock ϕ may be as low as 0.0001–0.02, while for some vuggy carbonates ϕ is as high as 0.4. Soils generally have higher porosities (see Table 1.1). Loamy soil typically has $\phi = 0.3$, while clays have ϕ in the range 0.4–0.85.

Table 10.1.　Physicochemical properties of rocks and soil.

Rock/Soil Type	Porosity (%)	Particle Density (kg/m^3)	Bulk Density (kg/m^3)	Permeability (m^2)
UNCONSOLIDATED				
Gravel	25–40	2,650	1,590–1,990	10^{-10}–10^{-5}
Sand	25–50	2,650	1,330–1,990	10^{-13}–10^{-9}
Loam	42–50	2,650	1,330–1,540	10^{-14}–10^{-10}
Silt	35–50	2,650	1,330–1,720	10^{-16}–10^{-12}
Clay	40–70	2,250	680–1,350	10^{-19}–10^{-16}
CONSOLIDATED				
Sandstone	5–30	2,650	1,860–2,520	10^{-17}–10^{-13}
Shale	0–10	2,250	1,980–2,250	10^{-20}–10^{-16}
Granite	0–5	2,700	2,570–2,700	10^{-20}–10^{-17}
Granite (fractured)	0–10	2,700	2,430–2,700	10^{-15}–10^{-11}
Limestone	0–20	2,870	2,300–2,870	10^{-16}–10^{-13}
Limestone (Karstic)	5–50	2,710	1,360–2,570	10^{-13}–10^{-9}
Basalt (permeable)	5–50	2,960	1,480–2,810	10^{-14}–10^{-9}

Table 10.2. Typical permeability values.

$\log_{10} K$ (m^2)	-7	-11		-16	-20
K(D)	100 000	10		0.0001	10^{-8}
Permeability	Pervious	Semi-pervious		Impervious	
Aquifer	Good	Poor		None	
Soils	Clean gravel	Clean sand or sand and gravel	Very fine sand, silt, loess, loam		
		Peat	Stratified clay	Unweathered clay	
Rocks			Oil rocks	Sandstone	Good limestone dolomite	Breccia, granite, shale

Source: Adapted from Bear (1972).

Table 10.3. Soil classification based on particle size.

Material	Particle Size (mm)
Clay	<0.004
Silt	0.004–0.062
Very fine sand	0.062–0.125
Fine sand	0.125–0.25
Medium sand	0.25–0.5
Coarse sand	0.5–1.0
Very coarse sand	1.0–2.0
Very fine gravel	2.0–4.0
Fine gravel	4.0–8.0
Medium gravel	8.0–16.0
Coarse gravel	16.0–32.0
Very coarse gravel	32.0–64.0

10.7. Estimating Permeability

It is easy to see that for most porous sedimentary rock K should be in the range of 10 mD–10,000 mD from Eq. (10.8). A porous rock or soil is only approximately modelled by a bundle of parallel tubes,

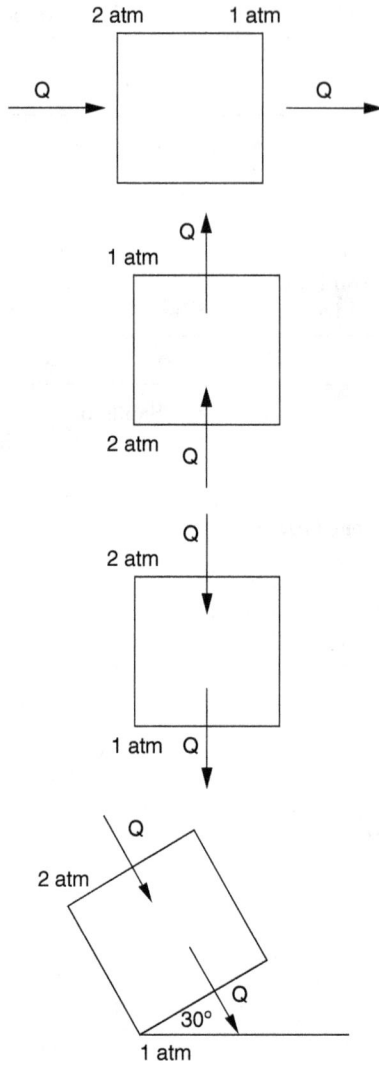

Figure 10.7. Diagram of fluid flow for the Darcy law exercise.

but to estimate K we may guess that r represents a typical throat size (10 μm–100 μm), while d, the distance between throats, is the grain diameter (100 μm–1000 μm), giving K in the range 0.1 D–10 D. If we account for diagenesis or compaction, which shrinks and closes some of the pore throats, leading to a tortuous confined pathway for fluid flow, then K may be orders of magnitude lower (as seen in deep oil reservoirs), whereas for unconsolidated gravel, the pore and grain size is much larger and permeabilities of thousands of Darcies are possible.

Permeability varies widely for different types of soil or rock. As can be seen in Tables 10.1–10.3, typical permeability values vary by 10 orders of magnitude from granite (where the fluid flows in small, poorly connected fractures) to clean gravel, with wide, well-connected pore spaces. Aquifers, consisting of sand of silt, normally have permeabilities between 1 D and 1,000 D. Bedrock and oil reservoirs may have permeabilities in the range from fractions of a mD to 1 D. Shales have permeabilities measured in nD to μD $(10^{-21}\,\mathrm{m^2}$ to $10^{-18}\,\mathrm{m^2})$.

10.8. Example Problem in Calculating Permeability

In the four situations shown in Fig. 10.7, the porous medium has a permeability of 1 D, a cross-sectional area of $1\,\mathrm{cm^2}$, and a length of 1 cm. The saturating fluid has a viscosity of 1 cp $(1\,\mathrm{cp} = 10^{-3}\,\mathrm{Pa\cdot s})$ and a density $10^3\,\mathrm{kg\cdot m^{-3}}$. For the inlet and outlet pressures shown, determine the total flow rate Q in $\mathrm{cm^3\cdot s^{-1}}$·$g$, the acceleration due to gravity, is $9.81\,\mathrm{m\cdot s^{-2}}$. You may take $1\,\mathrm{atm} = 10^5\,\mathrm{Pa}$ and $1\,\mathrm{D} = 10^{-12}\,\mathrm{m^2}$.

Chapter 11

Molecular Diffusion and Concentration

Darcy's law describes the flow of a single species in a porous medium — typically water or oil. Later we will return to study flow with multiple phases; however, before broaching this difficult subject, we will first discuss transport of species dissolved in a single phase.

Imagine that some pollutant is dissolved in the water, or consider a chemical constituent present in the oil. The pollutant (solute) not only follows the flow of the water, but also intermingles slowly with clean water because of molecular diffusion. In this section we write down an expression for the diffusive flux.

Imagine, as in Fig. 11.1, that we have a porous medium saturated with water which is not moving (in fact, we could do the same analysis if we considered a container just holding water with no porous medium at all). Initially on the left-hand-side the water is salty and on the right it is not very salty.

The salt (solute) particles have a random motion; this is due to thermal fluctuations that cause the particles to move constantly in random directions. After a long time this means that the particles will be evenly distributed in the system (Fig. 11.2). While the particle motion is random, on average particles will move from left to right simply because there are more particles on the left to begin with. An individual particle is just as likely to move to the right or the left.

This random motion of the solute is called molecular diffusion. It tends to smear out concentration gradients.

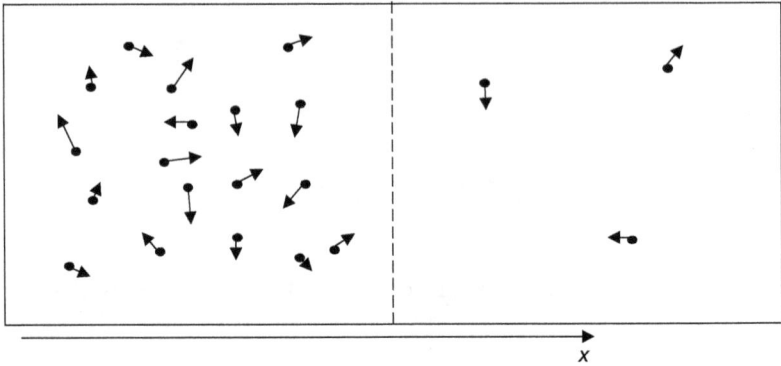

Figure 11.1. Diffusion in a porous medium. On the left (initially) there are more particles of a solute than on the right. The arrows represent the random velocities of the particles due to thermal motion. The dashed line is there for illustrative purposes.

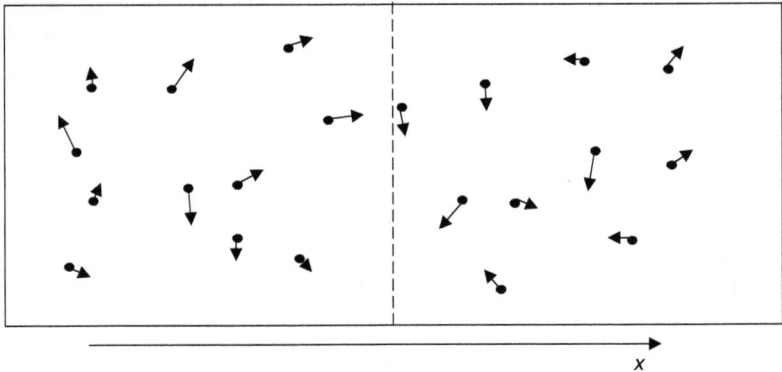

Figure 11.2. Diffusion in a porous medium. Eventually, the average concentration will be the same across the sample.

There is a flux of particles (solute) from high concentration to low concentration.

What is the flux? We will define a flux as the mass of solute moving per unit area per unit time — it has units of kg·m^{-2}s^{-1}.

Fick's law of diffusion (in a porous medium) states

$$J^\alpha = -\phi D^\alpha \nabla C^\alpha, \tag{11.1}$$

where J is the flux, D is the diffusion coefficient in a porous medium, and C is the concentration measured in mass per unit volume of water. The subscript α refers to the solute — different solutes have different diffusion coefficients. Note the minus sign — the flux is in the opposite direction to the concentration gradient. This is like Darcy's law, where the flow is in the opposite direction to the pressure gradient.

Why is there the factor of ϕ (the porosity)? This is simply a convention. Diffusion coefficients in porous media are lower than for bulk fluids, so the porosity factor corrects for this. In fact, the diffusion coefficient in Eq. (11.1) is still lower than in bulk — by a factor of 2 or 3 typically, because of the tortuosity of the pore space.

What are the units of the diffusion coefficient? It has units length^2/time or m$^2 \cdot$s^{-1} in SI units. The value of often very small — many low molecular weight solutes have diffusion coefficients at room temperature in the range 10^{-9} m$^2 \cdot$s^{-1} to 10^{-10} m$^2 \cdot$s^{-1}.

If we assume that the concentration gradient is in one direction — the x-direction in Figs. 11.1 and 11.2 — Eq. (11.1) simplifies to

$$J^\alpha = -\phi D^\alpha \frac{\partial C^\alpha}{\partial x}. \tag{11.2}$$

The last issue to discuss is how to relate this flux to situations when the water is also moving. If the water is moving, molecular diffusion still takes place. In this case the average motion of the water in the flow field is added to the random thermal motion of the molecules. The Darcy velocity q is the volume of fluid per unit area per unit time. If we multiply this by the concentration then qC^α is the mass per unit area per unit time, or a flux. This is the flux of solute ignoring diffusion. If we consider both diffusion and advection (flow) the total flux is

$$J^\alpha = qC^\alpha - \phi D^\alpha \frac{\partial C^\alpha}{\partial x}, \tag{11.3}$$

where q is given by Darcy's law, Eq. (10.6). In three dimensions we can write

$$J^\alpha = qC^\alpha - \phi D^\alpha \nabla C^\alpha. \tag{11.4}$$

Variations in permeability cause different portions of contaminant plume to follow different pathways through a porous medium. Molecular diffusion causes the contaminant to mix with clean water, causing a dilution of the plume. These two effects combined result in the formation of diluted, dispersed plumes that pollute large bodies of water, a concept that will be discussed in more detail later.

Chapter 12

Conservation Equation for
Single-phase Flow

Now we will derive an equation for the conservation of mass of a component α in a porous medium. This component is considered to be dissolved in water where C^α is the density of species α per unit pore volume. The porosity is ϕ. Then ϕC^α is the mass of α per unit volume of the soil or rock. We start by considering flow in one direction (the x-direction) through a small volume of length Δx along x with cross-sectional area A.

This is an important exercise, and I will expect you to be able to go through the steps yourself. Throughout this work we will derive conservation equations for different situations of interest and then solve these equations. I will also expect you to be able to derive — and solve — new equations describing new physical phenomena that you might not have met hitherto.

Referring to Fig. 12.1, the mass entering the volume per unit time is $AJ^\alpha(x)$, while the mass leaving the volume per unit time $= AJ^\alpha(x + \Delta x)$.

Expand the mass leaving as a Taylor series (A is a constant):

$$\text{mass leaving} = A\,J^\alpha(x) + A\Delta x \frac{\partial J^\alpha}{\partial x} + O(\Delta x^2). \qquad (12.1)$$

Hence, the difference in the mass in minus out is $-A\Delta x \frac{\partial J^\alpha}{\partial x}$. Now consider conservation of mass (mass in per unit time — mass out per

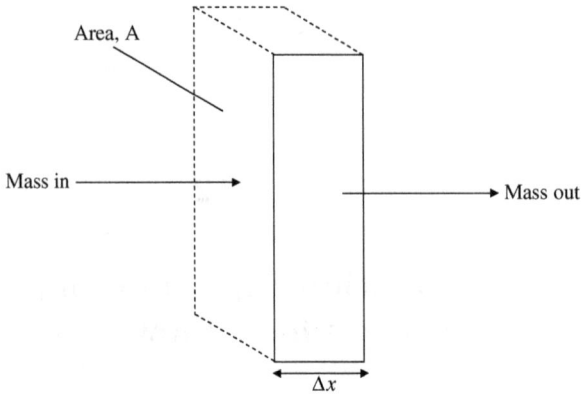

Figure 12.1. A schematic of transport in one dimension used to derive a conservation equation.

unit time = rate of change of mass in the volume $A\Delta x$):

$$-A\Delta x \frac{\partial J^\alpha}{\partial x} = A\Delta x \frac{\partial \phi C^\alpha}{\partial t}. \tag{12.2}$$

Hence, we can write (assuming that ϕ is constant in time):

$$\phi \frac{\partial C^\alpha}{\partial t} + \frac{\partial J^\alpha}{\partial x} = 0. \tag{12.3}$$

We can repeat this analysis for flow along the y- and z-axes. If q_x, q_y and q_z represent the flows in the x-, y- and z-directions respectively, then the 3D version of Eq. (12.3) is

$$\phi \frac{\partial C^\alpha}{\partial t} + \frac{\partial J^\alpha_x}{\partial x} + \frac{\partial J^\alpha_y}{\partial y} + \frac{\partial J^\alpha_z}{\partial z} = 0. \tag{12.4}$$

The 3D conservation equation can also be more elegantly and simply derived using vector calculus and Green's theorem. Consider an arbitrary volume V of the porous medium bounded by a surface S with a flux J^α through it. Then the equation of mass conservation is

$$\phi \int \frac{\partial C^\alpha}{\partial t} dV + \int \boldsymbol{J}^\alpha \cdot \boldsymbol{dS} = 0. \tag{12.5}$$

We use Green's theorem to convert the surface integral into a volume integral:

$$\phi \int \frac{\partial C^\alpha}{\partial t} dV + \int \nabla \cdot \boldsymbol{J}^\alpha dV = 0, \qquad (12.6)$$

and if this is true for an arbitrary volume, then the integrands must be related by

$$\phi \frac{\partial C^\alpha}{\partial t} + \nabla \cdot \boldsymbol{J}^\alpha = 0, \qquad (12.7)$$

which is the same as Eq. (12.4).

We can instead perform an overall volume balance. If we assume that the fluid is incompressible (i.e. the water density ρ_w is a constant), then the volume flowing into any arbitrary volume is the same as the volume flowing out. From the definition of the Darcy velocity, this leads to

$$\nabla \cdot \boldsymbol{q} = 0. \qquad (12.8)$$

We can then substitute Fick's law of diffusion and Darcy's law to find J^α, Eq. (11.4). If we assume that the diffusion constant is constant then Eq. (12.7) becomes

$$\phi \frac{\partial C}{\partial t} + q \cdot \nabla C = \phi D \nabla^2 C, \qquad (12.9)$$

where I have now dropped the superscript α for convenience. In one dimension,

$$\phi \frac{\partial C}{\partial t} + q \frac{\partial C}{\partial x} = \phi D \frac{\partial^2 C}{\partial x^2}. \qquad (12.10)$$

Equations (12.9) or (12.10) are called either the convection–diffusion equation or the advection–diffusion equation. The dissolved contaminant follows the overall flow of the fluid (water) by Darcy's law (convection or advection) as well as diffusing through the system by Fick's law. Later on we will extend this equation to multiphase flow.

In essence this conservation equation may be written in words as

$$\frac{\partial}{\partial t}(\text{Mass per unit volume}) + \frac{\partial}{\partial x}(\text{Mass flux}) = \text{Diffusion}. \quad (12.11)$$

It is possible to estimate the magnitude of the terms in Eq. (12.10) and so see whether diffusion or advection dominate. Imagine that the contaminant has spread over some distance $x = X$ in a time $t = T$. We're not going to attempt to solve any of these equations exactly — this comes later — but as a crude estimate we may estimate that the three terms in Eq. (12.10) have magnitude

$$\phi\frac{\partial C}{\partial t} + q\frac{\partial C}{\partial x} = \phi D\frac{\partial^2 C}{\partial x^2},$$
$$\frac{\phi C}{T} + \frac{\phi C}{X} \sim \frac{\phi D C}{X^2}. \quad (12.12)$$

This is a common way of analysing complex equations. We are going to compute the relative magnitude of each of the three quantities in Eq. (12.12). If advection dominates, then we equate the first two terms to find

$$\frac{X}{T} \sim \frac{q}{\phi}, \quad (12.13)$$

which we know already: X/T, or the typical flow speed is q/ϕ.

If diffusion dominates, then we equate the first and third terms of Eq. (12.12) to obtain

$$X \sim \sqrt{DT}. \quad (12.14)$$

Thus, for advective flow the pollutant spreads linearly with time, whereas if diffusion dominates, then the spread is proportional to the square root of time. The ratio of the diffusive to advective terms is

$$\frac{\text{Diffusion}}{\text{Advection}} \sim \frac{\phi D}{qX} = Pe. \quad (12.15)$$

This defines a dimensionless Peclet number, Pe. The length scale X is generally considered to be a representative scale in the porous medium, which is a mean grain (or pore) size: say around $100\,\mu m$

(0.1 mm). Typical flow speeds q/ϕ are of the order 1 m/day, or around 10^{-5} m·s^{-1}, and lower. The diffusion coefficient in water for most petroleum components (typical pollutants in groundwater) is around 10^{-9} m^2·s^{-1}. This gives a Peclet number of around 1; diffusion and advection are typically similar in magnitude at the pore scale.

Molecular diffusion is more significant over lengths X less than approximately 0.1 mm (corresponding to a time of around 10 s), but for large flows over hundreds of metres taking place over months and years, advection is by far the more important process.

We have derived differential equations that describe the flow of a dissolved contaminant in an incompressible fluid. These equations can either be solved analytically for simple cases, or numerically. Except for small-scale phenomena, advection (or flow computed using Darcy's law) is much more significant than molecular diffusion.

Also — rather interestingly — the same conservation equation pertains within a pore, if we ignore the porosity in the equations; we simply invoke conservation of mass in a volume of flowing fluid. We can also apply Fick's law (again without the porosity term), but the flow field is governed not by Darcy's law, but rather forms a solution of the Navier–Stokes equation, presented in Sec. 10. Primitively, the movement of a dissolved solute within the pore space is governed by the equation

$$\frac{\partial C}{\partial t} + v.\nabla C = D\nabla^2 C, \tag{12.16}$$

where v is the local flow velocity and D is, strictly, the molecular diffusion coefficient. v is given by the solution of the Navier–Stokes equation; it is this equation that can be used to describe transport at the pore scale.

Strictly speaking, Eqs. (12.9) and (12.10) are upscaled versions of Eq. (12.16) where now the Darcy velocity, q, is given by Darcy's law. The problem — as we discuss later — is the diffusive flux. This contains contributions not only from diffusion, but also from the random motion of the particles in a spatially heterogeneous flow field, which we have — so far — ignored.

12.1. Analytical Solution of the Advection-diffusion Equation

We will now present some solutions to Eq. (12.10). These solutions also apply to heat transport and can be found in the classic work by Carslaw and Jaeger (1946). In the end, there is no one correct way to arrive at a solution. Instead, we can use physical inference to find a functional form that is likely to work.

Figure 12.2 shows the solution to the equation for solute that is originally injected as a point source. Physically, we expect the plume to move with some average velocity $v = q/\phi$ and then spread out dependent on the degree of diffusion. This is what we see. We will use this insight to develop a possible mathematical form of the solution.

We can transform the governing partial differential equation into an ordinary differential equation using the following variable:

$$z = \frac{x - vt}{\sqrt{t}}, \tag{12.17}$$

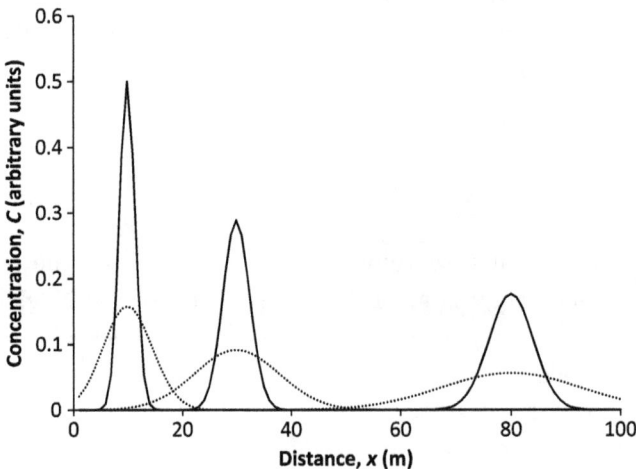

Figure 12.2. Solutions to the advection-diffusion equation using Eq. (12.26) with (solid lines) $v = 10^{-5}$ m/s, $D = 10^{-5}$ m^2/s and times of 1,000,000, 3,000,000 and 8,000,000 s. The dotted lines are with a diffusion coefficient, D, that is ten times higher.

where $v = q/\phi$ and we assume that the solution C can then be written as follows:

$$C = \frac{g(z)}{\sqrt{t}}, \qquad (12.18)$$

and we solve for the (unknown) function $g(z)$. There is — in principle — no reason why this has to be the case. All I need to demonstrate is that the solution obeys the governing partial differential equation and the boundary conditions. Near the end of this work, I will present another solution for a non-linear diffusion problem that uses a different set of variables, because the boundary conditions are different.

Then we define the following derivatives:

$$\frac{\partial g}{\partial t} = \left(-\frac{v}{\sqrt{t}} - \frac{z}{2t} \right) \frac{dg}{dz}, \qquad (12.19)$$

$$\frac{\partial g}{\partial x} = \frac{1}{\sqrt{t}} \frac{dg}{dz}, \qquad (12.20)$$

$$\frac{\partial^2 g}{\partial x^2} = \frac{1}{t} \frac{d^2 g}{dz}, \qquad (12.21)$$

and Eq. (12.10) becomes

$$g + z\frac{dg}{dz} + 2D\frac{d^2 g}{dz^2} = 0. \qquad (12.22)$$

This can be written

$$\frac{d}{dz} \left(2D\frac{dg}{dz} + gz \right) = 0. \qquad (12.23)$$

Hence, for an arbitrary constant c (not to be confused with concentration),

$$2D\frac{dg}{dz} + gz = c. \qquad (12.24)$$

We also know — from our schematic solution — that the concentration will be zero at large distances from the origin (large z); hence both g and dg/dz tend to zero for infinite z. This means that c in

Eq. (12.24) is zero and we can readily integrate:

$$4D \ln g = -z^2 + c',$$
(12.25)

for another constant c'. Then we can write

$$C(x,t) = \frac{M}{\sqrt{4Dt}} e^{-(x-vt)^2/4Dt},$$
(12.26)

where M is the initial mass (per unit area) of concentration. These are the solutions shown in Fig. 12.2.

This relationship, Eq. (12.26), makes use of the following identity to find the constant of integration:

$$\int_{-\infty}^{\infty} C(x,t)dx = M,$$
(12.27)

since

$$\int_{-\infty}^{\infty} e^{-z^2} dz = \sqrt{\pi}.$$
(12.28)

Note that Eq. (12.26) shows a mean (maximum) concentration that moves a distance $x = vt$, with a typical spread $x - vt = 2\sqrt{(Dt)}$, similar to our simple scaling analysis.

12.2.   Diffusion and Dispersion

The governing partial differential equation and its solution are well-known in the literature, as mentioned previously. However, they are a very poor approximation of what really happens in a porous medium, particularly the heterogeneous pore spaces we have shown earlier. The limitation in the derivation is that we assume a uniform flow field where the only flux associated with changes in concentration is due to molecular diffusion.

In reality — in porous media with tortuous pore spaces — the spreading and mixing of a solute are controlled by two factors: molecular diffusion that leads to a local (pore-scale) mixing of concentration, and variations in flow speed that allow the solute to follow different flow paths through the system. It is this second effect that normally dominates, with huge spreading of a dissolved plume of solute caused not so much by local-scale mixing, but governed by the variations in the flow field.

This is evident from an examination of the flow paths shown when we introduced Darcy's law. In particular, refer back to the images and flow fields shown in Sec. 10. To have an idea of what transport really looks like in heterogeneous porous media, Fig. 12.3 shows simulated concentration profiles — for an effective point source injection as described above — compared to measurements of water movement at the scale of a few mm to cm made using NMR techniques. For simplicity, the curves are shown in a dimensionless form. The y-axis represents concentration times average distance moved, while the x-axis is the distance divided by average speed times a dimensionless distance. A system with no diffusion (or dispersion) would then travel at unit speed on the graph. With simple diffusion — obeying Eq. (12.26) — we would see a Gaussian-type profile centred on 1. This is seen for the bead pack. Here the flow field is uniform and there is Fickian-type smearing controlled by molecular diffusion and small-scale heterogeneity in the flow field.

For the more heterogeneous porous media — the sandstone and carbonate — the behaviour is different: most of the solute resides in locally stagnant regions of the pore space and hardly moves, with a very long dispersed plume of solute in the faster flowing regions. There is no obvious concept of a typical speed with smearing about this average.

A full understanding of this phenomenon and how, properly, to describe dispersive transport is a rich topic of current research. An exploration of the ideas and how to describe transport mathematically in these cases is beyond the scope of this chapter. Needless to say an approach based on fluctuations in velocity and/or travel times is necessary, which does not fit neatly into an effective partial differential equation; it is evident from Fig. 12.3 that we cannot match the experiments simply by tweaking the effective diffusion (or dispersion) coefficient while leaving the functional form of the behaviour the same.

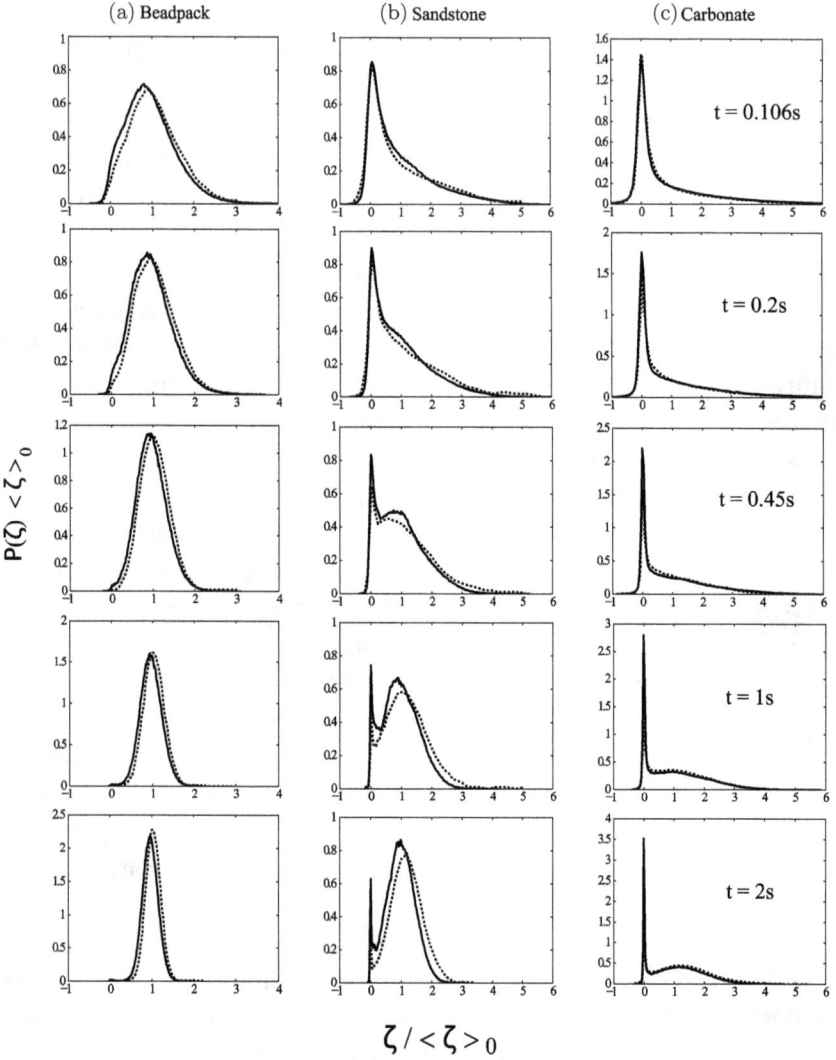

Figure 12.3. Predicted (solid lines) and measured (dotted lines) dimensionless concentration profiles for transport in a bead pack, sandstone and carbonate. Note that with the exception of the bead pack, the behaviour does not resemble a solution to the governing advection-diffusion equation, see Fig. 12.2. From Bijeljic *et al.* (2013a).

If — and in reality this is rarely, if ever, the case — we can indeed describe transport as the solution to an advection-diffusion-type equation in a heterogeneous medium, then traditionally we can write the following identical in functional form to Eq. (12.9):

$$\phi\frac{\partial C}{\partial t} + q\frac{\partial C}{\partial x} = \phi D\frac{\partial^2 C}{\partial x^2}, \tag{12.29}$$

where D is now the effective dispersion coefficient that accounts for both molecular diffusion and the random nature of the flow field. This equation is called the advection-dispersion equation. This coefficient D can be written as follows:

$$D = D_{dis} + D_m = \alpha v + D_m, \tag{12.30}$$

where now D_m is the molecular diffusion coefficient in the porous medium, and α is the dispersivity — it has the dimensions of a length and represents, physically, the typical scale of heterogeneity in the porous medium. However, natural geological media have structure over all length scales, and so the apparent dispersivity — and hence the spreading of a contaminant plume — appears to become more significant with scale, as more heterogeneity is encountered. Table 12.1 illustrates this phenomenon, where the dispersivity is, approximately, one-tenth of the scale of the system.

Equation (12.30) can be derived assuming that the solute experiences a series of random perturbations every time it moves a distance α; the velocity v in the dispersion coefficient represents the fact that the number of perturbations encountered in a given time is proportional to the flow speed.

In the plane of the aquifer (a nearly horizontal plane) the plume is therefore highly dispersed due to variations in permeability that cause huge fluctuations in the local flow rate. The permeability K will typically vary by several orders of magnitude or more over lengths of a few metres in heterogeneous formations. A contaminant plume does not flow at a constant rate and direction, but forms a ragged front due to variations in permeability. This causes the plume to spread. Small amounts of molecular diffusion mix the contaminant with clean water. The combined effects of permeability variation and diffusion

Table 12.1. Standard deviation and correlation scale of the natural logarithm of hydraulic conductivity or transmissivity (Gelhar, 1993).

Medium	Type*	σ_f	Correlation Scale (m)		Overall Scale (m)	
			Horizontal	Vertical	Horizontal	Vertical
Alluvial-basin aquifer	T	1.22	4,000		30,000	
Sandstone aquifer	A	1.5–2.2		0.3–1.0		100
Alluvial-basin aquifer	T	1.0	800		20,000	
Fluvial sand	A	0.9	>3	0.1	14	5
Limestone aquifer	T	2.3	6,300		30,000	
Sandstone aquifer	T	1.4	17,000		50,000	
Alluvial aquifer	T	0.6	150		5,000	
Alluvial aquifer	T	0.4	1,800		25,000	
Limestone aquifer	T	2.3	3,500		40,000	
Chalk	T	1.7	7,500		80,000	
Alluvial aquifer	T	0.8	820		5,000	
Fluvial soil	S	1.0	7.6		760	
Eolian sandstone outcrop	A	0.4	8	3	30	60
Glacial outwash sand	A	0.5	5	0.26	20	5
Sandstone aquifer	T	0.6	4.5×10^4		5×10^5	
Sand and gravel aquifer	A	1.9	20	0.5	100	20
Prairie soil	S	0.6	8		100	
Weathered shale subsoil	S	0.8	<2		14	
Fluvial sand and gravel aquifer	A	2.1	13	1.5	90	7
Homra red mediterranean soil	S	0.4–1.1	14–39		100	
Gravelly loamy sand soil	S	0.7	500		1,600	
Alluvial silty-clay loam soil	S	0.6	0.1		6	
Glacial outwash sand and gravel outcrop	A	0.8	5	0.4	30	30
Glacial lacustine sand aquifer	A	0.6	3	0.12	20	2
Alluvial soil (Yolo)	S	0.9	15		100	

Note: *Types of data: T, transmissivity; S, soils; A, 3D aquifer.

dilute the plume and disperse the contamination over a wide region of the aquifer. In virtually all circumstances the precise location of the contaminant cannot be predicted with any certainty unless the distribution of permeability in the subsurface is known; at every scale from the pore scale onwards, the distribution of contaminant is rarely, if ever, accurately predicted by an average displacement and some Gaussian-type variation about this mean.

Overall, the characterisation of dispersion as a diffusive process is flawed, since this does not — at the pore scale or the field scale — characterise transport in even a qualitative sense. The community does not — at present — have a good way to describe transport on all scales, although much of the mathematical and physical insight has been developed.

In some sense, multiphase flow, which we will now return to, is easier and we will revisit diffusive processes in this context — specifically capillary-controlled displacement — later.

Chapter 13

Capillary and Bond Numbers

Now we will return to multiphase flow in porous media and the concept of capillary pressure. In these notes, I presume that the fluid configurations are controlled by the Young–Laplace equation and contact angles. Fluid flow is slow and — as we show later — we presume that each phase flows independently.[1]

However, do capillary forces really dominate at the pore scale and what are the effects of buoyancy forces and flow rate?

If a representative pore radius is R, then a typical capillary pressure is of order σ/R. The viscous pressure drop over a length L is given from Darcy's law (where ΔP is a pressure drop — hence the removal of the minus sign):

$$q = \frac{K}{\mu}\frac{\Delta P}{L}; \quad \Delta P = \frac{q\mu L}{K}. \tag{13.1}$$

The ratio of a typical pressure drop to a capillary pressure is then

$$\text{ratio} = \frac{q\mu L R}{\sigma K}. \tag{13.2}$$

---

[1]This approximation can be relaxed and dynamic models of multiphase flow show that there is viscous coupling between the phases. However, in this volume, we will ignore these effects.

LR/K is a dimensionless ratio (consider the units — permeability has the units of length squared). In porous media this ratio is typically around 1,000 if L is a pore length. Typically, L/R is around 2–10.

Why isn't this ratio LR/K closer to 1? Recall the discussion in Sec. 10, which showed that permeability typically has a numerical value that is much lower than the square of a typical pore radius.

The capillary number is defined by

$$N_{cap} = \frac{q\mu}{\sigma}. \qquad (13.3)$$

Representative values for the capillary number for field-scale displacement are typically 10^{-8} to 10^{-6} or lower. If N_{cap} is around 0.001, then viscous and capillary forces are approximately equivalent at the pore scale — see Eq. (13.2).

In most natural flows in aquifers and oil reservoirs, capillary forces dominate at the pore scale: q is generally around 10^{-8}–10^{-5} m/s, μ is around 10^{-3} Pa·s (for water), while σ for oil/water systems is typically 0.05 N/m at ambient conditions and around half that number at oil-field temperatures. This leads to values of capillary number in the range 10^{-6} and lower. Viscous forces, however, dominate at the large (inter-well) scale; this can be seen by substituting $L = 100$ m–1,000 m (the inter-well scale) into Eq. (13.2).

We can perform a similar analysis for buoyancy. The pressure drop over a vertical distance L is $\Delta\rho g L$, where $\Delta\rho$ is the density difference between the phases. The ratio of buoyancy to capillary forces is given by

$$\text{Ratio} = \frac{\Delta\rho g L R}{\sigma}. \qquad (13.4)$$

Sometimes a Bond number B is defined by

$$B = \frac{\Delta\rho g L^2}{\sigma}. \qquad (13.5)$$

Once again using representative numbers — say a density difference of 200 kg·m^{-3} and a pore length of 10^{-4} m — the Bond number is of order 10^{-3}. Buoyancy forces are small in comparison with capillary forces at the pore scale.

The reason why this is important is that *if* viscous or buoyancy forces were to dominate at the pore scale, then two things happen:

1. The non-wetting phase (oil) is rarely trapped, since the invasion of water occurs through a flat connected front with essentially no snap-off and little bypassing.
2. Even if some ganglia of oil are stranded, viscous forces can push these blobs out of the pore space.

The result is that if the capillary number is around 0.001 or larger, then the residual oil saturation can be very low. Therefore, if we could increase the capillary number in this range (by increasing the flow rate q, or reducing the interfacial tension σ), then we would

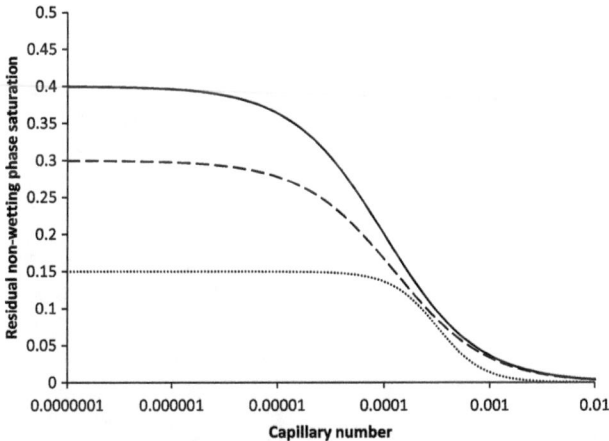

Figure 13.1. A schematic of residual saturation as a function of capillary number: solid line for a carbonate, dashed line for a sandstone and the dashed line for a sand pack. At capillary numbers higher than around 10^{-3} — meaning that viscous forces begin to dominate over capillary forces at the pore scale — the residual saturation can fall to very low values. This is the basis for surfactant flooding: if the interfacial tension is reduced to a sufficiently low value, then the capillary number is high enough to allow low residual saturations to be achieved, giving good oil recovery (Lake, 1989).

have a very efficient oil recovery process. This is the physics behind surfactant flooding. Here surfactants are added to the injected water, lowering the oil/water interfacial tension to 0.1 mN/m or smaller. If the capillary number increases to the range of 0.001 or above, then very high oil recoveries are observed.

The decrease in residual saturation as a function of capillary number is shown in Fig. 13.1 (based on Lake, 1989).

Chapter 14

Relative Permeability

We will now extend Darcy's law to cases when multiple fluid phases are flowing. We assume that each phase flows in its own sub-network of the pore space without affecting the flow of the other phases. This is applicable for flow at low capillary and Bond numbers, where capillary forces dominate at the pore scale.

For single-phase flow in one dimension, we have Eq. (10.11). Then the extension to multiphase flow of fluid p is (first proposed by Muskat and Meres in 1936; see Muskat's classic textbook from 1949)

$$q_p = -\frac{Kk_{rp}}{\mu}\left(\frac{\partial P_p}{\partial x} - \rho_p g_x\right),\qquad(14.1)$$

where k_{rp} is the relative permeability of phase p. It represents the mobility of the phase as a fraction of what it would be for single-phase flow. It is traditionally plotted as a function of saturation.

In this chapter, we will accept this characterisation of multiphase flow. For slow flow, dominated by capillary forces, this is a reasonable approximation. However, the relative permeabilities are not simply unique functions of saturation. As we found for capillary pressure, the relative permeabilities will also depend on wettability (and this can change during the course of a displacement, or series of displacements) and saturation history. Moreover, at higher flow rates, the relative permeabilities will also be functions of this flow rate, as well as viscosity ratio. Lastly, the flow of one phase can also affect the flow of the other, through viscous coupling at the

171

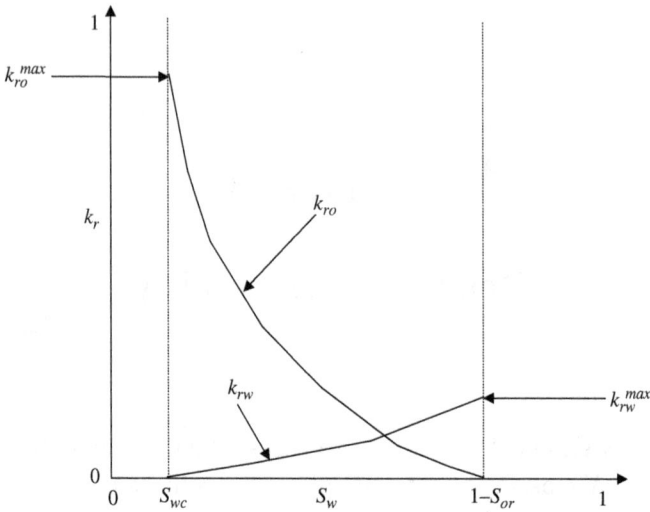

Figure 14.1. Typical relative permeability curves for waterflooding a water-wet medium. The points to note are the steep decrease in oil relative permeability with water saturation, the low water relative permeability (including its final value) and the high value of residual oil saturation S_{or}.

fluid interfaces. All of these effects may be significant, particularly when capillary forces are no longer dominant at the pore scale. However, in this treatment we will only consider the impact of pore structure, wettability and saturation path; this alone reveals a rich and important behaviour which is sufficient to explain and explore most displacement processes seen in oil fields and aquifers.

Figure 14.1 shows typical relative permeability curves for a water-wet medium. The curves are shown for waterflooding from the connate or irreducible water saturation.

The key features evident for a water-wet medium are as follows.

1. **Typical values of the maximum oil and water relative permeability.** At the beginning of water injection — marking the end of primary drainage — oil fills most of the pore space. Since wettability is altered only after oil invasion, the oil will reside in the larger regions of the pore space, confining water to the corners and the smallest pores. As a consequence, the

oil relative permeability at the beginning of waterflooding — the maximum value — is close to 1 (typically 0.8 or greater); the water relative permeability is zero, or close to zero, at the start of the displacement. When water is injected — for a water-wet system — it preferentially fills the narrower regions of the pore space, trapping oil in the larger pores. This means that the water always has poor connectivity and the relative permeability remains low; typically at the end of waterflooding the water relative permeability is only around 0.1 or lower. This is different — as described below — if the system is not strongly water-wet.

2. **The residual oil saturation is large.** This is related to the discussion for the previous point. Remember, in a water-wet system, water remains in the small pores, while oil is in the large pores. The oil can be trapped in these larger regions of the pore space by snap-off, leading to a large immobile or residual saturation at the end of waterflooding; typically S_{or} is in the range 0.2–0.5 for a water-wet system.

3. **Why the sum of the relative permeabilities is less than 1 for all saturations, and why the sum is much less than when the curves cross.** Any interface between oil and water, across a pore space, prevents flow. Hence, the more interfaces between the phases, the more flow is restricted. This is particularly true when water invades by snap-off, cutting off oil flow through the largest pores, while remaining poorly connected itself. Hence, two phases in combination have a much lower conductance than for single-phase flow (one or more of the relative permeabilities is typically always very low). When the relative permeabilities cross is usually when there is most phase interference and both values are likely to be small.

In the following subsection, we will amplify these points through experimental and modelling results. Relative permeability is very difficult to measure accurately; sophisticated apparatus is used to measure this important quantity. A discussion of experimental techniques is, however, beyond the scope of this course. By default

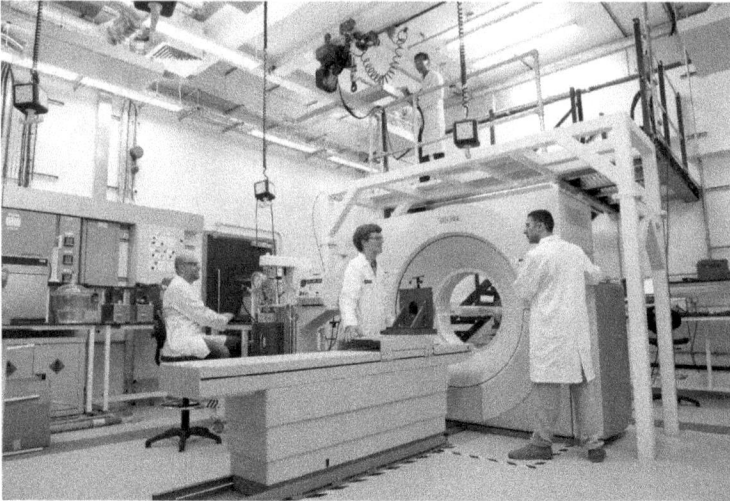

Figure 14.2. A photograph of the apparatus at Imperial College used to measure relative permeability.

the data refer to water injection, since commonly this is the most important process involving multiphase flow in oil reservoirs. Figure 14.2 shows some of our apparatus used to measure relative permeability at Imperial College. There is an adapted medical X-ray scanner (with a resolution of around 1 mm) that can monitor fluid movement within rock cores several cm across and up to 1 m long.

The key controls on relative permeability are wettability and the connectivity of the pore structure, discussed in more detail later in this section.

14.1.   Relative Permeabilities for Sandstones and Predictions Using Pore-scale Modelling

The points above can be illustrated in the relative permeability curves shown in Fig. 14.3. The experimental results are part of a classic series of measurements made on Berea sandstone for both two- and three-phase (oil, water and gas flowing — discussed later) by Oak *et al.* (1990). These measurements have served as a benchmark for analysis in the literature, because Berea is a standard quarry

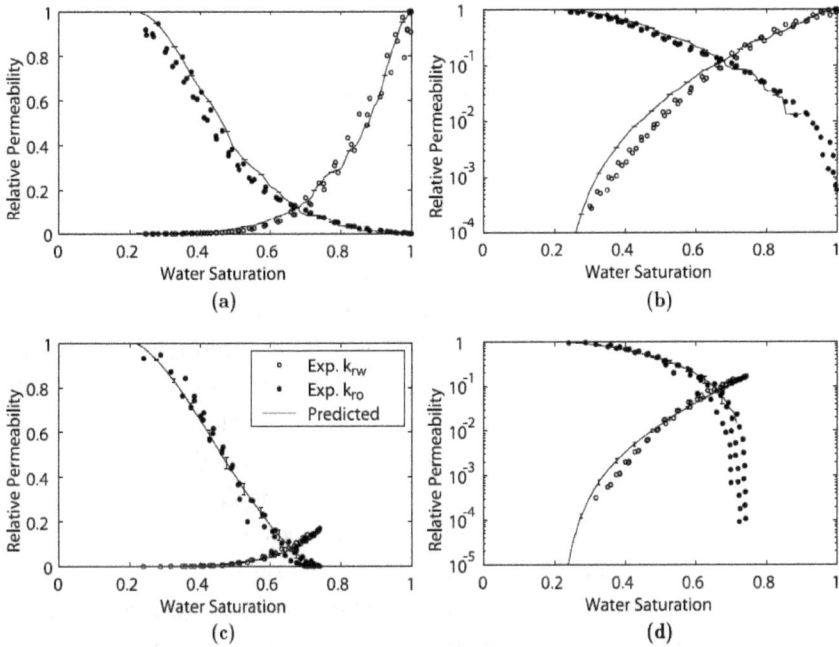

Figure 14.3. Measured (points) and predicted (lines) relative permeability curves in a water-wet Berea sandstone. The curves are shown on both linear and logarithmic axes. The primary drainage curves are shown at the top, and the waterflooding curves at the bottom. Network modelling — capturing the connectivity of the pore space using a random lattice of pores and throats, combined with an accurate assessment of pore-scale displacement processes — can predict the behaviour accurately. From Valvatne and Blunt (2004).

sandstone used by many researchers and the raw data is available in spreadsheet form. The results of three sets of experiments are shown.

Also shown are network modelling predictions performed by Valvatne and Blunt (2004). These predictions use the network for Berea presented earlier in these notes and employ the pore-scale displacement processes we have described. Primary drainage is an invasion percolation process; it is assumed that the wetting phase (water) is strongly wetting and the contact angle is zero. We make good predictions of the measured data. For waterflooding, there is an uncertainty. As discussed in Sec. 4, the (advancing) contact angle is larger in this case, mainly due to the roughness of the solid surface. In network modelling we assign an effective contact angle that accounts for roughness and the converging/diverging nature of the pores; this angle is around 60°. With this larger contact angle — which tends to suppress snap-off in some pores — we predict the relative permeability curves accurately.

The measured data and the predictions show the features mentioned above. Note that the residual saturation is around 0.3 and the maximum water relative permeability is only about 0.1. Blocking the flow in the largest 30% of the pore space reduces the water conductivity by a factor of 10. This is an important observation; small changes in fluid configuration that prevent flow through a few large channels have a big impact on relative permeability.

We can also study the effect of different displacement paths on the relative permeability. The data in the curves in Fig. 14.4 are taken from Akbarabadi and Piri (2013). Here carbon dioxide is the non-wetting phase and is injected into a Berea core to different initial saturations. Then brine is injected to displace the carbon dioxide, resulting in different curves (relative permeability hysteresis) and different amounts of trapped non-wetting phase, as discussed previously.

We can continue our study of relative permeability with more results where carbon dioxide is the non-wetting phase (the application here is carbon dioxide storage in aquifers). The graphs in Fig. 14.5 show primary drainage capillary pressure (Sec. 6), relative permeability and trapping curves (Sec. 7) on Berea sandstone, comparing the results of Krevor *et al.* (2012) with those of other researchers. This is the combination of multiphase properties that control fluid movement and recovery in the sub-surface.

14.1.1.   *Effect of Wettability in Sandstones*

We can also consider the effects of wettability on the relative permeability curves. This will be presented in more detail later in the context of carbonate rocks. In general, carbonates tend to experience a stronger wettability alteration in contact with crude oil than sandstones. So, while we see the whole range of behaviour from water-wet to strongly oil-wet in sandstones, more typically in carbonates we see mixed-wet to oil-wet properties. Figure 14.6 shows

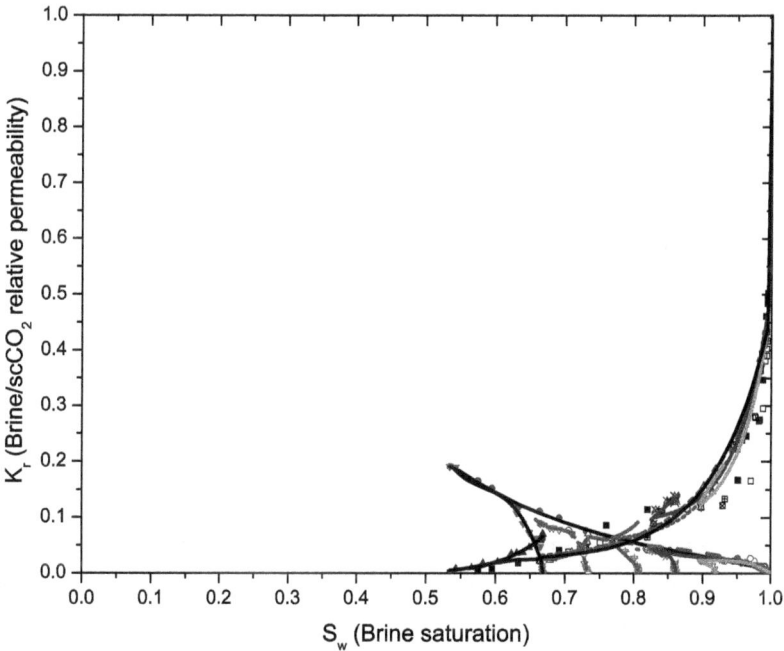

Figure 14.4. Measured relative permeability on Berea sandstone. Here carbon dioxide is the non-wetting phase and is injected to different initial saturations before brine injection. The points are the data, while the lines are simply curve fits. We see the relative permeability hysteresis; notice the different residual saturations, dependent on the initial saturation, as discussed previously. From Akbarabadi and Piri (2013).

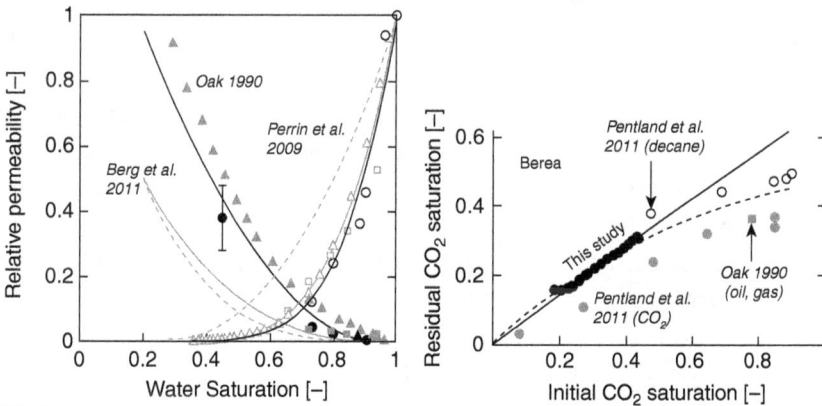

Figure 14.5. Measured primary drainage relative permeability and trapping curves for carbon dioxide injection in Berea sandstone. From Krevor *et al.* (2012).

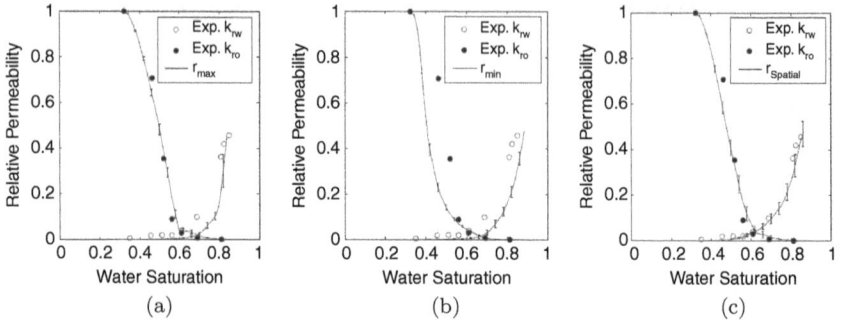

Figure 14.6. Predicted and measured relative permeability for a mixed-wet reservoir sandstone. The predictions (lines) use different assumptions as to which pores become oil-wet after primary drainage; the results are relatively insensitive to this assignment. From Valvatne and Blunt (2004).

predicted waterflood relative permeabilities for a mixed-wet reservoir sandstone compared to the data. The predictions use different models to assign wettability; there is some discussion in the literature over whether large or small pores are more likely to undergo a significant wettability change. The results are also taken from Valvatne and Blunt (2004).

In a mixed-wet system the key features are a low residual oil saturation (noted previously for capillary pressure, Sec. 9) and low oil and water relative permeabilities. The low residual is due to the connectivity and slow drainage of oil layers. The low oil relative permeability is also easy to explain: where the system is oil-wet, the oil resides in the smallest pore spaces and in layers that, while interconnected, have a very low conductance. The water relative permeability is also low; this is a significant feature that has a major impact on waterflood oil recovery at the field scale and is discussed further later. When water is first injected, it preferentially fills the water-wet regions: the smallest water-wet pores and throats. The water saturation increases, but the connectivity of the water is poor and so the relative permeability remains low. Then water fills the oil-wet regions, and the largest oil-wet pores first. Again, to begin with, the connectivity remains low; it is only at the highest water saturations that water becomes well connected through the larger

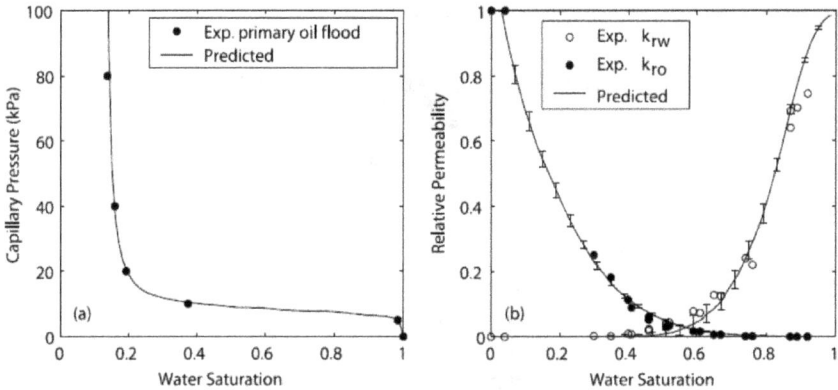

Figure 14.7. Predicted and measured primary drainage capillary pressure and waterflood relative permeability for an oil-wet reservoir sandstone. Good predictions are made. Note the low oil relative permeability at higher water saturations — this is the oil layer drainage regime, where oil can be displaced to a very low residual saturation (less than 10%). Once the water becomes well connected through the pore space occupying the larger pores, its relative permeability rises quickly. From Valvatne and Blunt (2004).

regions of the pore space and the relative permeability rises rapidly. Its maximum value is higher than for water-wet systems, since the residual oil saturation is lower (there is more water in the rock) and the water preferentially occupies the larger oil-wet pores.

The final set of sandstone curves are for an oil-wet reservoir rock. Again good predictions can be made. In Fig. 14.7, the layer drainage regime is evident; this is where the oil relative permeability is low, but allows flow down to a very low residual saturation. The water relative permeability can reach high values, once water is well connected through the pore space in the larger pores.

14.2. Imbibition and Oil Recovery Processes

There are two distinct recovery processes in oil fields when water is injected to displace brine. The first is direct displacement, shown in Fig. 14.8: water is injected and essentially pushes out oil. At the pore scale, we know that capillary forces dominate and the

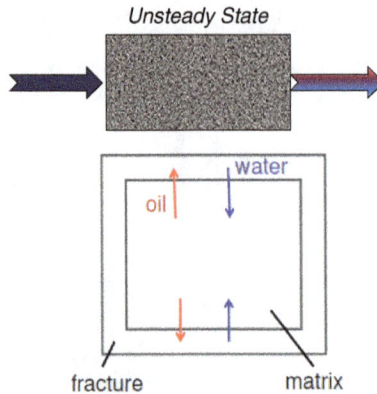

Figure 14.8. A schematic of the two types of recovery process in reservoirs. The top picture shows displacement, where water is injected to displace oil. This occurs in most unfractured reservoirs. However, where there is extensive fracturing, and these fractures provide high permeability paths for flow, the behaviour is different, as shown in the bottom picture. Here recovery occurs by imbibition of water into the matrix (normal unfractured rock).

recovery is controlled by the residual saturation. How fast this recovery occurs is controlled by the relative permeabilities; the ideal is a low water relative permeability that holds water back and a high oil relative permeability, allowing oil to flow rapidly and be displaced ahead of the water. This is discussed later through a rigorous mathematical treatment of the governing flow equations, but qualitatively, a low residual saturation says how much oil can be recovered in theory (a low residual indicates high recovery) while the relative permeability gives the rate at which recovery occurs.

The second process is imbibition. This is simple to imagine. It is the same as placing a piece of rock in water: the water spontaneously enters the rock under the influence of capillary pressure. This is the dominant recovery process in fractured reservoirs — typically seen for brittle rocks, such as carbonates. Here the injected water, rather than forcing out the oil, flows rapidly along the high permeability fractures. Then water enters the matrix (normal unfractured rock) by imbibition. This is a process controlled by capillary forces, with some help due to the density difference between water and oil that helps push the water into the bottom of a matrix block. Again this process is shown schematically in Fig. 14.8. In this case the recovery is controlled by how much oil remains after spontaneous imbibition; from our discussion of capillary pressure, this saturation (when the capillary pressure is zero) is much lower than the residual saturation for mixed-wet systems. The rate of recovery is controlled by the water relative permeability, typically at low saturation, since this is the saturation range of interest and limits how fast the oil can be displaced.

Both of these problems can be analysed through analytical solutions (in one dimension) to the governing flow equations; these are presented later (Secs. 17 and 18). Here we explain the results physically in terms of the relative permeabilities and capillary pressures.

In imbibition, the recovery as a function of time has a characteristic behaviour that has been studied by many authors. Shown in Fig. 14.9 is a compilation of 48 datasets in the literature, compiled by Schmid and Geiger (2012) in a classic paper that also presents a closed-form analytic solution to the flow equations for this problem (which is presented later, Sec. 18).

In Fig. 14.9, the recovery is plotted as a function of dimensionless time. A full discussion giving an analytical solution is provided in Sec. 18; however, we can use physical principles to estimate likely time-scales. We will show that this is a diffusive problem mathematically, so we can readily examine the likely scaling of the displacement.

The driving force is capillary pressure, which is the interfacial tension divided by a typical pore radius. As before (Sec. 8) we can relate this to the square root of the permeability divided by the porosity. Imagine that the wetting phase has invaded a distance x into the porous medium. Hence, the pressure gradient driving flow can be written

$$\frac{\partial P}{\partial x} = \frac{\sigma}{x}\sqrt{\frac{\phi}{K}}. \tag{14.2}$$

Then from the multiphase Darcy law, Eq. (14.1), assuming that flow is limited by the water relative permeability, we can find the flow

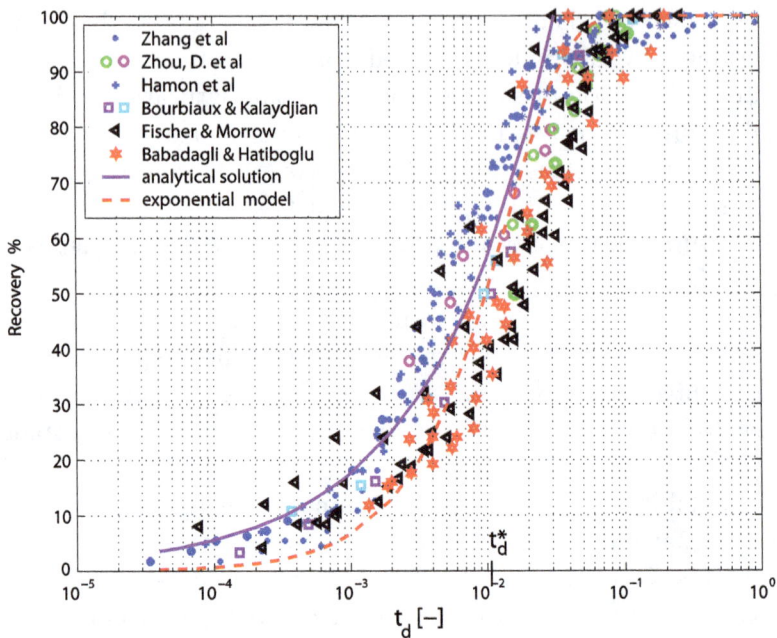

Figure 14.9. Recovery — as a percentage of the final recovery — as a function of dimensionless time for spontaneous imbibition for 48 experiments in the literature compiled by Schmid and Geiger (2012). The exponential model uses Eq. (14.6) with $\alpha = 0.05$.

rate, which determines how fast the distance x changes with time:

$$q = \phi \frac{dx}{dt} \sim \frac{\sigma K k_{rw}}{\mu_w} \sqrt{\frac{\phi}{K}} \frac{1}{x}. \qquad (14.3)$$

The porosity term for dx/dt converts a Darcy velocity into a speed. Equation (14.3) has the solution

$$x(t) = \sqrt{At}, \qquad (14.4)$$

where

$$A = \frac{\sigma k_{rw}}{2\mu_w} \sqrt{\frac{K}{\phi}}. \qquad (14.5)$$

Note that the distance travelled (and hence recovery) scales — at early time, before the imbibing front reaches the ends or boundaries of the system — as the square root of time. This mathematically and physically is a diffusive process, as opposed to recovery by direct displacement where recovery and front movement increases linearly with time.

Eventually, the wetting front reaches the end of the system (say a distance $x = L$); from then on recovery is much slower. Empirically — simply a match to the compilation of recovery results shown previously — we find that the recovery can be written as

$$R = R_\infty \left(1 - e^{-\alpha t_D}\right),\tag{14.6}$$

where R is the oil recovery, R_∞ is the ultimate recovery, α is a constant used to match the data and t_D is a dimensionless time. This is the analytical match shown in Fig. 14.9. In our analysis, it would be given by

$$t_D = t\frac{\sigma k_{rw}}{\mu_w L^2}\sqrt{\frac{K}{\phi}}.\tag{14.7}$$

However, this is not necessarily accurate, as this was a simplistic analysis; a more complex but analytically correct expression that accounts for the flow of both water and oil is found in Schmid and Geiger (2012); this is presented later in Sec. 18.

We can use Eq. (14.7) though to estimate time-scales for imbibition recovery. What is a typical imbibition time for a water-wet rock of size 1 cm? This is the real time necessary to have t_D around 1 in Eq. (14.7). Using $\sigma = 0.04\,\text{N/m}$, $\mu_w = 10^{-3}\,\text{Pa}\cdot\text{s}$, $K = 10^{-14}\,\text{m}^2$ (10 mD) and $\phi = 0.2$, we find, for a typical end-point relative permeability value of 0.1 (a water-wet rock), times around 100 s; imbibition is typically quite quick for small systems. What about a matrix block 10 m across? Notice that the time-scale increases as length squared, and so in this case imbibition takes a million times longer, or around 3 years.

A further complexity arises when we consider mixed-wet systems. Shown in Fig. 14.10 is a comparison of waterflood and imbibition

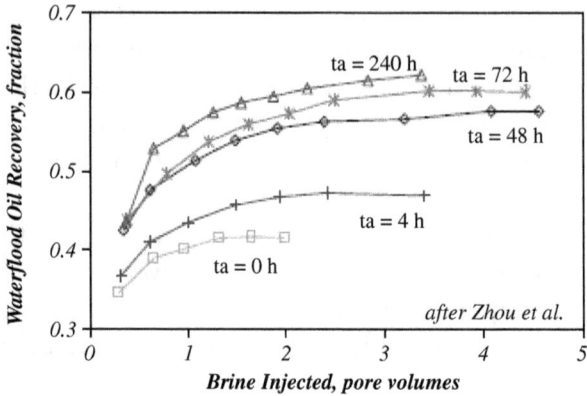

(a)

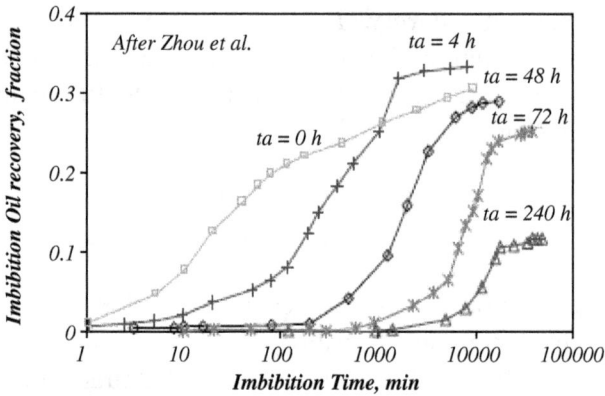

(b)

Figure 14.10. Waterflood recovery (a) and imbibition recovery (b) for sandstone cores aged for the length of time (in hours) indicated on the graphs. The more the core is aged in crude oil (essentially soaked in crude for different amounts of time), the more mixed-wet in character it becomes. No ageing is least favourable for waterflooding, because of the high residual oil saturation in this water-wet case, but is favourable and fastest if recovery is controlled by imbibition. From Behbahani and Blunt (2005) based on the experiments of Zhou *et al.* (2000).

recoveries as a sandstone core becomes more mixed-wet in character. Imbibition becomes less favourable as more of the pore space becomes oil-wet, since there is no recovery from these regions. Furthermore, recovery is much slower, since the water relative permeability is very low, as discussed above; recall that if there is little imbibition, then the water flow is governed by displacement at low water saturations. This makes an enormous difference to recovery rates and may make recovery uneconomic in a field setting; if we consider our previous example, but with, say, a 1 m block size and a representative but low matrix block permeability (say 1 mD) and relative permeability (say 10^{-4}) in Eq. (14.7), we find an imbibition time of 100 years, which is uneconomic for field-scale recovery.

In contrast, the waterflood recovery improves as the system becomes more mixed-wet. This is a consequence of the lowered residual oil saturation. Also, as discussed in more detail later, the low water relative permeability holds back the injected water, allowing oil to escape and providing — in this case — a favourable displacement efficiency.

14.3. Analysis of Relative Permeability in Mixed-wet Carbonates

In this section, we will go through a network analysis of relative permeability, to show how we predict multiphase flow properties, their behaviour and how this relates to field-scale recovery. We will also compare the results against experimental data in the literature. The emphasis in this section will be on mixed-wet systems, which comprise the vast majority of carbonate rocks, which in turn contain most of the world's remaining reserves of conventional oil, mainly in the Middle East. Most of the analysis is taken from Gharbi and Blunt (2012).

14.3.1. *Pore Structure and Connectivity*

We start by showing images of the carbonates that we will study and the networks extracted from these images (Fig. 14.11). This forms the basis of the modelling.

A detailed description of the extracted networks is provided in Table 14.1. The samples cover a wide range of average coordination numbers: ME1 and Portland are poorly connected with coordination numbers of approximately 2.5 whereas Guiting and Mount Gambier

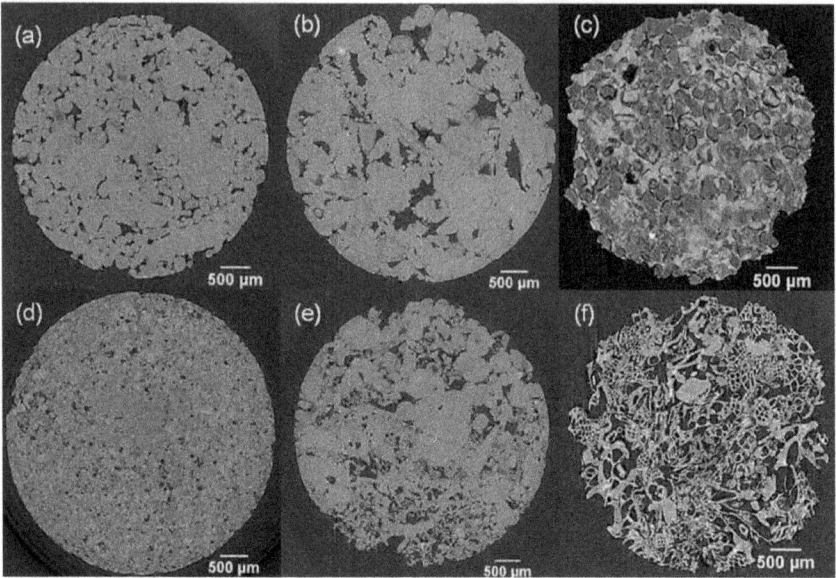

Figure 14.11. 2D cross-sections of 3D micro-CT images of different carbonate samples. (a) Portland limestone; (b) Indiana limestone; (c) Guiting carbonate; (d) Middle Eastern Carbonate 1, a carbonate sample from a deep, highly saline Middle Eastern aquifer; (e) Middle Eastern Carbonate 2, a second sample from a deep, highly saline Middle Eastern aquifer; (f) Mount Gambier limestone.

are highly connected with average coordination numbers of 5.1 and 7.4, respectively. As we show later, the average coordination number (average number of throats connected to a single pore) is a key determinant of relative permeability and residual saturation. It is derived from the network extraction analysis and is an indicator of the connectivity of the void space.

The pore and throat distributions of the networks are presented in Figs. 14.3 and 14.4.

Capillary controlled displacement is simulated using the pore network model developed by Valvatne and Blunt (2004). Initially, the medium is assumed to be filled with the wetting phase (brine) and oil is then injected. After oil invasion, we alter the wettability of the pore spaces in direct contact with oil to represent mixed-wet conditions. Waterflooding is then simulated and relative permeability curves are generated.

Table 14.1. Description of the extracted networks.

	ME1	Portland	Indiana	ME2	Guiting	Mount Gambier
Voxel resolution (μm)	7.7	9	7.7	7.7	7.7	9
Number of voxels	380^3	320^3	330^3	320^3	350^3	350^3
Physical volume (mm^3)	25.05	23.89	16.41	14.96	19.57	31.26
Number of pores	55828	6129	5653	10855	25707	22665
Number of throats	70612	7939	8539	20071	66279	84593
Total number of elements	126440	14068	14192	30926	91986	107258
Average coordination number	2.50	2.53	2.97	3.64	5.11	7.41
Min pore radius (μm)	7.7	9	7.7	7.7	7.7	9
Max pore radius (μm)	51.52	93.51	99.48	107.82	74.09	119.88
Average pore radius (μm)	8.44	14.89	10.17	10.90	11.16	18.17
Average aspect ratio	1.87	2.28	1.88	2.08	2.00	2.59
Porosity (%)	14.37	9.32	13.05	18.60	29.79	56.27
Permeability (m^2)	3.23×10^{-14}	1.37×10^{-13}	5.69×10^{-13}	9.40×10^{-13}	3.72×10^{-13}	2.20×10^{-11}

Average coordination number is the average number of throats connected to each pore. The average aspect ratio is the average of the ratio of the pore radius to the mean radius of the throats connected to it. The permeability is computed from a flow simulation through the network.

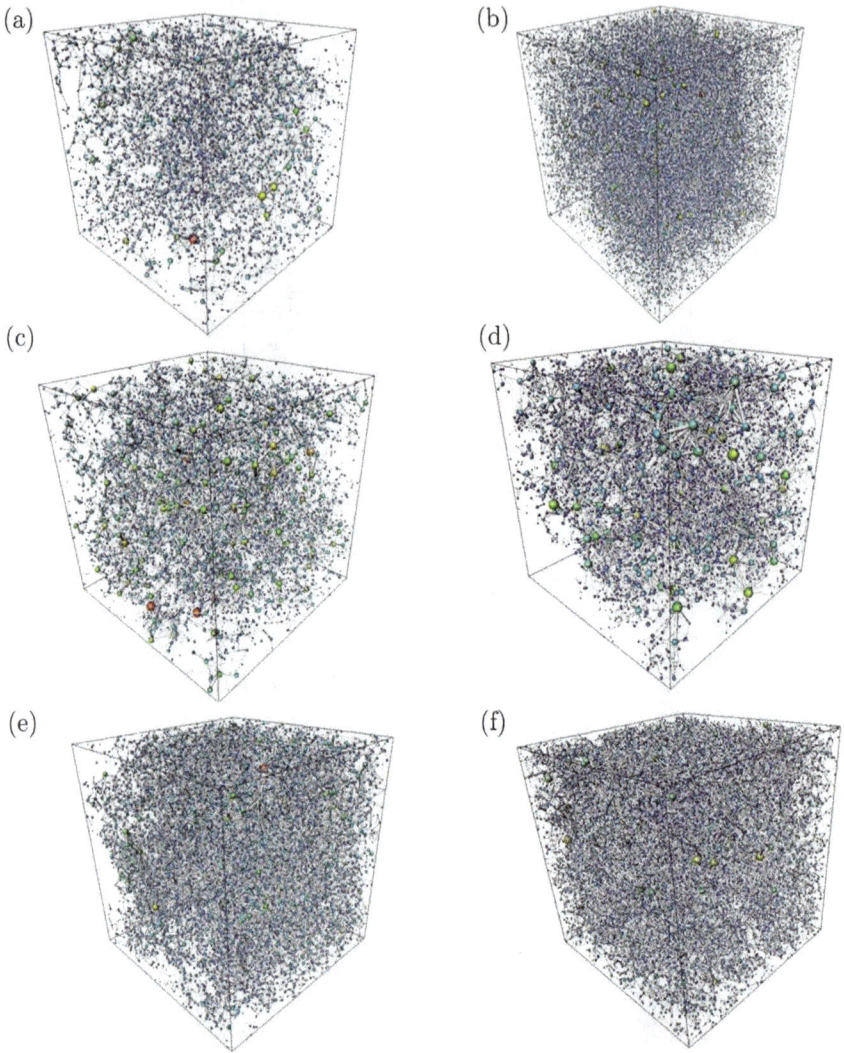

Figure 14.12. Pore networks extracted from the images shown in Fig. 14.11. The pore space is represented by a lattice of pores (represented by spheres) and throats (represented by cylinders); in cross-section each pore and throat is a scalene triangle.

We study the impact of wettability in mixed-wet media where some fraction, f, of the pore space occupied by oil is made oil-wet and a fraction $1 - f$ remains water-wet. We vary the oil-wet fraction from zero (a strongly water-wet case) to 1 (strongly oil-wet rock). In

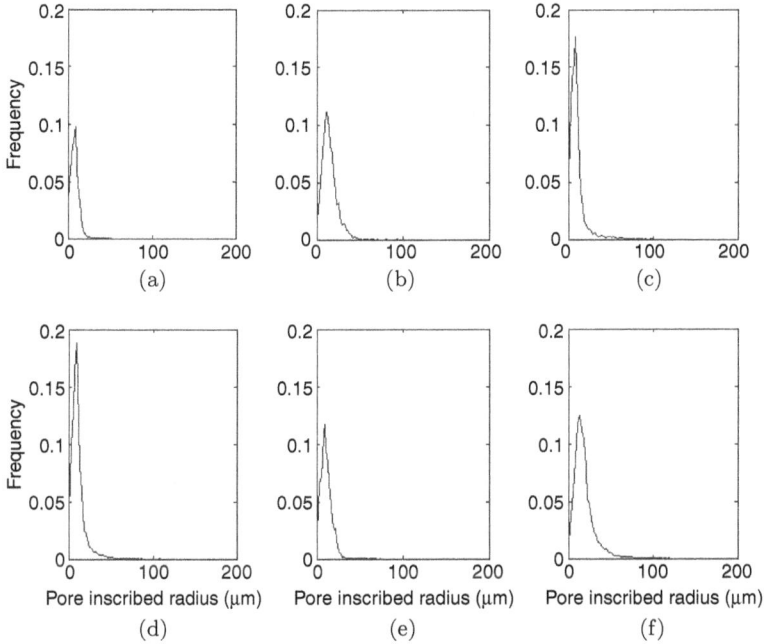

Figure 14.13. Pore inscribed radius distributions for (a) Middle Eastern sample 1; (b) Portland limestone; (c) Indiana limestone; (d) Middle Eastern sample 2; (e) Guiting carbonate; and (f) Mount Gambier limestone. In this and subsequent figures, samples are presented in order of increasing coordination number: from a low connectivity sample (a) to a very high connectivity sample (f). From Gharbi and Blunt (2012).

addition to modelling mixed-wet media, this methodology reproduces wettability alteration which is due to asphaltene deposition/precipitation in carbonates. This alteration, governed by oil composition, brine salinity and rock mineralogy is difficult to predict *a priori*.

Where oil has been in contact with the carbonate surface (pores and throats), random contact angles with no spatial correlation are assigned with different distributions (given in Table 14.2) for the water-wet and oil-wet pores and throats.

The 3D networks are composed of individual elements (pores and throats) with circular, triangular or square cross-sectional shapes. Using square or triangular shaped networks elements allows for the explicit modelling of wetting layers where non-wetting phase occupies

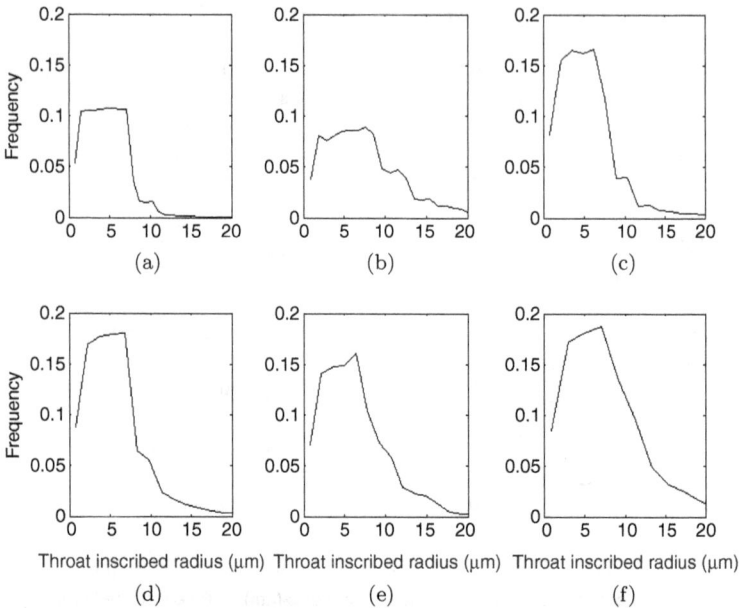

Figure 14.14. Throat inscribed radius distributions for (a) Middle Eastern sample 1; (b) Portland limestone; (c) Indiana limestone; (d) Middle Eastern sample 2; (e) Guiting carbonate; (f) Mount Gambier limestone. Samples are presented in order of increasing coordination number. From Gharbi and Blunt (2012).

Table 14.2. Input parameters for relative permeability computations.

Input Parameters	
Initial contact angle (degrees)	0
Interfacial tension (mN/m)	48.3
Water-wet contact angles (degrees)	0–60
Oil-wet contact angles (degrees)	100–160
Oil viscosity (mPa·s)	0.547
Water viscosity (mPa·s)	0.4554

the centre of the element and wetting phase remains in the corners. The pore space in carbonates is highly irregular with water remaining in the grooves and crevices after primary oil flooding due to capillary forces. The wetting layers might not be more than a few microns in thickness, with little effect on the overall saturation or flow. Their contribution to wetting phase connectivity is, however, of vital importance, ensuring low residual wetting phase saturation by preventing trapping. Wetting layers of water are always present in the corners, while layers of oil sandwiched between water in the corners and water in the centre can be observed in oil-wet regions. Layer drainage is when oil flows in these layers, allowing, slowly, very low saturations to be reached.

14.3.2. *Effect of Fractional Wettability on Relative Permeability*

Five wettability distributions are studied: $f = 0$, $f = 0.25$, $f = 0.5$, $f = 0.75$ and $f = 1$. For the water-wet case ($f = 0$), as expected, water remains in the smallest portions of the pore space, giving very low water relative permeability and significant trapping of oil in the larger pores at the end of waterflooding, mainly caused by snap-off (see Fig. 14.15). In the case of poorly connected carbonates (ME1, Portland and Indiana limestones), up to 75% of the pore space can be trapped. However, for the better connected networks, namely ME2, Guiting and Mount Gambier, the water relative permeability is higher and there is less trapping (there are more pathways for the oil to escape), although the residual saturation is around 40% or higher in all cases.

For a mixed-wet case with $f = 0.25$ (Fig. 14.16), the small fraction of oil-wet pores tends to increase the amount of oil trapping, particularly in the less connected networks where now there is little or no range of saturation when two phases flow simultaneously, except very slow flow in wetting layers. The water phase connectivity is reduced and the water relative permeability is in general lower than the strongly water-wet case. The water-wet regions fill first in a capillary-controlled displacement at the pore scale. These are the small pores and poorly connected; however, they surround most of the oil-wet pores that are then trapped. These pores cannot then be displaced during forced water injection, which explains the increase in residual oil saturation. Here again, for the highly connected networks, the water relative permeability is higher since the water has more

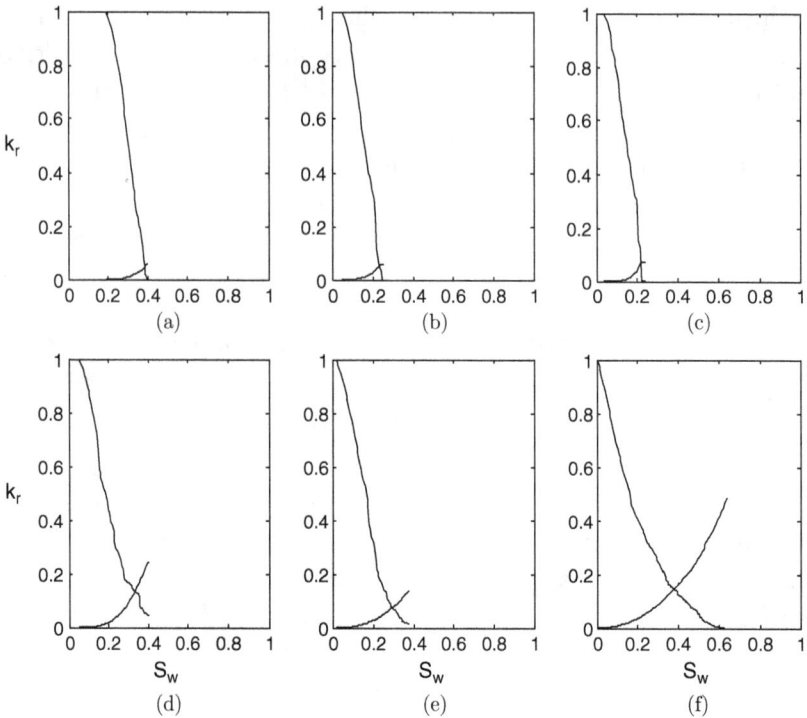

Figure 14.15. Waterflood relative permeability for the strongly water-wet case ($f = 0$). Curves are presented in order of increasing connectivity: (a) Middle Eastern sample 1; (b) Portland limestone; (c) Indiana limestone; (d) Middle Eastern sample 2; (e) Guiting carbonate; (f) Mount Gambier limestone. From Gharbi and Blunt (2012).

possible pathways through the system and there is both spontaneous and forced displacement by water.

When the fractional wettability is 0.5 (Fig. 14.17), an equal mix of water-wet and oil-wet pores, at low water saturations, a similar behaviour is observed regardless of the connectivity of the pore space. At the beginning of the waterflooding, the water is still poorly connected and flows only through the smallest water-filled pores and thin wetting layers of the pore space; therefore the water relative permeability is low. However, in an equal mix of water-wet and oil-wet fractions of the pore space, depending on the connectivity, an important increase in the water relative permeability is noticeable.

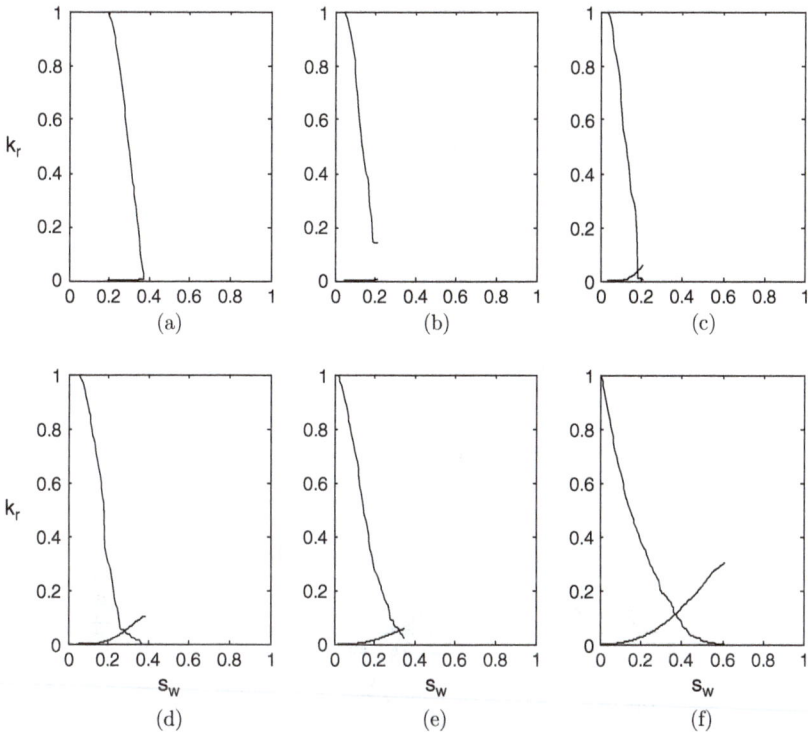

Figure 14.16. Waterflood relative permeability for the mixed-wet case ($f = 0.25$). Curves are presented in order of increasing connectivity: (a) Middle Eastern sample 1; (b) Portland limestone; (c) Indiana limestone; (d) Middle Eastern sample 2; (e) Guiting carbonate; (f) Mount Gambier limestone. From Gharbi and Blunt (2012).

After spontaneous imbibition, a significant forced displacement of oil occurs as the oil-wet pores and throats connect through the network. The residual oil saturation is generally lower since oil remains connected in the oil-wet region in layers. This effect is noticeable in the shape of the oil relative permeability for the well-connected samples, which show a long region where the oil relative permeability is very low, but there is still displacement; this behaviour is controlled by slow flow in oil layers. The poorly connected samples still show a water-wet controlled behaviour, where there is a sharp decrease in the oil relative permeability and significant trapping. Here there is little connectivity of the oil-wet regions and as a consequence layer

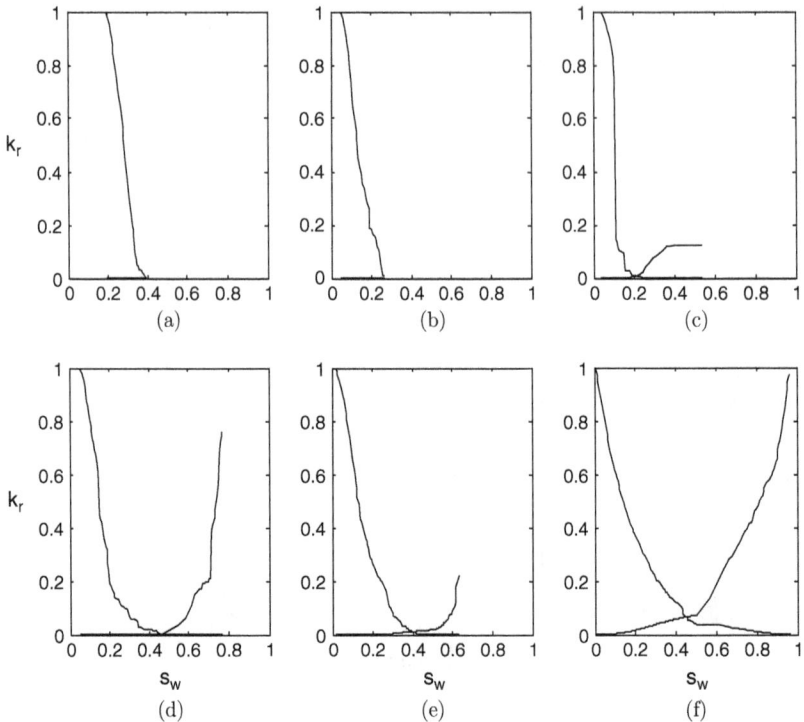

Figure 14.17. Waterflood relative permeability for the mixed-wet case ($f = 0.5$). Curves are presented in order of increasing connectivity: (a) Middle Eastern sample 1; (b) Portland limestone; (c) Indiana limestone; (d) Middle Eastern sample 2; (e) Guiting carbonate; (f) Mount Gambier limestone. From Gharbi and Blunt (2012).

drainage is unable to achieve low residual saturations. In addition, the maximum water relative permeability varies from very low to very high values dependent on the degree of trapping and the connectivity of the water phase. Where the residual saturation is low, water can fill most of the pore space — and the larger pores in the oil-wet regions — and has a high end-point value. A wide range of behaviour is seen in this case dependent on the pore structure of the medium.

When the oil-wet fraction is higher, $f = 0.75$ (Fig. 14.18), the residual saturation is now very low as the oil remains connected in layers throughout the displacement. The water relative permeability can rise to high values in all cases as the water fills the centres of

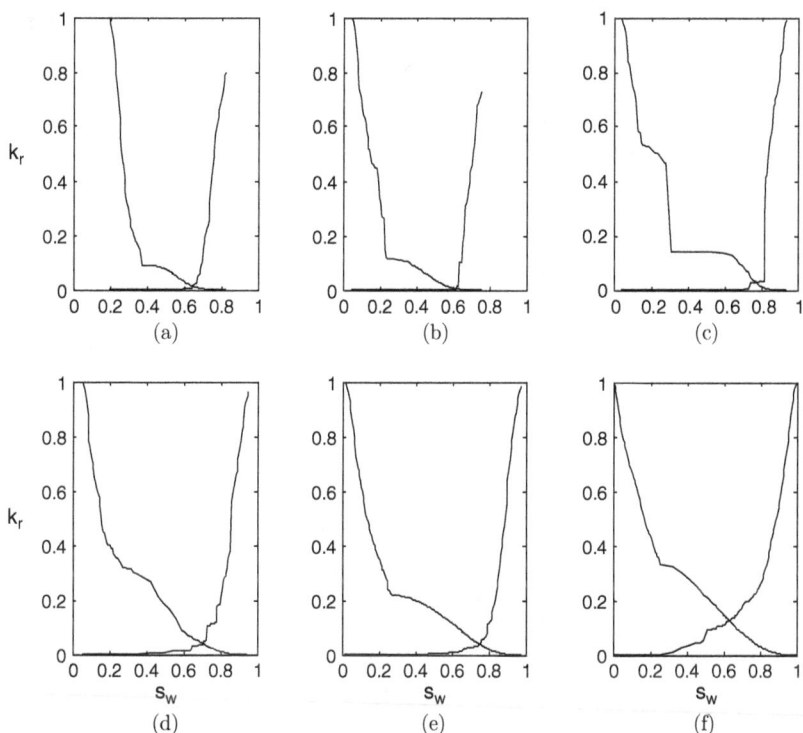

Figure 14.18. Waterflood relative permeability for a mixed-wet case ($f = 0.75$). Curves are presented in order of increasing connectivity: (a) Middle Eastern sample 1; (b) Portland limestone; (c) Indiana limestone; (d) Middle Eastern sample 2; (e) Guiting carbonate; (f) Mount Gambier limestone. From Gharbi and Blunt (2012).

the larger regions of the pore space. This is a sign of a more typical oil-wet behaviour with displacement over a wide saturation range and low relative permeabilities of both oil and water at low saturations of their respective phases, controlled by wetting layer flow. This behaviour is generically similar to network modelling calculations for sandstones (Valvatne and Blunt, 2004; Zhao *et al.*, 2010). The jumps in some of the curves reflect the relatively small size of the networks studied; improvements in imaging should soon allow larger networks to be constructed.

For the fully oil-wet case ($f = 1$) the behaviour is generally quite similar to that observed for the mixed-wet case ($f = 0.75$): very low

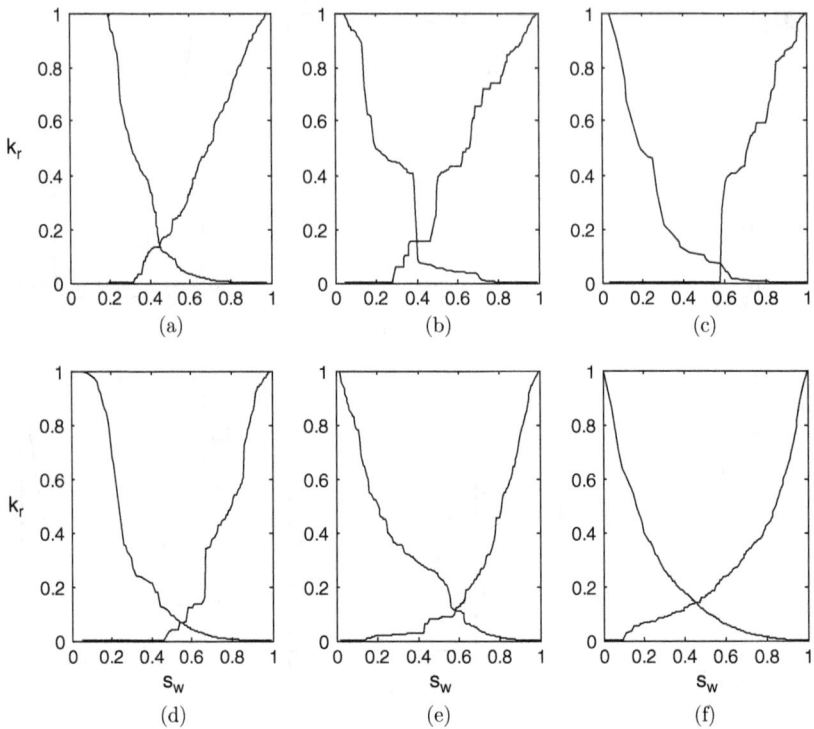

Figure 14.19. Waterflood relative permeability for the strongly oil-wet case ($f = 1$). Curves are presented in order of increasing connectivity: (a) Middle Eastern sample 1; (b) Portland limestone; (c) Indiana limestone; (d) Middle Eastern sample 2; (e) Guiting carbonate; (f) Mount Gambier limestone. From Gharbi and Blunt (2012).

residual oil saturation, a prolonged layer drainage regime (low oil relative permeability at low oil saturation) and high end-point water relative permeability (Fig. 14.9).

To summarise the previous description: we analyse the impact of wettability and average coordination number on the relative permeability behaviour. The evolution of residual oil saturation with the fractional wettability shows that the residual oil saturation reaches a maximum for the fractionally wet case with $f = 0.25$, and then decreases sharply to very low saturations as the medium becomes more oil-wet. Waterflooding gives a high local displacement

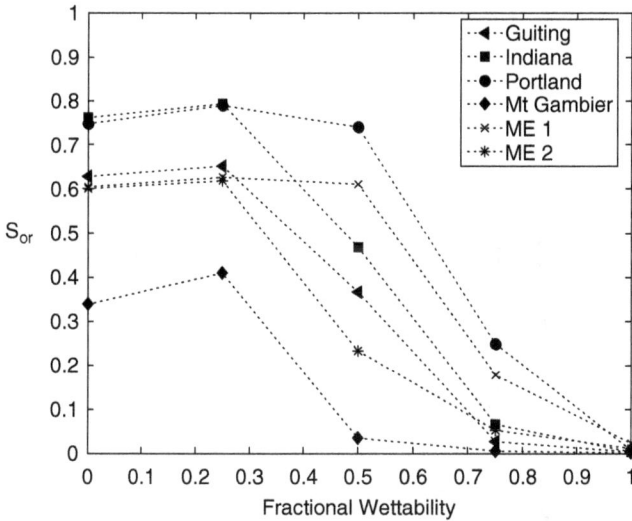

Figure 14.20. Residual oil saturation as a function of fractional wettability for: Guiting (triangles), Indiana (rectangles), Portland (circles), Mount Gambier (diamonds), Middle Eastern sample 1 (crosses), and Middle Eastern sample 2 (stars). From Gharbi and Blunt (2012).

efficiency for the cases $f = 0.75$ and $f = 1$, where the behaviour is controlled by oil layers.

The impact of connectivity on the residual oil saturation is shown in Figs. 14.20 and 14.21. The residual oil saturation tends to decrease with increasing connectivity, regardless of wettability.

One indication of waterflood displacement efficiency that is used to characterise the wettability is the water saturation value at which the oil and water relative permeabilities are equal (S_w where $k_{rw} = k_{ro}$) (Craig, 1971). For water saturations higher than the crossover saturation, waterflooding becomes less efficient, since (for equal viscosities) more water flows than oil. The water saturation at the crossover as a function of wettability for the different carbonate samples is shown below. In most cases, the water saturation is highest for the mixed-wet case $f = 0.75$. This confirms that waterflooding is most effective for mixed-wet carbonates that have preference to an oil-wet behaviour. The smallest water saturation at the crossover point is reached for the water-wet and weakly mixed-wet cases

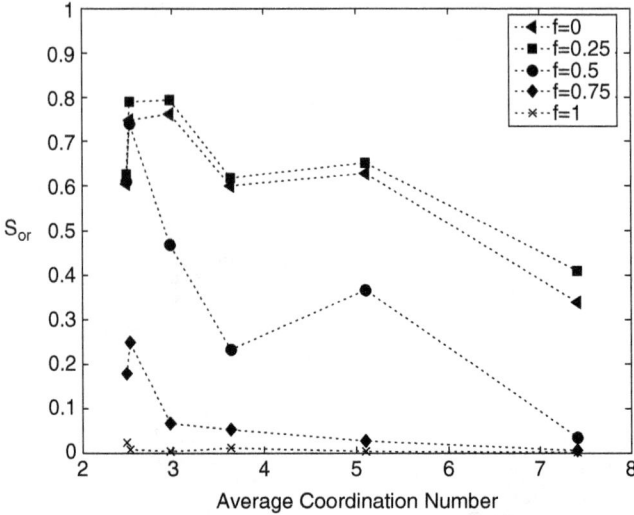

Figure 14.21. Residual oil saturation as a function of the average coordination number for: Guiting (triangles), Indiana (rectangles), Portland (circles), Mount Gambier (diamonds), Middle Eastern sample 1 (crosses), and Middle Eastern sample 2 (stars). From Gharbi and Blunt (2012).

($f = 0.25$); these are least efficient for waterflooding. This contrasts with traditional analyses of relative permeability which suggests that the crossover point is at more than 50% water saturation for water-wet cases and less than 50% water saturation for mixed-wet or oil-wet samples (Craig, 1971). We only see this trend in the near oil-wet region; this rule does not apply in general because of the low estimated water relative permeability.

14.4. Comparison of Network Model Results with Experimental Data

We will now compare our computations to measurements found in the literature on reservoir carbonate samples. The approach is not necessarily genuinely predictive as scans of the reservoir samples and an independent measurement of wettability are not available; we simply make an assessment whether the estimated connectivity and wettability are plausible for the experimental sample studied. Also, the objective of this comparison is not to have a perfect match between the laboratory measurements and the results of the network modelling by fine-tuning the oil-wet fraction or the contact angles; rather, the goal is to determine if our calculated behaviour is supported by the available experimental evidence and discuss the impact of wettability and pore structure on field-scale recovery.

We study three sets of waterflood relative permeabilities measured on Middle Eastern carbonate reservoir samples. A summary of the petrophysical and geological description of the samples is provided in Table 14.3.

Case 1. Al-Sayari (2009) measured steady-state waterflood relative permeability on an aged (restored state) reservoir carbonate sample from the Middle East. Through analysis of thin sections, mercury injection capillary pressure and NMR response, the reservoir sample was described as having a well-connected pore structure with a relatively low fraction of micro-porosity (Fig. 14.22).

A similar relative permeability to that measured can be observed for the case of $f = 0.25$ for the well-connected Guiting and Mount Gambier networks. The relatively low residual oil saturation and the shape of the oil relative permeability curve indicate a mixed-wet behaviour. For Guiting, the discrepancies in the water relative permeability can be explained by the unresolved micro-porosity.

Case 2. Meissner *et al.* (2009) performed detailed measurements on several samples from the Arab-D reservoir of the Dukhan field, onshore Qatar. They reported the results of several steady-state relative permeability tests for oil/brine and gas/oil systems. Results were reported for both native and the restored state cores. The results were reported in terms of normalized saturations and relative permeabilities:

$$S_{wn} = \frac{S_w - S_{wi}}{1 - S_{wi} - S_{or}}, \tag{14.8}$$

Table 14.3. A summary of the petro-physical and geological descriptions of the reservoir samples found in the literature.

	Wettability	Wettability Measurement	Geology	Lithology	NMR Description
Al-Sayari (2009)	Mixed-wet	N/A	Kharaib formation	Dual pore system	Multi-modal, micro-porosity
Meissner et al. (2009)	Mixed-wet preference to oil	USBM	Arab-D reservoir	Lime grainstone	Complex multi-modal pore structures, micro-porosity
Meissner et al. (2009)	Mixed-wet preference to oil	USBM	Arab-D reservoir	Lime mudstone	
Meissner et al. (2009)	Mixed-wet preference to oil	USBM	Arab-D reservoir	Lime grainstone	
Meissner et al. (2009)	Mixed-wet preference to oil	USBM	Arab-D reservoir	Lime grainstone	
Okasha et al. (2007)	Neutral to slightly water-wet	Amott	Arab-D reservoir Haradh area	N/A	N/A
Okasha et al. (2007)	Generally oil-wet to intermediate-wet	Static imbibition Amott USBM	Arab-D reservoir Utmaniyah area	N/A	N/A

Note: USBM stands for US Bureau of Mines and is a somewhat cumbersome method to measure wettability: it measures the ratio of the area under the capillary pressure curves for spontaneous and forced displacement.

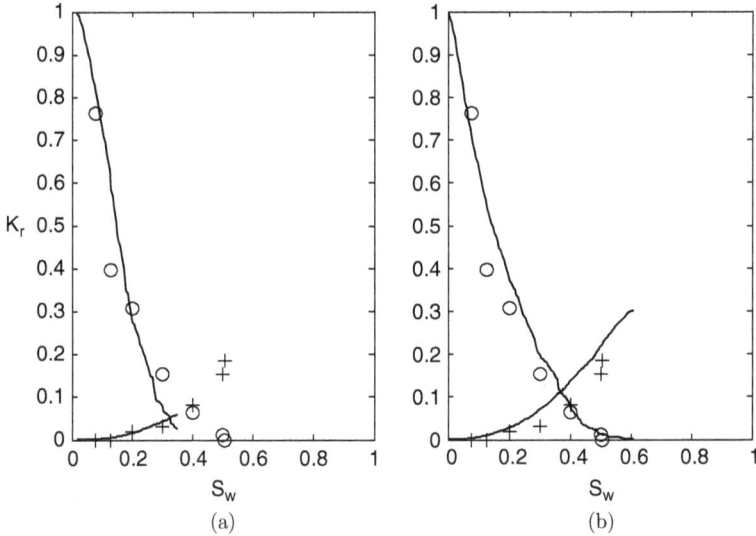

Figure 14.22. Comparison between relative permeability measurements from a Middle Eastern reservoir (oil relative permeability, circles; water relative permeability, crosses; Al-Sayari, 2009) with (a) Guiting limestone and (b) Mount Gambier limestone for a fractional wettability of $f = 0.25$. From Gharbi and Blunt (2012).

where S_{wi}, the initial water saturation, is determined after primary drainage and S_{or} is the residual oil saturation determined by extrapolation of the oil relative permeability as it asymptotically approaches zero.

In this case, to introduce an initial water saturation, we set the maximum primary drainage capillary to be equal to 690 kPa (approximately 100 psi). This value is chosen based on the different capillary pressure measurements that showed a sharp increase in the pressure for an average pressure of around 100 psi.

Figure 14.23 shows a comparison between the four measurements reported of water/oil relative permeability on the native state sub-surface cores with the relative permeability generated for the strongly oil-wet case $f = 1$ for ME1. The suggestion here is that the reservoir is strongly oil-wet with a structure similar to that observed in the sub-surface sample from which we extracted a network.

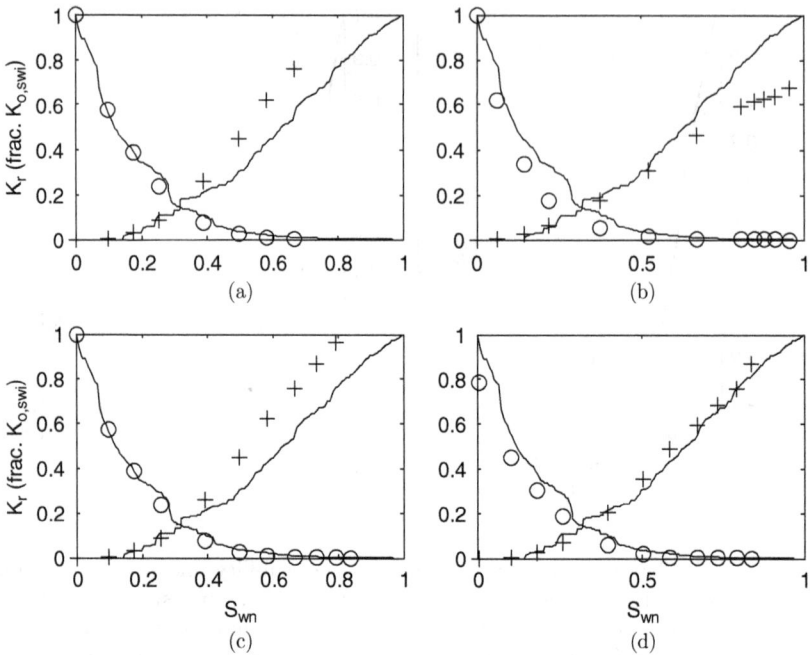

Figure 14.23. A comparison between the waterflood relative permeability for the Middle Eastern sample 1, for a strongly oil-wet case $f = 1$ with measurements on native state sub-surface reservoir cores (oil relative permeability, circles; and water relative permeability, crosses) obtained from Meissner *et al.* (2009). From Gharbi and Blunt (2012).

Case 3. Okasha *et al.* (2007) reported unsteady-state relative permeability measurements on carbonate reservoir samples from the Arab-D reservoir of the Ghawar field in Saudi Arabia. This is the world's largest conventional oil field. Three data sets were presented for three samples obtained from different areas of the Ghawar field: Utmaniyah, Hawiyah and Haradh. Here, since the measured values are presented in a non-normalised form, we simply compare with the data without changing the initial water saturation.

Figure 14.24 shows a good agreement between the measurements and the relative permeability generated by network modelling for the mixed-wet Mount Gambier network ($f = 0.25$) for one of the three samples. Note that we suggest that in this field the wettability and pore structure are different from the sub-surface Middle Eastern

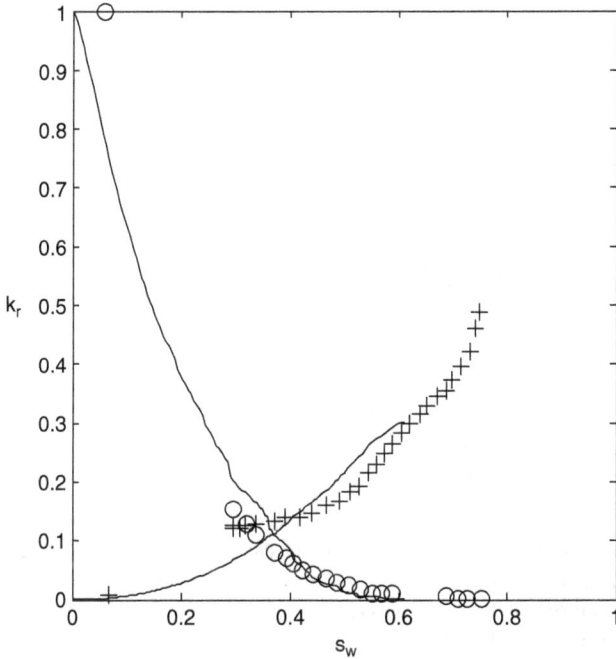

Figure 14.24. Mount Gambier waterflood relative permeability for the mixed-wet case with an oil-wet fraction of $f = 0.25$ (solid) compared to measurements on a reservoir sample obtained from Okasha *et al.* (2007) (oil relative permeability, circles; water relative permeability, crosses). From Gharbi and Blunt (2012).

sample shown previously. Figure 14.25 shows good agreement for the second measured sample with low connectivity carbonates, i.e. Portland and ME1 for a strongly oil-wet case. The difference of the wettabilities is evidence of local variations of wettability within the reservoir.

Good agreement was not obtained for the third sample which had high connate water saturation.

14.5. Impact of Relative Permeability on Field-scale Recovery

In waterflooding, for oil and water of similar viscosity, the saturation at which the relative permeabilities cross — as discussed

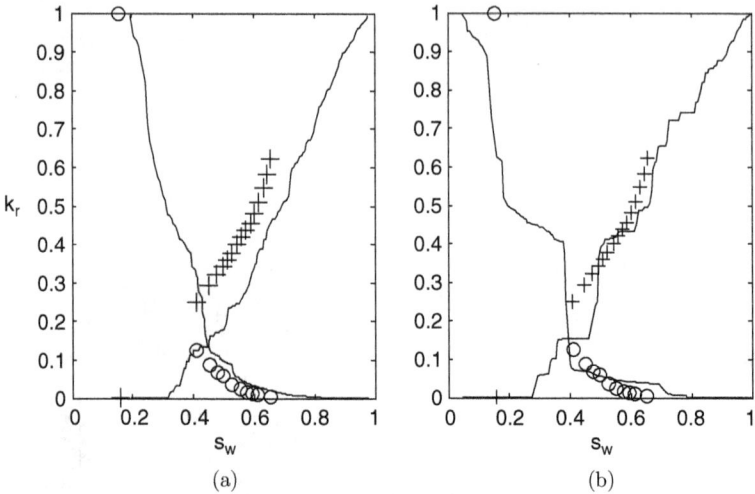

Figure 14.25. Middle Eastern sample 1 (b) and Portland limestone (a) waterflood relative permeability for the strongly oil-wet case with an oil-wet fraction of $f = 1$ (solid) compared to measurements on reservoir samples obtained from Okasha *et al.* (2007) (oil relative permeability, circles; water relative permeability, crosses). From Gharbi and Blunt (2012).

below — gives a useful and simple indicator of the recovery. For water saturations beyond the crossover point, more water will be produced than oil — beyond this point, oil production becomes increasingly uneconomic. Hence, a rough guide to recovery can be derived from the change in saturation from its initial value to when the relative permeabilities cross.[1] Later, we show how to perform this analysis rigorously and predict — for given relative permeabilities and fluid viscosities — the amount of oil recovered as a function of the amount of water injected.

Our simulations indicate that the optimal waterflood efficiency is observed for a mixed-wet system with a large fraction of oil-wet pores, around 0.75. The highest waterflood efficiency is implied for

---

[1]Please note that this change in saturation is not a recovery factor. For this, we need to compute the volume of oil produced, convert it to surface conditions and then divide by the total volume of oil initially in the reservoir (also measured at surface conditions). This is an approximate physically motivated assessment of recovery and no excuse for not doing a proper analysis, as described later.

the less well-connected samples, since in these cases the waterflood relative permeability is very low and this holds back the movement of water, allowing oil to be displaced. For better connected samples, there is less sensitivity to wettability and overall a lower crossover saturation, indicating less favourable recoveries.

This is a somewhat surprising conclusion and implies that waterflooding in mixed to oil-wet carbonates of poor pore-space connectivity may be an effective process. This behaviour stands in contrast to sandstones, where network modelling studies indicate that more neutrally-wet conditions provide optimal recovery (Øren *et al.*, 1998; Valvatne and Blunt, 2004). Moreover, experimental measurements presented by Jadhunandan and Morrow (1995) have shown that oil recovery by waterflooding in sandstones reach a maximum at close to neutral wettability.

As mentioned at the beginning of this section there are two distinct recovery processes in carbonates, depending on whether or not fractures dominate the flow. If they do not, then viscous forces are significant for displacement through the porous matrix and local recovery is determined by the relative permeabilities. It is possible to perform a Buckley–Leverett analysis to compute, analytically, recovery for a homogeneous 1D displacement from the relative permeabilities; the method to do this is presented later in these notes (Sec. 17). However, as mentioned previously, the likely local waterflood displacement efficiency can be estimated rapidly from direct inspection of the relative permeability curves. Imagine that the reservoir-condition oil and water viscosities are the same. Then, if the saturation near the production well is where the relative permeabilities cross, then the sub-surface ratio of oil to water production will be 1:1. Wells are abandoned when the cost of recycling and processing the produced water exceeds the economic benefit of the oil produced; this is normally when the oil/water ratio is between 1:2 and 1:10. On the other hand, the oil viscosity is typically greater than that of water, and the flow rate is determined by the ratio of relative permeability to viscosity. Hence, in most cases, production ceases close to where the relative permeabilities cross — between the producers, where water has displaced oil, the saturations will be higher, but this very simple trick allows a quick comparative study of recovery trends. Hence, waterflooding is quite favourable in the less well-connected carbonate samples. Most of the moveable pore volume is displaced, and the residual saturation is low. The reason for this is that the poorly connected water phase holds back water advance, allowing the efficient displacement of oil. For better-connected samples, the water rapidly finds a pathway of large pores through the system. This allows water to bypass oil at the pore scale, leading to less favourable waterflood recovery.

Now consider a reservoir where flow is dominated by fractures. In this case the fractures effectively short circuit the flow field and it is not possible to impose a substantial viscous pressure drop across the matrix. Recovery is mediated by capillary and gravitational forces. Imagine that water quickly invades the fractures surrounding a region of matrix (a so-called matrix block, although it does not have to be exactly, or even remotely, cuboidal in shape). Then recovery will occur by spontaneous imbibition — i.e. recovery will occur until the capillary pressure is zero. Figure 14.26 shows the capillary pressures for three carbonate samples. These are not the same samples as already discussed, with the exception of Mount Gambier, with a well-connected pore space. Ketton and Estaillades are both more poorly connected and are — for the sake of this discussion — similar in behaviour to the other low coordination number samples: Portland, Indiana and ME1.

In our examples in the figures above this means that only around 25% of the moveable pore volume is recovered. Furthermore, the rate of recovery is limited by the rate at which water can advance into the pore space — i.e. the relative permeability in the low water saturation range where the capillary pressure is positive. In this case the more favourable system is now the Mount Gambier — the water relative permeability is higher, indicating a more rapid displacement, while the degree of spontaneous imbibition is larger, since the well-connected pore space allows all the water-wet regions of the rock to be accessed easily; in contrast the poorly connected Ketton and Estaillades have a lower water relative permeability and

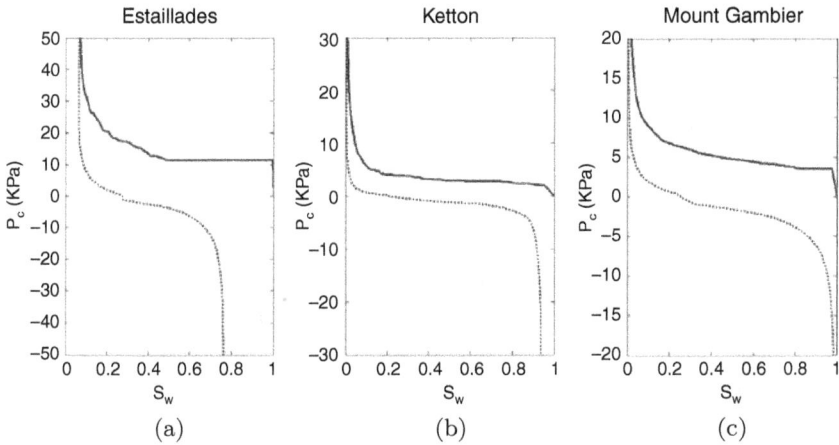

Figure 14.26. Primary drainage (solid line) and waterflood (dotted line) capillary pressures predicted using pore-scale network modelling. Here the oil-wet fraction f is 0.75. If recovery occurs by imbibition — in a fractured reservoir — the final recovery is controlled by the saturation when the capillary pressure is zero, not the residual saturation. From Gharbi and Blunt (2012).

not all the water-wet regions of the pore space are interconnected, leading to less displacement at a positive capillary pressure.

Gravitational forces can also play an important role in the displacement. If water floods a vertical fracture then oil, being less dense, is preferentially produced from the top of the matrix, and the weight of water in the fracture acts as a driving force. If we assume that the capillary pressure in the fractures is very small and is equal to zero at the top of a matrix block, then the capillary pressure at the base is $\Delta\rho g h$, where $\Delta\rho$ is the density difference between water and oil and h is the effective height of the matrix block. The capillary pressure is negative; the water has a higher pressure than oil. This allows forced displacement to a lower oil saturation. If we take typical values, $g = 9.81\,\mathrm{m\cdot s^{-2}}$; $\Delta\rho = 300\,\mathrm{kg\cdot m^{-3}}$ and, say, $h = 2\,\mathrm{m}$, then the negative capillary pressure that can be reached is around $-6\,\mathrm{kPa}$. Reading off the graph in Fig. 14.26, we can see that this driving force displaces a further 15% of the oil for the lowest permeability sample, Estiallades. Even if we consider lower permeability rocks (the capillary pressure approximately increases as $1/K^{1/2}$, where K is the permeability — this is a consequence of Leverett J-function scaling, Sec. 8), there is likely to be significant displacement with this driving force and this demonstrates how both capillary and gravitational forces mediate recovery in field settings.

Gravity also determines the initial water saturation before water-flooding. As is apparent from the capillary pressures, for the lowest permeability sample, Estaillades (which is still high permeability compared to most reservoir rocks), an effective matrix block height of around 10 m would be required to displace all the oil to close to residual saturation. There is a corollary to this; it also indicates that the initial saturation determined by capillary-gravity equilibrium (based, typically, on the primary drainage capillary pressure) has a transition zone — with varying saturation above the irreducible value — of height around 10 m–100 m for rocks with permeabilities between 1 mD and 100 mD (using the $1/K^{1/2}$ scaling mentioned above). The initial water saturation affects both the wettability (at high saturation less of the rock is contacted directly by oil and, as the imposed capillary pressure is lower, the wettability alteration is likely to be less strong) and the starting point for waterflooding. There is often a wettability trend from water-wet near the oil–water contact, through mixed-wet in most of the reservoir with more oil-wet conditions at the crest, as discussed previously. Usually, core-flood measurements are made from samples near the top of the reservoir; this could suggest oil-wet conditions and unfavourable waterflood recovery, when the reality is a much more efficient displacement in most of the reservoir column. Pore-scale modelling, allowing the prediction of relative permeabilities as a consistent function of initial water saturation, has enormous potential to improve the characterisation of such reservoirs.

This rather simple analysis already leads to some interesting and surprising conclusions. For the same wettability, in a reservoir where flow is not fracture-dominated, local waterflood recovery is higher in the lower-permeability, less well-connected sample, since the low water relative permeability holds back the water advance. On the other hand, if the reservoir is extensively fractured, the better-connected sample gives faster and better recovery, since there is a greater degree of spontaneous imbibition allowed. This is a clear indication that both the nature of the reservoir — fractured or unfractured — and the multiphase flow properties are crucial for any reasonable assessment of recovery.

I call this conundrum over recovery — mixed-wet systems are good for displacement, but bad for imbibition — the "trillion barrel question" since it will determine the recovery of most of our remaining conventional reserves of oil, mainly in the Middle East. While I do not have a simple answer — and it is unlikely that there is a simple answer — it does underscore the importance of this topic and how important it is to have good measurements (and predictions) of relative permeability.

Chapter 15

Three-phase Flow

Oil, water and gas may all flow together in reservoirs. Examples include gas injection, including carbon dioxide injection, solution gas drive (when the reservoir pressure is dropped below the bubble point), gas cap expansion and steam injection. In environmental settings, when a non-aqueous phase pollutant migrates downwards towards the water table in a moist soil, there are three mobile fluid phases: the pollutant, water and air.

First we will consider oil, water and gas at the pore scale, and how they are arranged. We use this insight to discuss wettability and relative permeability, as well as oil recovery.

15.1. Spreading, Wetting and Oil Layers

What happens when I place a drop of oil on water?

Consider the spreading coefficient, which is defined by

$$C_s = \sigma_{gw} - \sigma_{go} - \sigma_{ow}. \tag{15.1}$$

If $C_s > 0$, then the oil spreads on water. Light alkanes and many alkane mixtures — and indeed most crude oils — spread; the arrangement shown in Fig. 15.1 is not stable and it will be energetically favourable for the oil to cover the interface between gas and water.

If $C_s < 0$, then the oil does not spread on water. Dense (denser than water) non-aqueous phase liquids (such as chlorinated solvents)

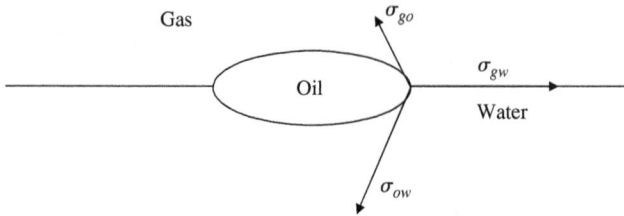

Figure 15.1. The arrangement of a small droplet of oil floating on water in the presence of a gas.

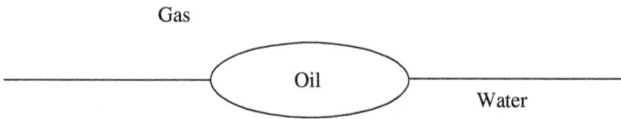

Figure 15.2. The equilibrium arrangement of a small droplet of oil floating on water in the presence of a gas. Here the spreading coefficient is negative and no oil film forms between the gas and water.

and long chain alkanes (such as decane, dodecane, etc.) do not spread on water.

In thermodynamic equilibrium, there are then three things that can happen to the drop of oil.

1. $C_s < 0$ and the drop is stable (Fig. 15.2).
2. $C_s > 0$ but if we define an equilibrium spreading coefficient when the gas/water interface is covered by a molecular film of oil (Fig. 15.3). We can define a spreading coefficient, C_s^e, using Eq. (15.1) when the phases are in equilibrium, coated by films. In this case, the oil film reduces the effective gas/water tension, $C_s^e < 0$ and the drop is stable.
3. $C_s > 0$ and $C_s^e = 0$ (Fig. 15.4). The oil film swells without limit as more oil is added to the system. When the oil film is sufficiently thick, the effective interfacial tension between the gas and water is the sum of the interfacial tensions between oil and water, and gas and oil. This means that the equilibrium spreading coefficient is zero.

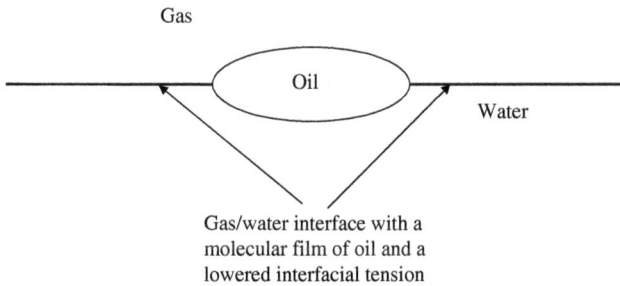

Gas/water interface with a
molecular film of oil and a
lowered interfacial tension

Figure 15.3. The equilibrium arrangement of a small droplet of oil floating on water in the presence of a gas. Here the initial spreading coefficient is positive; an oil film forms, generating a new effective gas/water interface with a negative equilibrium spreading coefficient.

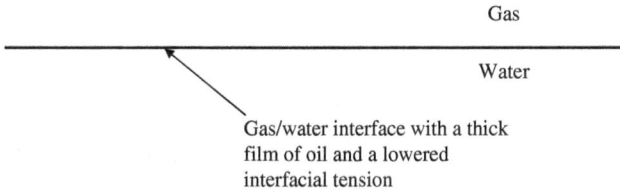

Gas/water interface with a thick
film of oil and a lowered
interfacial tension

Figure 15.4. The third and final possibility for the drop of oil is that it spreads without limit forming a thick film. Effectively the interfacial tension across the interface containing the film is simply the sum of the oil/water and gas/oil interfacial tensions, giving an equilibrium spreading coefficient of zero.

In thermodynamic equilibrium, $C_s^e \leq 0$; if the equilibrium spreading coefficient were positive then oil would continue to spread until its value was zero — case 3 .

Now we will discuss what this implies about the arrangement and flow of three phases in the pore space. The typical arrangement of the phases is shown in Fig. 15.5.

The oil can form layers in a (water-wet) pore space, sandwiched between water in the corners and gas in the centre. This formation of the oil layer is favoured by having a low (or zero) equilibrium spreading coefficient, as this controls the contact angle between the gas and oil.

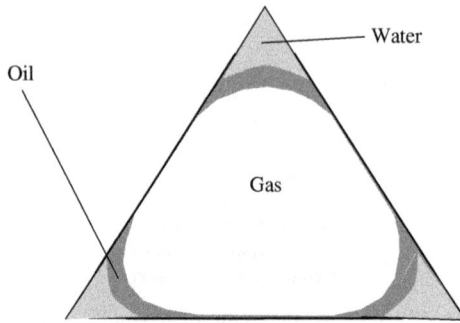

Figure 15.5. The arrangement of oil, water and gas in a water-wet pore of triangular cross-section. Note that the oil resides as a layer sandwiched between the water in the corners and gas occupying the centre of the pore.

However, for an oil layer to form (i.e. to be able to draw a layer in an angular pore space)[1]:

$$\alpha + \theta_{go} < \pi/2. \tag{15.2}$$

Spreading oils (with an equilibrium spreading coefficient of zero) have $\theta_{go} = 0$ since, at the microscopic level, there is no angle of contact between the gas and oil. θ_{go} increases as the spreading coefficient becomes more negative, as we will show below.

Consider the Young equation on a flat surface, with different combinations of fluids, shown in Fig. 15.6. Rearranging the equations leads to an important equality in three-phase flow, known as the Bartell–Osterhof (1927) equation:

$$\sigma_{gw}\cos\theta_{gw} = \sigma_{go}\cos\theta_{go} + \sigma_{ow}\cos\theta_{ow}. \tag{15.3}$$

Here we assume that the contact angles and interfacial tensions are measured in thermodynamic equilibrium. This relationship provides a constraint between the contact angles and the interfacial tensions; there are only two independent contact angles. Conventionally

---

[1]This is a necessary but not sufficient condition for the formation of a layer. Strictly, we need to consider an energy balance for the displacement. Conceptually this is straightforward, as we consider the change in interfacial energy associated with different possible fluid configurations, but can be complex when dealing with three-phase flow.

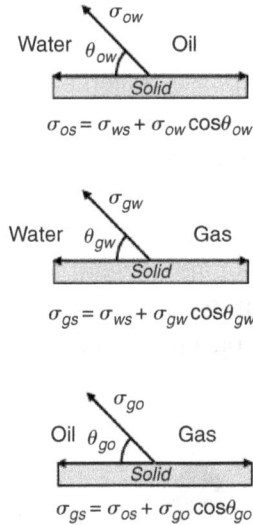

Figure 15.6. The Young equations for different combinations of fluids on a solid surface. From this an important constraint between contact angles and interfacial tensions can be derived, Eq. (15.3).

we consider that the wettability controls θ_{ow}, while spreading controls θ_{go}.

If the system is strongly water-wet with $\theta_{gw} = \theta_{ow} = 0$, then the gas/oil contact angle θ_{ow} is simply

$$\cos\theta_{go} = 1 + \frac{C_s^e}{\sigma_{go}}, \tag{15.4}$$

using our definition of spreading coefficient, which has a negative (or zero) value.

Now consider that we have residual oil surrounded by water and that gas then enters the system. Examples of this include lowering the water table in a soil with non-aqueous phase pollutants present, gas injection in oil reservoirs, solution gas drive (the primary production mechanism that occurs when the pressure in an oil field drops below the bubble point) and gravity drainage through gas cap expansion. In a water-wet system, this oil will occupy the centres of the larger pore spaces.

When the gas phase is introduced, oil spreads in the porous medium between water (coating the solid surfaces) and gas (which as the most non-wetting phase preferentially fills the centres of the largest pores). Oil layers form that occupy the crevices and corners of the pore space between water and gas — it is now connected wherever there is gas, and the oil can flow. We can drain to essentially zero saturation. Remaining oil saturations as low as 0.1% have been observed after gravity drainage in sand packs (Sahni *et al.*, 1998).

A full discussion of three-phase flow rapidly becomes rather complicated, beyond the key concept of oil layers; as an example, Fig. 15.7

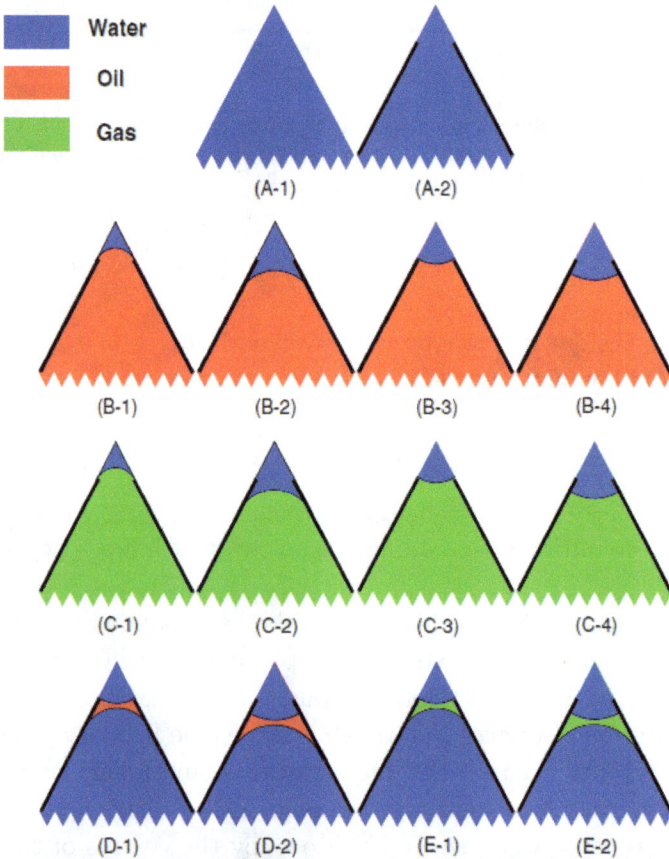

Figure 15.7. Some of the possible configurations of three phases — oil (red), water (blue) and gas (green) — in a corner pore space. From Piri and Blunt (2005a).

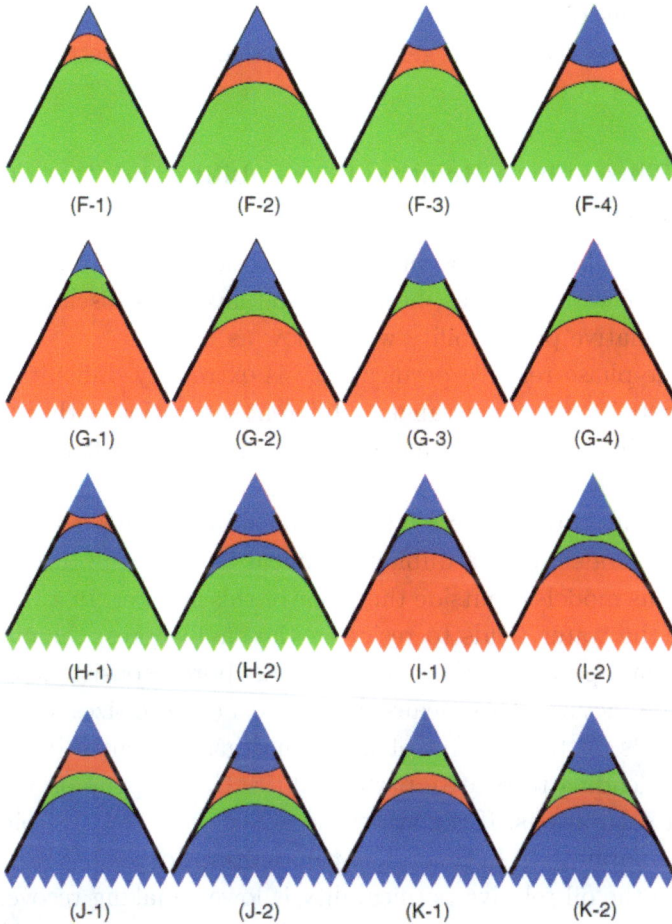

Figure 15.7. (*Continued*)

shows (just some of) the configurations of two and three phases in a pore space, dependent on saturation path and wettability. All of these fluid configurations simply use the concepts of wetting, spreading, contact angle and the Young–Laplace equation. Rather than go through all the details, we will present briefly a discussion on wettability, pore-scale configurations (specifically layers) and recovery later in this section.

We can also see oil layers in mixed wet and oil-wet media; gas is always non-wetting to oil, and so can reside in the centre of a pore, with an oil layer (as in two-phase flow) present in the corners. However, as we show later, gas is not necessarily non-wetting to water in an oil-wet system.

15.2. Three-phase Relative Permeability and Trapped Saturations

We expect to find very low values of the residual oil saturation in the presence of water and gas, because of oil layers, as discussed above, but the relative permeability will be low as well.

Three-phase relative permeability is extremely difficult to measure. There is also a huge range of different saturation paths that may be taken during a displacement that all may have different relative permeabilities.

Normally three-phase relative permeabilities are predicted using empirical models with a dubious physical basis. A full discussion of the various models is outside the scope of this chapter. In a water-wet system, the water tends to reside in the small pores, the oil in the intermediate pores and the gas in the large pores, consistent with our pore-scale picture. This means that the exact pore sizes seen by the oil depends on the amount of water and gas present in the porous medium, leading to relative permeabilities that depend on two independent saturations. If the system is mixed-wet then this picture is further complicated. In general, in three-phase flow — in the presence of gas — the oil relative permeability is lower, making recovery low, but the residual oil can also be very low, leading to high ultimate recoveries, because of the drainage of oil layers. The gas relative permeability — in a water-wet system — is high, since gas resides in the larger pores, leading to early breakthrough and poor overall recovery when gas is injected. On the field scale, the main design criterion is how to keep the injected gas in the reservoir, allowing the oil to flow to low saturation. It is less easy to make general statements about recovery, as the system is now much more complex, and we have to rely on field-scale simulation models to assess recovery efficiency.

To help advance a more physically based picture of three-phase flow, micro-CT images of trapped phases in three-phase flow are presented in Fig. 15.8. When water is injected into a porous medium

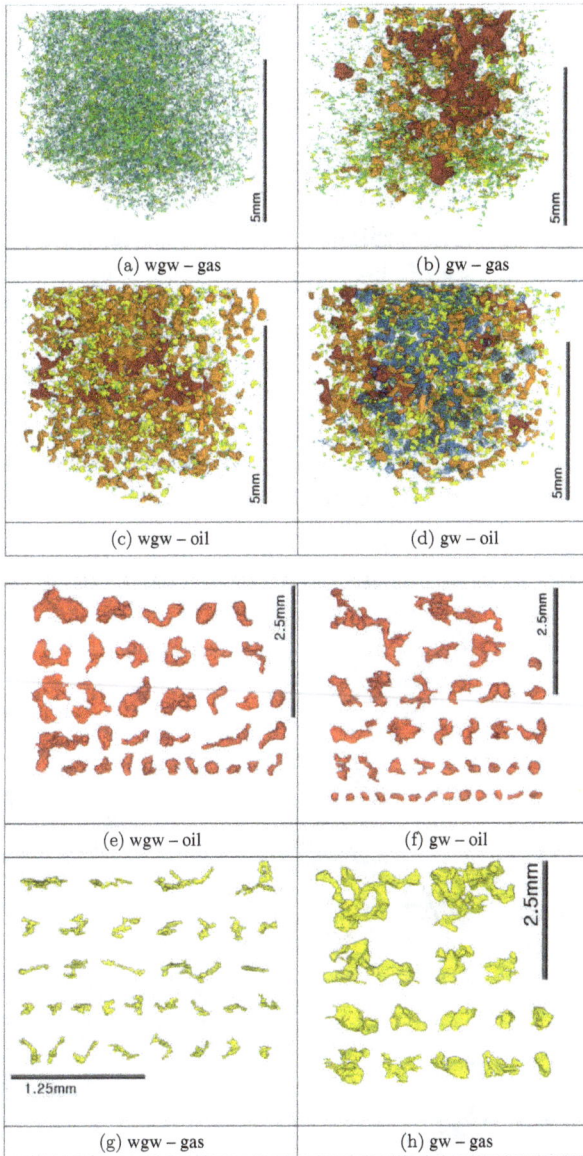

(a) wgw – gas

(b) gw – gas

(c) wgw – oil

(d) gw – oil

(e) wgw – oil

(f) gw – oil

(g) wgw – gas

(h) gw – gas

Figure 15.8. Trapped oil and gas imaged in a water-wet sandstone (lglauer *et al.*, 2013). Initially the core is full of water. Then oil is injected — this is primary drainage. Then two displacement sequences are considered. The first is gas injection, followed by water injection (gw), while the second is waterflooding, followed by gas injection followed by a further waterflood (wgw). The gw sequence leads to considerably more trapping of gas in the pore space.

containing gas and oil, both gas and oil can be trapped. If the system is water-wet, then snap-off can strand ganglia of both hydrocarbon phases, as shown.

The amount of trapping — of both gas and oil — is dependent on the displacement sequence, with less trapped if we waterflood the reservoir before gas injection (this is called tertiary injection, as opposed to gas injection straight away, which is secondary gas injection). The morphology of the trapped clusters is also different, with smaller clusters of gas seen for the wgw sequence. This is an active topic of research and we do not have, as yet, a full understanding of recovery and displacement processes in three-phase flow.

One important observation is that — in water-wet systems — more gas can be trapped in a three-phase displacement than if displaced only by water. This could be used to design gas injection to retain the gas in the reservoir, while mobile oil is produced. Experimental evidence for this is shown in Fig. 15.9 for experiments on sand packs; more gas can be trapped in a three-phase displacement involving oil and water than when gas is displaced by water alone.

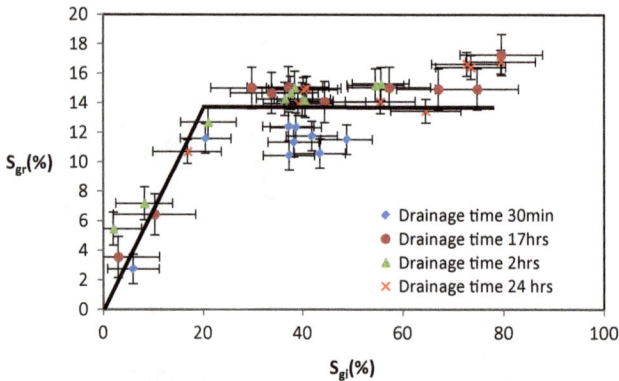

Figure 15.9. Trapped gas saturation as a function of initial gas saturation (determined by the time period of gravity drainage indicated on the graph) for three-phase displacements in a water-wet sand-pack. More gas can be trapped in the presence of oil than when gas is displaced by water alone (the solid black lines). From Ameachi *et al.* (2014).

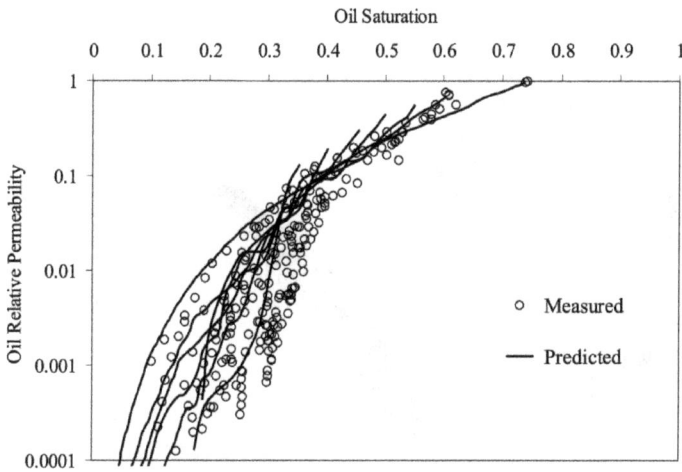

Figure 15.10. Measured and predicated oil relative permeabilities for water-wet Berea sandstone. The different points refer to different displacement sequences in the experiments. In Figs. 15.10–15.12, the data comes from Oak *et al.* (1990), while the predictions are based on the work of Piri and Blunt (2005b).

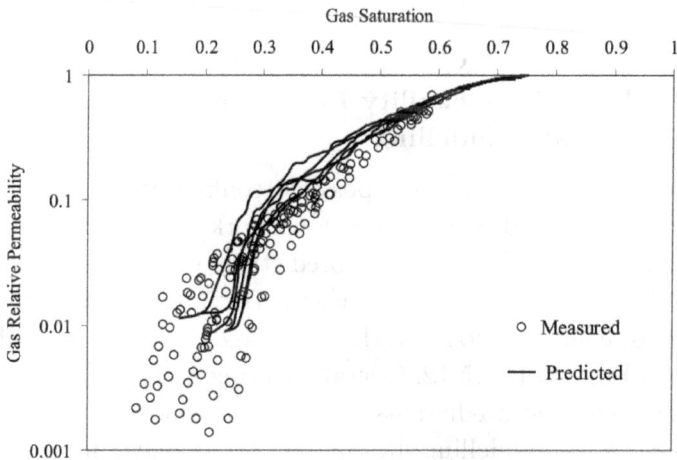

Figure 15.11. Measured and predicted gas relative permeabilities. The different points refer to different displacement sequences in the experiments. From Piri and Blunt (2005b).

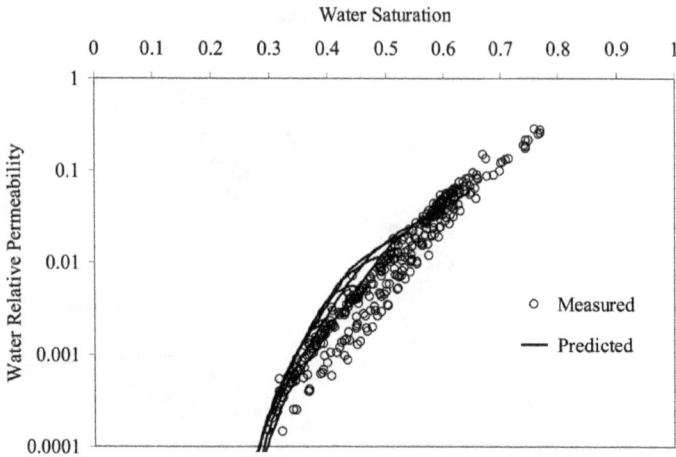

Figure 15.12. Measured and predicted water relative permeabilities. The different sets of points refer to different displacement sequences in the experiments. From Piri and Blunt (2005b).

In general, there is more trapping of oil and gas combined than in two-phase flow, and more trapping of gas alone, but less trapping of oil alone.

15.3. Relative Permeability Predictions Using Pore-scale Modelling

If we consider all the various possible configurations of phases in the pore space, and have a good network representation of the rock, then it is possible to make predictions of three-phase relative permeabilities. The dataset is the classic Berea measurements of Oak *et al.* (1990) compared to network model predictions of Piri and Blunt (2005b) in Figs. 15.10–15.12. Overall, bearing in mind the complexity of the problem, the predictions shown are a good test of the ability of pore-network modelling to predict the behaviour of complex systems.

15.4. Layer Drainage and Wettability

We will now make some statements concerning wettability, layers and recovery. Figure 15.13 shows the results of gravity drainage experiments (gas enters a long sand column, while oil and water drain

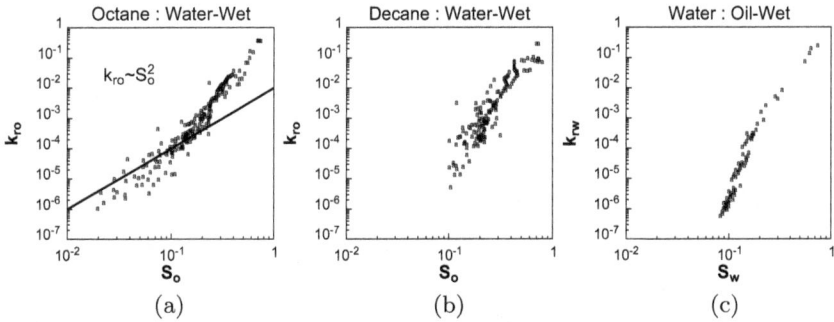

Figure 15.13. Measured relative permeabilities for gas gravity drainage in sand packs. From (a) to (c) the oil relative permeability in a water-wet medium, when octane is the oil phase; the same experiment but with decane as the oil; the water relative permeability in an oil-wet system. From DiCarlo *et al.* (2000).

out of the bottom of the column under gravity; the same process occurs in an oil reservoir, if gas is introduced to the crest of the field, or a natural gas cap expands). The relative permeability is shown for three cases: octane as the oil in a water-wet system; decane as the oil in a water-wet system; and the water relative permeability for an oil-wet case.

The behaviour is different in each case and we can understand this and discuss the implications for recovery using our discussion of spreading coefficient and the Bartell–Osterhof relation, Eq. (15.3).

Octane spreads — or almost spreads — on water, with an effective gas/oil contact angle in a water-wet system close to zero. Hence, oil layers readily form. If gas is injected, then oil layers form and allow drainage down to very low saturation — below 1% in the sand pack studied. At low oil saturation, the flow is dominated by this layer drainage; the oil saturation is simply proportional to the area of oil open to flow. The conductance scales as the square of the area — this is important — rather than proportionally to it. This is a direct consequence of the Navier–Stokes equation (consider Poiseuille flow where the flow rate is proportional to the fourth power of radius (second power of area)). Physically, this is because there is no flow at a solid boundary. If we increase the area to flow, then the flow speed in the centre of the channel can increase, as it is further away from the walls. This, combined with the fact that the area is greater, is what leads to the quadratic dependence on area. This result contrasts with electrical conductance, which scales linearly with area, as discussed in Sec. 10, Eq. (10.7).

The oil relative permeability is just the fractional flow conductance of the oil and so this discussion leads to the prediction

$$k_{ro} \sim S_o^2. \tag{15.5}$$

This is seen in the experimental results shown, and in other measurements.

Decane has a higher interfacial tension with water than octane does and does not spread on water. Its contact angle in the presence of gas is non-zero and this non-spreading oil does not form oil layers in the pore space. Hence there is no oil layer drainage regime, the oil relative permeability drops rapidly at low saturation and we see significant trapping. This is observed in Fig. 15.13 (b).

It is considered likely that in reservoir settings, most oils are spreading, and so high recoveries are potentially possible. However, in environmental applications many non-aqueous phase liquids, particularly chlorinated solvents, do not spread on water and therefore can remain trapped even in the presence of gas (air).

15.5. Why Ducks Don't Get Wet

If instead the porous medium is oil-wet, then perhaps we can simply swap phases. That is, the water relative permeability in an oil-wet system is the same as the oil relative permeability in a water-wet system. Figure 15.13 (c) shows that this is not the case if the oil is spreading. The water relative permeability drops sharply and has an irreducible saturation. Why is this, and what has it got to do with ducks?

Figure 15.14 explains this. If we rearrange the Bartell–Osterhof equation to find the gas/water contact angle for a spreading oil ($\theta_{go} = 0$) in a strongly oil-wet system ($\theta_{ow} = 180°$), we find

$$\cos\theta_{gw} = \frac{\sigma_{go} - \sigma_{ow}}{\sigma_{gw}}. \tag{15.6}$$

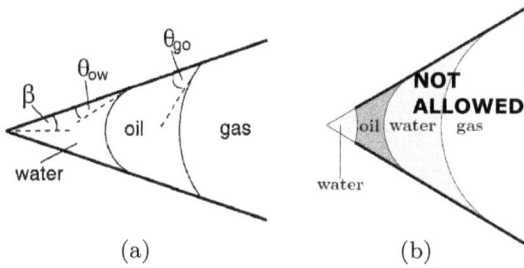

Figure 15.14. Diagram showing oil layers in the pore space an oil-wet systems: this is not allowed for normal values of interfacial tensions as the gas/water contact angle will be greater than 90°.

The interfacial tension between oil and water is always larger than between gas and oil. Hence, the gas/water contact angle has a negative cosine and is greater than 90°. Water cannot spread on oil in the presence of gas. Indeed, in a strongly oil-wet system water is the most non-wetting phase and can be trapped by gas and water.

This is why ducks don't get wet. Their feathers are covered in oil and form an oil-wet porous medium. This makes water the most non-wetting phase: you have to force water into the feathers — they prefer to be surrounded by air — keeping the duck dry and insulated. This is also why water beads and runs off an oily surface. It's also why sea-birds often die if crude oil (from a spill) is washed off their feathers using soap; now their feathers are water-wet, they imbibe water and the birds soon die of hypothermia.

While this observation is the topic of a children's book ("Ducks Don't Get Wet" by Augusta Goldin), the concept still struggles to be accepted by petroleum engineers, where it is still widely assumed that gas "must" be the most non-wetting phase.

The consequences of wettability for recovery in three-phase flow are still not fully understood. We do expect good ultimate recoveries if oil layers form — as they do in a spreading system, regardless of wettability (oil is always more wetting that gas). We can also trap and suppress the movement of gas in mixed-wet media, which is favourable for recovery.

Chapter 16

Conservation Equation for Multiphase Flow

The approach so far in this volume has been physically motivated, with a minimum of equations. However, to provide some rigour and background to our comments on recovery, we will now derive flow equations for multiple phases in a porous medium.

16.1. 1D Flow

We will consider a conservation equation for a case where we have multiple phases. This is an extension of the derivations for single-phase flow in Sec. 12. The approach will be very slightly different, but is straightforward. Here we consider the transport of saturation, rather than concentration. Consider conservation of mass of one phase (water) in Fig. 16.1.

The mass of water that enters the box — the shaded region shown in Fig. 16.1 — in a time $\Delta t = A\Delta t \rho_w q_w(x)$, where q_w is given by the multiphase Darcy law, Eq. (14.1). A is the cross-sectional area to flow. Similarly, the mass that leaves is given by $A\Delta t \rho_w q_w(x + \Delta x)$. The mass of water in the box is given by $A\Delta x \phi S_w$. The mass in minus the mass out is the change in mass:

$$A\Delta t \rho_w \left(q_w\left(x \right) - q_w\left(x + \Delta x \right) \right)$$
$$= A\Delta x \rho_w \phi \left(S_w\left(t + \Delta t \right) - S_w\left(t \right) \right), \qquad (16.1)$$

$$\phi \frac{S_w\left(t + \Delta t \right) - S_w\left(t \right)}{\Delta t} + \frac{q_w\left(x + \Delta x \right) - q_w\left(x \right)}{\Delta x} = 0, \qquad (16.2)$$

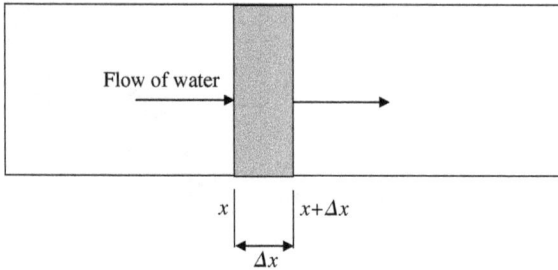

Figure 16.1. An illustration of conservation of water for 1D flow. Here we will allow other phases — oil and/or gas — to be flowing as well.

where we have assumed that the density and porosity are constant (assuming, as in Sec. 12, that the flow is incompressible). Then we take the limit of small Δx and Δt to obtain a differential equation:

$$\phi \frac{\partial S_w}{\partial t} + \frac{\partial q_w}{\partial x} = 0. \qquad (16.3)$$

This simple form of the conservation equation can be rearranged by substituting in Darcy's law, Eq. (14.1). This takes some algebra and ends with an equation that can be solved analytically for saturation. We start with Eq. (14.1) for 1D flow of the water phase:

$$q_w = -\frac{K k_{rw}}{\mu_w} \left(\frac{\partial P_w}{\partial x} - \rho_w g_x \right). \qquad (16.4)$$

Similarly for oil:

$$q_o = -\frac{K k_{ro}}{\mu_o} \left(\frac{\partial P_o}{\partial x} - \rho_o g_x \right), \qquad (16.5)$$

and $P_c = P_o - P_w$ is the capillary pressure. k_{ro}, k_{rw} and P_c are known as functions of S_w.

We can write a similar conservation equation to Eq. (16.3) for oil:

$$\phi \frac{\partial S_o}{\partial t} + \frac{\partial q_o}{\partial x} = 0. \qquad (16.6)$$

Add the two conservation Eqs. (16.3) and (16.6):

$$\phi \frac{\partial (S_w + S_o)}{\partial t} + \frac{\partial (q_w + q_o)}{\partial x} = 0. \qquad (16.7)$$

We define $q_t = q_w + q_o$ as the total velocity. Then Eq. (16.7) is (the saturation term is zero as for two-phase flow; the sum of the oil and water saturation is one, a constant)

$$\frac{\partial q_t}{\partial x} = 0. \tag{16.8}$$

The total velocity is constant in space (it can vary over time; $q_t(t)$) for 1D flow.

From the multiphase Darcy equations for oil and water, Eqs. (16.4) and (16.5), and writing the expression in terms of the water pressure only:

$$q_t = q_w + q_o = -\frac{K k_{rw}}{\mu_w} \left(\frac{\partial P_w}{\partial x} - \rho_w g_x \right)$$
$$- \frac{K k_{ro}}{\mu_o} \left(\frac{\partial P_w}{\partial x} + \frac{\partial P_c}{\partial x} - \rho_o g_x \right). \tag{16.9}$$

Then, defining mobilities by $\lambda_w = k_{rw}/\mu_w$ and $\lambda_o = k_{ro}/\mu_o$, with the total mobility given by $\lambda_t = \lambda_w + \lambda_o$, Eq. (16.9) becomes

$$q_t = -K\lambda_t \frac{\partial P_w}{\partial x} + K g_x (\rho_w \lambda_w + \rho_o \lambda_o) - K\lambda_o \frac{\partial P_c}{\partial x}. \tag{16.10}$$

Then we substitute $\frac{\partial P_w}{\partial x}$ from Eq. (16.10) in q_w in the Darcy Equation (16.4):

$$q_w = \frac{\lambda_w}{\lambda_o} q_t - K \frac{\lambda_o \lambda_w}{\lambda_t} \rho_o g_x - K \frac{\lambda_w^2}{\lambda_t} \rho_w g_x$$
$$+ K\lambda_w \rho_w g_x - K \frac{\lambda_o \lambda_w}{\lambda_t} \frac{\partial P_c}{\partial x}. \tag{16.11}$$

Rearrange terms to find

$$q_w = \frac{\lambda_w}{\lambda_o} q_t + K \frac{\lambda_o \lambda_w}{\lambda_t} (\rho_w - \rho_o) g_x + K \frac{\lambda_o \lambda_w}{\lambda_t} \frac{\partial P_c}{\partial x}. \tag{16.12}$$

In words, the water Darcy velocity = pressure gradient + capillary pressure + gravity.

16.1.1.   *Fractional Flow*

We can write the conservation equation (16.3), in terms of the water fractional flow (the fraction q_w/q_t):

$$\phi \frac{\partial S_w}{\partial t} + q_t \frac{\partial f_w}{\partial x} = 0, \tag{16.13}$$

where f_w is the water fractional flow defined by $q_w = f_w q_t$:

$$f_w = \frac{\lambda_w}{\lambda_t} \left[1 + K \frac{\lambda_o}{q_t} \left(\frac{\partial P_c}{\partial x} + (\rho_w - \rho_o) g_x \right) \right]. \tag{16.14}$$

The fractional flow has three terms representing the three physical forces that impact the fluid movement: advection (governed by the pressure gradient), capillary pressure and gravity (buoyancy).

16.1.2.   *Note About Nomenclature*

Many authors define mobility as $\lambda_w = K k_{rw}/\mu_w$ with an extra factor of K. Also we use Q (volume per unit time) and q (Darcy velocity) rather than q and v respectively, as in some books. Often you see conservation equations with an explicit area A. Our equations are per unit area: $Q = qA$.

16.2.   Richards Equation

The conservation equations we have developed are for an oil/water displacement, where water displaces oil, as occurs in a hydrocarbon reservoir.

   We will now address some special cases, where simplifications to the equations can be made. While we cannot solve the full equation directly, we can explore solutions in a variety of different limits. One case will be considered in this section.

   If we have gas/water flow, then the mobility of the gas phase can be considered to be much larger than for the water. The Richards equation describes the transport of water in this case.

   In Eqs. (16.13) and (16.14) λ_o is replaced by λ_g (for gas) and $\lambda_g \gg \lambda_w$ and so $\lambda_t = \lambda_g$. Then the fractional flow can be written as

$$q_t f_w = K \lambda_w \left(\frac{\partial P_c}{\partial x} + (\rho_w - \rho_g) g_x \right), \tag{16.15}$$

where the first (advection) term in Eq. (16.14) is now considered to be negligible. Then the conservation equation, Eq. (16.13), is

$$\phi \frac{\partial S_w}{\partial t} + K \frac{\partial}{\partial x} \left[\lambda_w \left(\frac{\partial P_c}{\partial x} + (\rho_w - \rho_g) g_x \right) \right] = 0. \qquad (16.16)$$

More usually this equation is written in terms of the pressure head $p = P_w/\rho_w g + z$. Write $\psi = P_w/\rho_w g$. If the gas (air) density is considered to be negligible and the air pressure constant at atmospheric ($P_c = P_{\text{atm}} - P_w$), then Eq. (16.16) becomes for vertical flow ($g_x = g$)

$$\phi \frac{\partial S_w}{\partial t} = K_H \frac{\partial}{\partial x} \left[k_{rw} \left(\frac{\partial \psi}{\partial x} - 1 \right) \right], \qquad (16.17)$$

using the standard definition for hydraulic conductivity, K_H (see Sec. 10). Often rather than seeing the relative permeability and capillary pressure written as a function of pressure, the saturation and relative permeability are written as a function of scaled capillary pressure (ψ).

Equation (16.17) is the standard transport equation in hydrology to describe the movement of water under gravity and capillary pressure: it is called the Richards equation.

Chapter 17

Fractional Flow and Analytic Solutions

We will now begin to construct an analytical solution for multiphase flow in one dimension. The derivation can be somewhat cumbersome and is helped if we first develop the concept of the fractional flow: the fraction of the total flow of oil and water that is taken by water alone.

Consider a reservoir with constant dip and linear flow, as shown in Fig. 17.1. Using the conservation equation, Eq. (16.13), we can write the fractional flow equation, Eq. (16.4), as

$$f_w = \frac{\lambda_w}{\lambda_t} \left[1 + K \frac{\lambda_o}{q_t} \left(\frac{\partial P_c}{\partial x} + \Delta \rho g \sin \theta \right) \right], \quad (17.1)$$

where θ is the angle to the horizontal and the density difference is written as $\Delta \rho$. $\theta > 0$ represents downwards flow, while $\theta < 0$ is flow uphill.

How important is the capillary pressure term at the field scale? While it dominates at the pore scale, as discussed previously in this volume, it is small over scales of hundreds of metres to kilometres; the effect of capillary pressure is encapsulated in the relative permeabilities. This was discussed in Sec. 13 in the context of capillary and Bond numbers.

We will estimate the relative contribution of viscous forces (advection), capillary pressure and buoyancy to the fractional flow by

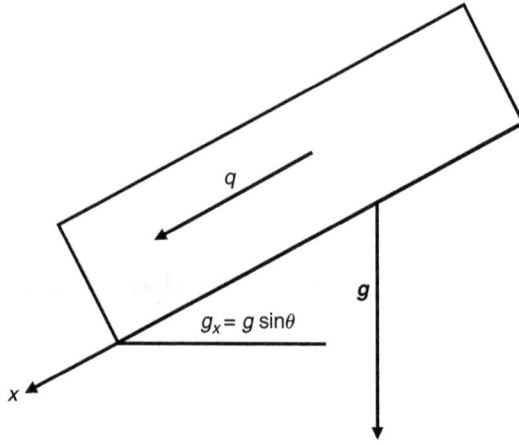

Figure 17.1. A Schematic showing 1D flow in a tilted reservoir.

considering the magnitude of each of the three terms in the brackets of Eq. (17.1). By definition the viscous term is 1. For q_t of the order of 10^{-5} m·s^{-1} (around 1 m/day), K of 10^{-13} m^2 (100 mD), a viscosity of 10^{-3} Pa·s and a relative permeability of order 1, the $K\frac{\lambda_o}{q_t}$ term in Eq. (17.1) is around 10^{-5} m·Pa^{-1}. Then for a typical capillary pressure of 10^4 Pa varying over a typical distance between wells (say 100 m), the capillary pressure term in Eq. (17.1) is of order 10^{-3}. The gravitational term, for, say, a density difference of 300 kg·m^{-3}, is around 0.03. This indicates that buoyancy forces are generally small (but not negligible) in comparison to advection, while the effect of capillary pressure — at the field scale — is tiny. Another way of seeing this is to consider the pressure drop between injection and production wells — generally a few MPa — compared to capillary pressures of 0.01 MPa–0.1 MPa.

So, to construct an analytical solution applicable at large scales, we ignore the capillary pressure (later, in Sec. 18, we will solve for the opposite limit — where capillary pressure dominates — which applies for imbibition in fractured media) and write Eq. (17.1) as

$$f_w = \frac{\lambda_w}{\lambda_t}\left[1 + K\frac{\lambda_o}{q_t}\Delta\rho g \sin\theta\right]. \qquad (17.2)$$

Define a gravity number N_G (this is different from the Bond number introduced in Sec. 13) as

$$N_G = \frac{K \Delta \rho g}{\mu_o q_t}. \tag{17.3}$$

Also, traditionally, we write

$$\frac{\lambda_w}{\lambda_t} = \frac{1}{1 + \frac{\lambda_o}{\lambda_w}} = \frac{1}{1 + \frac{k_{ro}\mu_w}{k_{rw}\mu_o}}. \tag{17.4}$$

Then the fractional flow Eq. (17.2) becomes

$$f_w = \frac{1 + N_G k_{ro} \sin\theta}{1 + \frac{k_{ro}\mu_w}{k_{rw}\mu_o}}. \tag{17.5}$$

Let's look at typical fractional flow curves, shown in Fig. 17.2, for $\theta = 0$ as a function of the endpoint water/oil mobility ratio,

$$M = \frac{k_{rw}^{\max} \mu_o}{k_{ro}^{\max} \mu_w}. \tag{17.6}$$

The curves have a characteristic S shape, with a point of inflection when we consider horizontal flow (without gravity).

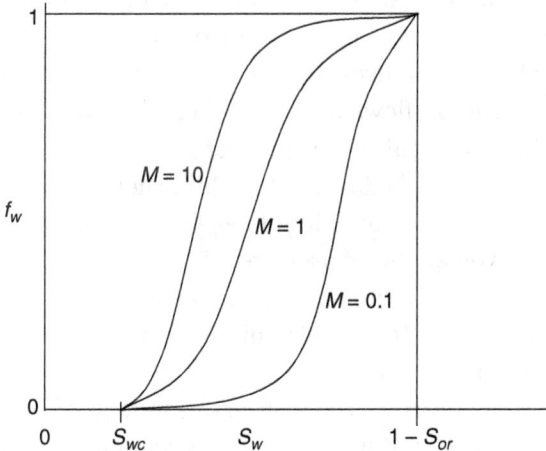

Figure 17.2. Example fractional flow curves for a horizontal system with different end-point mobility ratios, M, Eq. (17.6).

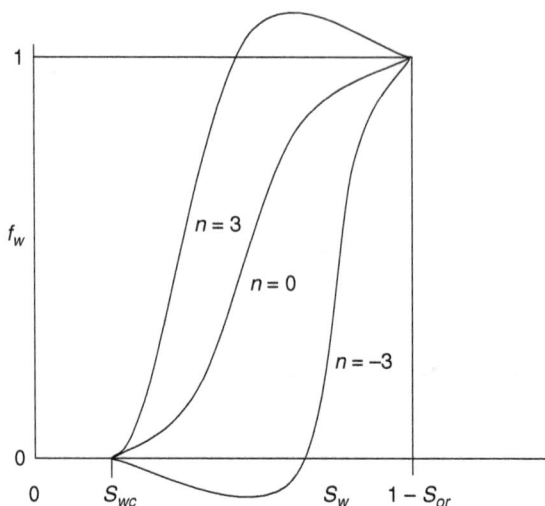

Figure 17.3. Example fractional flow curves including the effect of gravity. If gravitational effects are sufficiently strong, we can have — for some range of saturation — counter-current flow where the oil and water flow in opposing directions. In this case, the water fractional flow is either greater than 1 or negative.

If we include gravity, we can study the behaviour schematically for $M = 1$, where $n = N_G k_{ro}^{\max} \sin\theta$ (see Fig. 17.3). Here the fractional flow curves can be greater than one, or less than zero. This represents, physically, counter-current flow, where water moves downwards while oil moves up; if the overall flow direction is downwards and the oil flows upwards, then the water flow is greater than q_t and the fractional flow is greater than one. Conversely, if the flow is uphill and water is flowing downhill, then the water fractional flow is negative. Mathematically, $f_o + f_w = 1$. Thus, when $f_w > 1$, $f_o < 0$ and vice versa; the phases are flowing in opposite directions. Hence, if gravity is sufficiently strong, it can generate counter-current flow of oil and water. To repeat: this means that the water moves downhill, while oil moves uphill.

We have now derived the conservation equation and plotted different typical fractional flow curves. For reference, we will continue with a specific example, shown in Fig. 17.4. The water and oil relative permeabilities are written as follows (this is a common

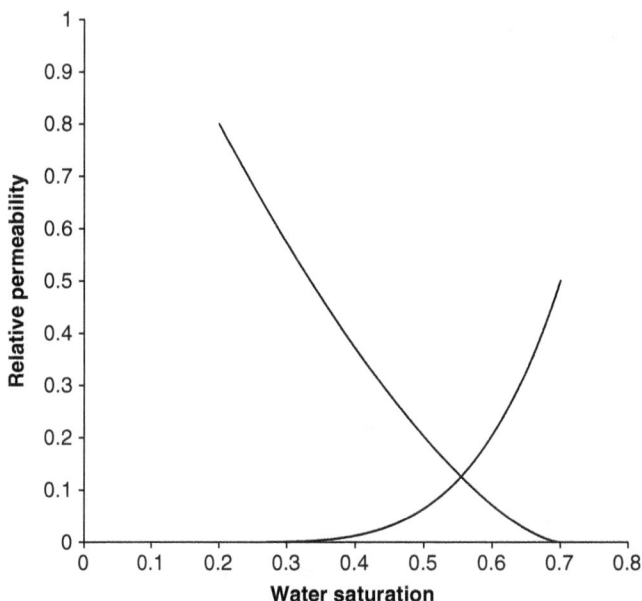

Figure 17.4. Relative permeabilities used for our example test case. This represents a weakly water-wet system.

parameterisation of these curves and normally fit to somewhat scattered experimental data)[1]:

$$k_{rw} = k_{rw}^{max} \frac{(S_w - S_{wc})^a}{(1 - S_{or} - S_{wc})^a}, \tag{17.7}$$

$$k_{ro} = k_{ro}^{max} \frac{(S_o - S_{or})^b}{(1 - S_{or} - S_{wc})^b}. \tag{17.8}$$

In this specific case we take a maximum water relative permeability of 0.5, a maximum oil relative permeability of 0.8, $a = 4$ and $b = 1.5$ with $S_{wc} = 0.2$ and $S_{or} = 0.3$. For the fractional flow the water

---

[1]These are sometimes called Corey curves and the power-laws Corey exponents. In reality, the original paper was a little more specific and proposed a physical justification (which has no foundation) for presenting equations for relative permeability and did not directly present these simple forms, but in any event Corey (or Brooks and Corey) were the first authors to suggest fitting relative permeability data to a power-law form.

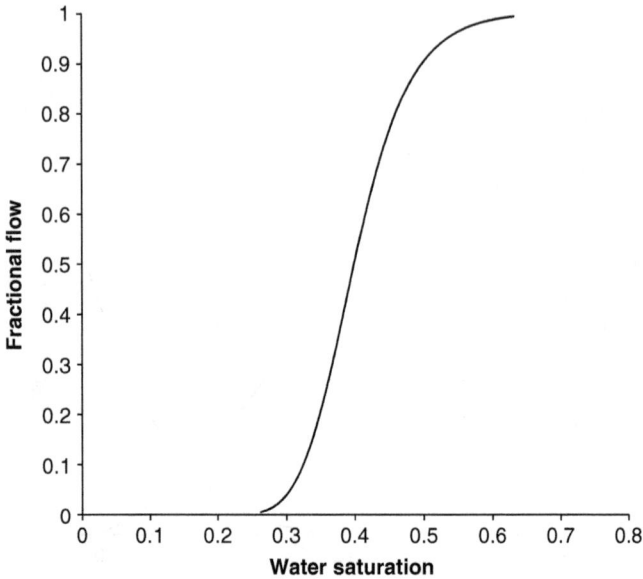

Figure 17.5. The water fractional flow corresponding to the relative permeabilities shown in Fig. 17.4.

and oil viscosities are 0.001 Pa $\cdot$ s and 0.03 Pa $\cdot$ s respectively; the relative permeabilities and fractional flow are shown below. From Eq. (17.6) the mobility ratio $M = 18.75$ and $n = 0$ (horizontal flow). The fractional flow is shown in Fig. 17.5.

17.1. Buckley–Leverett Solution

We will now show how to solve the conservation equation, Eq. (16.13) with the fractional flow given by Eq. (17.5).

We can write the saturation conservation equation, Eq. (16.13), as

$$\frac{\partial S_w}{\partial t} + v \frac{\partial f_w}{\partial x} = 0, \tag{17.9}$$

where $v = q_t/\phi$ is an interstitial velocity. This can be written as

$$\frac{\partial S_w}{\partial t} + v \frac{df_w}{dS_w} \frac{\partial S_w}{\partial x} = 0. \tag{17.10}$$

We are interested in water injection from a well into a reservoir containing some initial (usually irreducible) water saturation. We consider 1D flow with an injection well placed at $x = 0$ and a producer at $x = L$. Then the initial condition is $t = 0$, $S_w(x, 0) = S_{wi}$, while the boundary condition at the well is $x = 0$, (well) $S_w(0, t) = S_{w0}$.

We control rates (f_w) at wells, not saturation. Thus, we find S_{w0} that has given $f_w(0, t)$. Normally we have $S_{wi} = S_{wc}$ and inject 100% water; hence $f_w(0, t) = 1$, and $S_{w0} = 1 - S_{or}$.

We now define dimensionless variables:

$$x_D = \frac{x}{L}. \tag{17.11}$$

This is the fractional distance between wells.

We also define a dimensionless time. This requires some more thought — t_D is the pore volumes of water injected. This gives an indication of how much water has entered the system, in comparison with the total capacity of the reservoir. It is defined as

$$t_D = \int_0^t \frac{v}{L} dt = \int_0^t \frac{q_t}{\phi L} dt = \int_0^t \frac{Q}{\phi A L} dt = \frac{1}{V_p} \int_0^t Q dt, \tag{17.12}$$

where Q is the total flow rate and $V_p = \phi A L$ is the pore volume. If the flow rate is constant, then the integral simply becomes Qt; however, this definition does allow varying flow rates to be accommodated in the analysis. A useful relationship, which we will use later, is to convert dimensional speeds to a dimensionless quantity. If we have a speed $v = x/t$, then using Eqs. (17.11) and (17.12), $v = q_t/\phi \, v_D$, where $v_D = x_D/t_D$.

Then transformation of variables means that the conservation equation, Eq. (17.10), becomes

$$\frac{\partial S_w}{\partial t_D} + \frac{df_w}{dS_w} \frac{\partial S_w}{\partial x_D} = 0. \tag{17.13}$$

We will now solve this equation by the method of characteristics (MOC). What this means is that we will find the dimensionless velocity with which a given saturation moves. This means that we find the solution as a function of dimensionless velocity. At first, this can be a confusing concept. However, for a given time, the profile of saturation as a function of velocity is the same shape as saturation as a function of distance: the profile elongates linearly with time. It is also straightforward to convert between dimensionless and real variables, with care.

Write Eq. (17.13) as a function of $v_D = x_D/t_D^2$:

$$\frac{\partial S_w}{\partial t_D} = \frac{dS_w}{dv_D}\frac{dv_D}{dt}\bigg|_x = -\frac{v_D}{t_D}\frac{dS_w}{dv_D}, \qquad (17.14)$$

$$\frac{\partial S_w}{\partial x_D} = \frac{dS_w}{dv_D}\frac{dv_D}{dx}\bigg|_t = \frac{1}{t_D}\frac{dS_w}{dv_D}. \qquad (17.15)$$

Thus Eq. (17.13) becomes

$$\frac{dS_w}{\partial v_D}\left(v_D - \frac{df_w}{dS_w}\right) = 0. \qquad (17.16)$$

One solution is a so-called constant state — i.e. a saturation that does not change with dimensionless wavespeed. The non-trivial solution is

$$v_D = \frac{df_w}{dS_w}. \qquad (17.17)$$

Let's look at the example fractional flow shown previously in Fig. 15.5; since $v_D = df_w/dS_w$, we find the derivative which resembles the curve shown in Fig. 17.6.

Rearranging the plot to find water saturation as a function of speed (which — at a fixed time — would represent water saturation as a function of distance), we arrive at something that does not make sense: we have multiple solutions, as shown in Fig. 17.7.

The solution to this conundrum is to introduce the concept of a shock, or a discontinuity in saturation that moves with a distinct speed; it is not possible to construct a solution where the saturation varies smoothly with dimensionless speed (or distance).

---

[2]An entirely equivalent approach is to assume that the solution for water saturation is a function of $z = x - vt$ only and find v. This makes it even more explicit that the solutions are profiles that move with some characteristic speed.

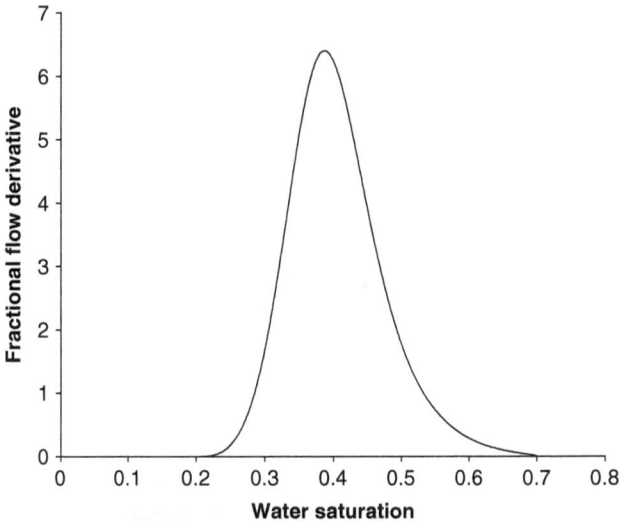

Figure 17.6. A schematic of the derivative of the fractional flow. Note that the derivative — indicating dimensionless wavespeed — has a maximum at intermediate water saturation.

17.2. Shocks

Shocks are discontinuities in saturation, which means that we can't use a differential equation to describe them. Shocks are encountered in other physical situations, such as a bomb blast, the flash of light after a nuclear explosion, or traffic jams.

Consider the situation in Fig. 17.8, with a shock between two saturations — a left state and a right state. We are only showing the region at the shock — the saturations can vary smoothly with distance away from the shock.

Imagine a shock moving at speed v_{sh}, as shown in Fig 17.8. Similar to our derivation of the conservation equation, consider the change in mass as the shock moves in a time dt. Just as before, when deriving a differential equation, flux in − flux out = rate of change of mass:

$$\rho_w q_t (f_w^L - f_w^R) = \rho_w v_{sh} \phi (S_w^L - S_w^R), \tag{17.18}$$

$$v_{sh} = \frac{q_t}{\phi} \frac{(f_w^L - f_w^R)}{(S_w^L - S_w^R)} = \frac{q_t}{\phi} \frac{\Delta f_w}{\Delta S_w}, \tag{17.19}$$

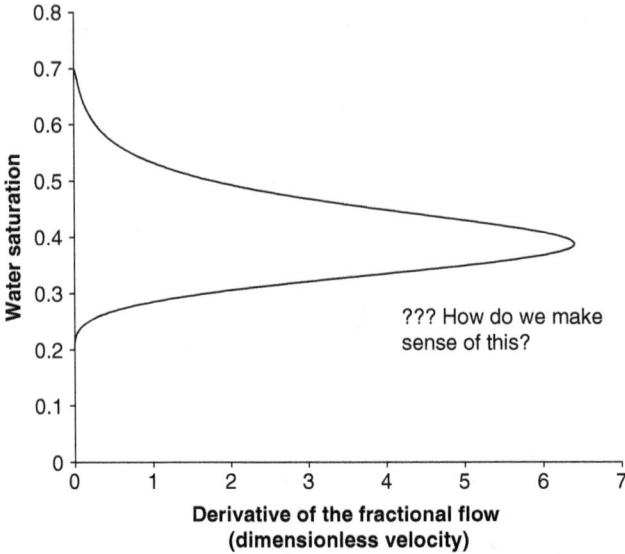

Figure 17.7. A schematic of the saturation as a function of dimensionless wavespeed (v_D). This appears to give two values of saturation for a given speed (or, if we consider some fixed time, for a given distance between the wells). This does not make physical sense. What happens in reality is that a shock, or discontinuity in saturation, builds up, as described next.

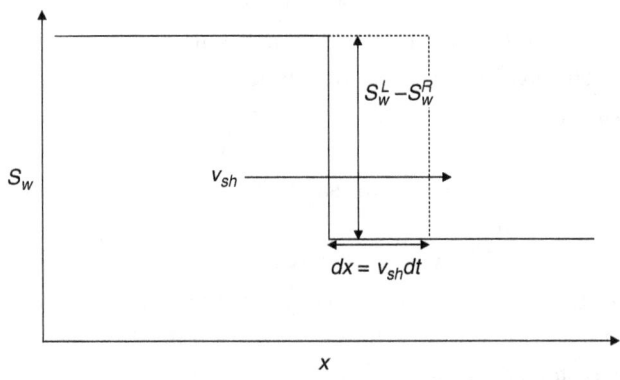

Figure 17.8. A diagram illustrating conservation of volume across a shock front. A left state of saturation is displacing a right state at a characteristic speed $\mathbf{v}_{sh}$.

again assuming incompressible flow (constant density and porosity). In dimensionless form,

$$v_{shD} = \frac{\Delta f_w}{\Delta S_w}. \tag{17.20}$$

This is the difference form of the governing equation for speed. Notice that if there is no shock, but a smooth change in wavespeed, then the speed reduces to Eq. (17.17). Indeed, this is a more elegant and rapid derivation to find the wavespeed than the cumbersome derivation of a partial differential equation for volume conservation. In the next section, I show how to find the correct shock and its speed using a graphical construction.

17.3. Welge Construction

The solution for saturation must represent a monotonic decrease in water saturation from $1 - S_{or}$ ($f_w = 1$) for $v_D = 0$ to S_{wc} ($f_w = 0$) for large v_D (large distance for a given time). The solution can be a constant saturation, a smooth variation (called a rarefaction) obeying Eq. (17.17) or a shock that obeys Eq. (17.20). Mathematically, there are many ways to do this, but only one that makes physical sense — i.e. a solution that is the correct physical limit when capillary pressure (which smooths out the shock) becomes small. It is possible to find this physically correct solution graphically, as shown in Fig. 17.9. This is the Welge construction: a line is drawn from the initial condition ($S_w = S_{wc}$; $f_w = 0$) that is tangent to the fractional flow curve. The shock is from the initial condition (this is the right state of the shock) to the saturation (and fractional flow) where the line hits the fractional flow curve (the left state). The slope of the line is the change in fractional flow divided by the change in saturation, and so represents the dimensionless shock speed. Mathematically, this can be written as

$$v_{shD} = \left.\frac{df_w}{dS_w}\right|_{S_w^L} = \frac{\Delta f_w}{\Delta S_w}. \tag{17.21}$$

The solution for water saturation as a function of dimensionless speed is then as shown schematically in Fig. 17.10 for our example

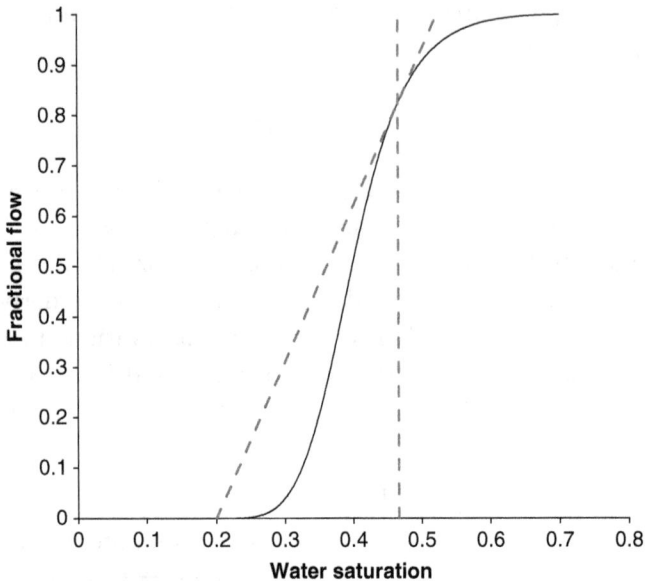

Figure 17.9. A diagram illustrating the Welge construction to find the correct shock for the Buckley–Leverett solution. The sloping dashed line is tangent from the initial conditions to the fractional flow curve. Where this line hits the fractional flow curve represents the left state of the shock (the saturation indicated by the dashed vertical line) — the right state is the initial water saturation (0.2 in this case). The slope of the tangent is the dimensionless shock speed.

case. The smooth part of the profile — the rarefaction — is found by computing the slope of the fractional flow curves for saturations higher than the shock front, as a function of saturation. This can be done analytically (from a closed-form expression of the fractional flow) or — with care — graphically.

The solution shown in Fig. 17.10, obeys conservation of volume and the boundary conditions. Other possible shocks either give unphysical solutions (multiple values of saturation for a given speed) or are unstable physically. The correct shock is self-sharpening. To explain this, consider the real situation where there is some capillary pressure. The capillary pressure has a diffusive effect and tends to smear out the shock. At the leading edge — small saturations — the wavespeed is slower than the shock speed, so this saturation is caught up by the saturation behind it. To the left side of the shock,

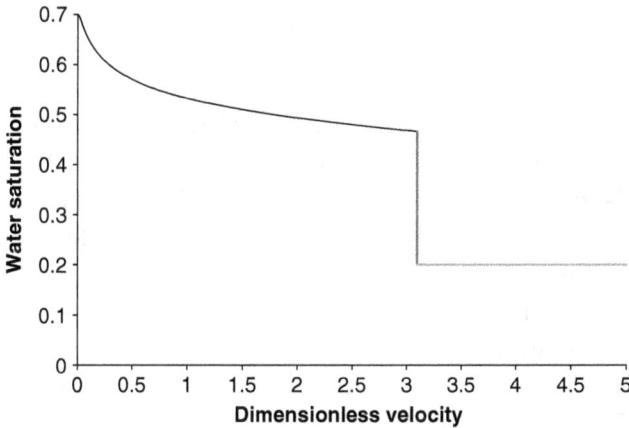

Figure 17.10. The Buckley–Leverett solution, featuring constant states (saturation constant), a rarefaction where the saturation varies smoothly with velocity and a shock (vertical line), constructed as explained in Fig. 17.9.

any smearing appears to slow down the saturation, but here, the natural wavespeed will be higher than the shock speed and so this water will speed up. The net result is a shock that is stable against perturbations due to capillary pressure. Any other possible solution (and to avoid confusion I will not present them here) will lead to a shock that will decompose and rearrange as the stable shock we have just described.

This solution is called the Buckley–Leverett solution after the authors who first presented it; the conservation equation, Eq. (17.10) and its variants are often called the Buckley–Leverett equation.

17.4. Wave, Particle Speeds and Definitions

Remember that the wavespeed IS NOT the same as the particle (tracer) velocity:

$$v_{pD} = \frac{f_w}{S_w}. \tag{17.22}$$

The particle speed is the speed with which a single molecule of water moves through the pore space. This can be explained simply by considering conservation of volume in a system where the saturation is constant (or over a length where the change in saturation is small). Imagine that, say, blue water is injected to displace red water. Then if we inject a volume $q_w = f_w q_t$ per unit area per unit time, this blue water will fill a volume $f_w q_t / (\phi S_w)$ of the porous medium — in dimensionless form this is the speed given by Eq. (17.22).

This is distinct from the wavespeed. This is the speed with which a given saturation moves, not a single particle. Why is this different? One way to consider this is through analogy with traffic flow. Travelling in a car on a motorway, you are always moving forwards, or stationary. But now imagine yourself in a helicopter flying above the motorway (and, for a more graphic illustration this is the M25 during rush hour, so there are many traffic jams). You can see waves of traffic density — this is analogous to saturation. If there is an accident, for instance, the cars at the crash site are stationary — the particle speed is zero. However, a wave of stationary cars moves *backwards* up the motorway; the wavespeed associated with a dense packing of (unmoving) cars is negative.[3] So, here, the wavespeed for saturation is a speed that a particular value of saturation moves, but this is not the same as the speed of a particular water molecule (the particle speed).

Now we briefly present some definitions useful for understanding the terminology used to describe these solutions. A *spreading wave* or *rarefaction* is when the wave becomes more diffuse with time — smooth changes in saturation.

A *sharpening wave* is when the wave becomes less diffuse and sharpens to form a shock.

Indifferent is the case when the wave neither spreads nor sharpens.

A *constant state* is a fixed saturation with distance (or velocity). This is also an acceptable solution to the equations.

---

[3]This is why driving when there is an accident (particularly in poor visibility) is so dangerous. You might think that you have sufficient time (and space) to stop before the car in front comes to a halt, but in reality a wave of stationary traffic is moving *towards* you — this is less safe than there being a brick wall just out of sight.

17.5. Effect of Gravity

We can also consider the impact of gravity on our analytical solutions, as shown in Figs. 17.11 and 17.12 with some schematic curves. If the water flow is uphill, then the water flow is held back, resulting in a higher shock saturation moving more slowly than a corresponding case with no gravity. Even if the fractional flow becomes negative, the shock jumps across this region and so we do not see explicit counter-current flow for water injected at the bottom of the formation.

Note that it is possible that the wavespeed at $1 - S_{or}$ is not zero (as in the previous examples) but finite — this is the derivative of the fractional flow at $S_w = 1 - S_{or}$. In this case, we have a constant state from $v_D = 0$ to the computed value. As mentioned previously, this is a perfectly acceptable solution of the conservation equation.

If we flow downhill, then water moves faster and we see a faster-moving, shallower water shock. If the water fractional flow at $v_D = 0$ is 1, then again we cannot observe strictly counter-current

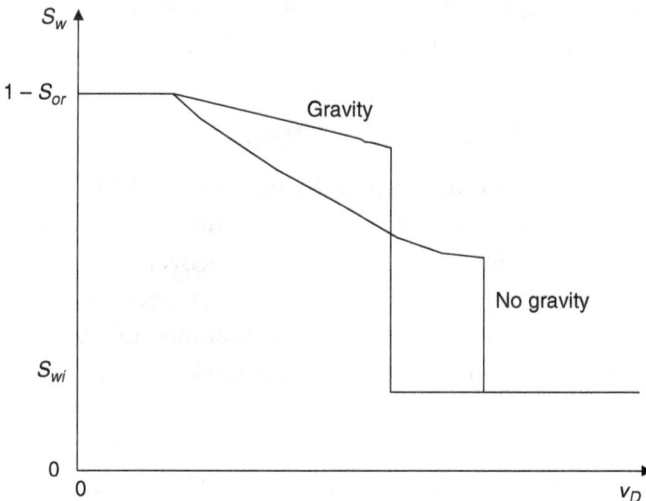

Figure 17.11. The Buckley–Leverett solution showing the effect of gravity for water flowing uphill. Buoyancy here leads to a higher, slower-moving water shock.

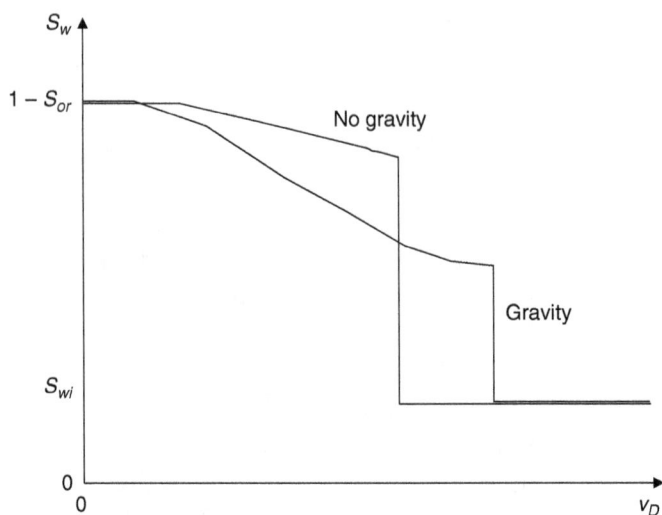

Figure 17.12. The Buckley–Leverett solution showing the effect of gravity for water flowing downhill. Buoyancy in this case leads to a lower, faster-moving water shock.

flow. However, it is possible to construct solutions where we have a backwards moving shown or rarefaction, representing the portion of the fractional flow curve which is greater than 1.

17.6. Average Saturation and Recovery

The final stage in the analysis is to use the analytical solution for saturation as a function of speed (and hence, for a given time, as a function of distance) to compute recovery. In our previous discussions, we have always discussed how recovery is a function of how much water is injected (the pore volumes of water injected). The way to derive at a recovery calculation is first to consider the average saturation. The recovery is proportional to the change from the initial (usually connate) water saturation to this average value: specifically, the average saturation in a domain after breakthrough — i.e. where the shock front has already reached the production well. Consider the schematic saturation profile shown in Fig. 17.13.

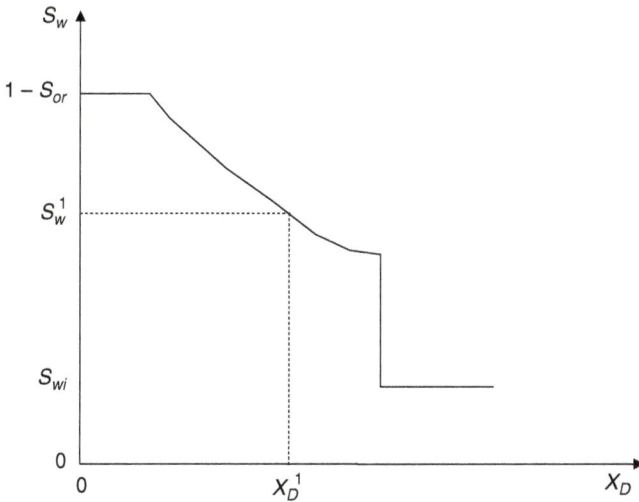

Figure 17.13. The Buckley–Leverett solution as a function of dimensionless distance employed to compute the average saturation and hence recovery. x_D^1 represents the production well ($x_D = 1$) and we compute the average saturation between this and the injector at $x_D = 0$.

The average saturation behind x_D^1 is easy to define, but the mathematics is somewhat tedious to produce a tractable solution:

$$\bar{S}_w(t_D) = \frac{1}{x_D^1} \int_0^{x_D^1} S_w dx_D. \tag{17.23}$$

Integrate by parts:

$$\bar{S}_w(t_D) = \frac{1}{x_D^1} \left([x_D S_w]_0^{x_D^1} - \int_{1-S_{or}}^{S_w^1} x_D dS_w \right). \tag{17.24}$$

We can write $x_D^1 = \left. \frac{df_w}{dS_w} \right|_{S_w=S_w^1}$, and thus Eq. (17.24) becomes

$$\bar{S}_w(t_D) = S_w^1 - \frac{t_D}{x_D^1} \int_{1-S_{or}}^{S_w^1} \frac{df_w}{dS_w} dS_w = S_w^1 - \frac{t_D}{x_D^1} \int_1^{f_w^1} df_w$$

$$= S_w^1 + \frac{t_D}{x_D^1} (1 - f_w^1) = S_w^1 + \frac{1 - f_w^1}{f_w^{1'}}. \tag{17.25}$$

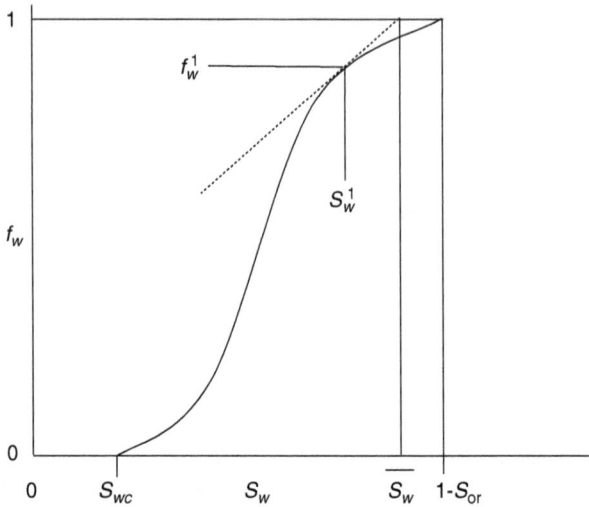

Figure 17.14. A schematic of the construction used to find the average saturation when the saturation at the well ($x_D = 1$) is some arbitrary value S_w^1.

Notice the prime on the final equation, denoting the derivative. Graphically this can be represented as the extension of the tangent of the curve at some location on the rarefaction to $f_w = 1$ giving the average saturation behind the front. This construction is shown schematically in Fig. 17.14. From a practical perspective, choose any saturation above the shock-front value, draw a tangent and find the saturation value when this tangent reaches $f_w = 1$.

We now use this construction to compute the recovery — pore volumes of oil produced — as a function of time, or pore volumes of water injected: N_{pD} vs. t_D. We know that by definition, the average saturation is $S_{wi} + N_{pD}$, where N_{pD} are the pore volumes of oil produced.

Before breakthrough, the reservoir volume of oil produced is equal to the reservoir volume of water injected and so $N_{pD} = t_D$. Breakthrough occurs when the shock moving at dimensionless speed v_{shD} reaches $x_D = 1$; this is a dimensionless time $t_D = 1/v_{shD}$. So, the first step in constructing a recovery curve is to draw a straight line of unit slope from $N_{pD} = t_D = 0$ to $N_{pD} = t_D = 1/v_{shD}$.

After breakthrough, the recovery curve has a slope less than one, indicating the production of both oil and water. What is the maximum recovery? This is $1 - S_{or} - S_{wc}$; you cannot produce more oil than this. When will this occur? If the wavespeed for $S_w = 1 - S_{or}$ is zero, then this happens at infinite time, so the maximum recovery is met asymptotically at infinite time. If the wavespeed has a finite minimum values, say v_{Dmin}, then maximum recovery is met at a time $1/v_{Dmin}$. Now construct one or two points in between. Choose a value of saturation in the rarefaction. Find the tangent through this saturation value on the fractional flow curve. The slope of the tangent is the wavespeed, v_D, while the intersect when $f_w = 1$ is the average saturation. The recovery is the average saturation minus the initial value, while the dimensionless time $t_D = 1/v_D$.

The resultant plot for our example case is shown in Fig. 17.15. In words, the procedure does seem a little bewildering — the only way to learn this is through performing the exercise yourself.

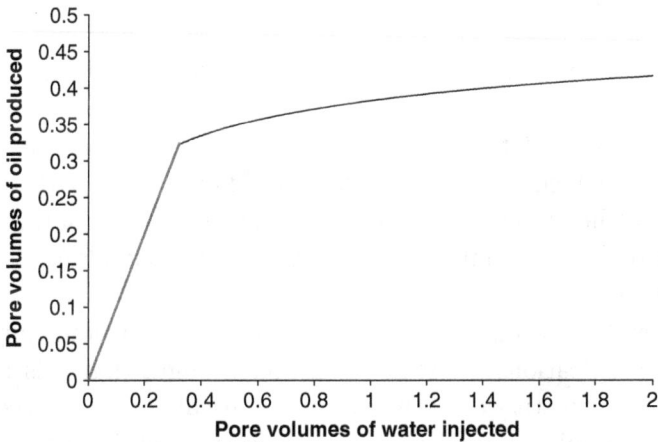

Figure 17.15. Pore volumes produced as a function of pore volumes of water injected for our example case. The straight line has unit slope and occurs before breakthrough. After breakthrough recovery was constructed using the methodology described in the text. The ultimate recovery is 0.5, but this is only reached asymptotically once an infinite number of pore volumes are produced. Rarely is more than pore volumes injected in a reservoir waterflood and so there is no point showing recovery beyond, say, 2 pore volumes, as here.

To recap, the steps to do a complete Buckley–Leverett analysis are as follows:

1. Given relative permeability curves, you first compute the fractional flow and plot this as a function of saturation.
2. Perform the Welge construction to find the shock front saturation and shock speed.
3. Plot the saturation as a function of dimensionless velocity; in the rarefaction the speed is the slope of the fractional flow curve.
4. Compute the dimensionless recovery (pore volumes produced) as a function of dimensionless time (pore volumes injected). Before breakthrough, since we have assumed that the oil and water are incompressible, these two quantities are the same. After breakthrough of water (when the shock front reaches the production well), the pore volumes of oil produced are less than the pore volumes of water injected, since some water is produced as well. You find the recovery from choosing points in the rarefaction and extrapolating a tangent on the fractional flow curve to $f_w = 1$, as described above.

This is an important exercise, and is performed by any reservoir engineer to assess waterflood recovery. In the end the behaviour at the field scale is determined by both this analysis and the geology of the field, which defines the preferential flow paths for the injected water, and the placement of the wells — an assessment of this usually requires reservoir simulation methods and lies outside the scope of this volume.

We will now discuss different types of behaviour physically and the implications for recovery. Note, though, that this rigorous analysis can substitute for the rather empirical approach used previously and should be used to assess recovery when both the relative permeabilities and viscosities are known. There are generically three types of solutions:

1. **Classic Buckley–Leverett.** This is the case we have shown with a rarefaction and a shock. This is observed for most power-law relative permeabilities and reproduces the behaviour of most

weakly water-wet or mixed-wet systems. Note though that the minimum wavespeed is not necessarily zero (as in our example) and so the rarefaction may be preceded by a constant state.

2. **All shock.** For strongly water-wet media, or if the water mobility is extremely low (say polymer flooding) the shock may extend all the way to $1 - S_{or}$; indeed it is not possible to construct a tangent and the fractional flow curve is entirely concave. This is the simplest solution with a shock moving with speed $1/(1 - S_{or} - S_{wc})$ and a recovery that reaches its maximum at breakthrough. Physically this is why waterflooding a water-wet system is favourable (despite the high residual saturation) since the water is held back in the pore space, and the oil moves ahead, leading to a sharp front.

3. **No shock.** It is possible, for oil-wet systems and/or with a very unfavourable viscosity ratio (the oil is much more viscous than the water), for there to be no shock, in that the maximum wavespeed $v_{D\text{max}}$ occurs for $S_w = S_{wc}$. In this case, the fractional flow is convex and the solution only a rarefaction. Breakthrough occurs at a dimensionless time $1/v_{D\text{max}}$. Note that the maximum wavespeed is never infinite — this does not make physical sense.

The last point to note is that a Buckley–Leverett-style analysis is not confined to a 1D analytical analysis; it is also useful to compare field-scale recovery (either real data or simulation predictions) to Buckley–Leverett predictions. It is straightforward (with a very careful scrutiny of the definitions) to convert data for oil produced at surface conditions into pore volumes produced, and real time into pore volumes of water injected. Since the real reservoir is heterogeneous, the recovery will always lie below that from an idealized Buckley–Leverett analysis, but it does serve as a very useful basis of comparison. This is also helpful in field management: the closer the predicted (or real) recovery to Buckley–Leverett, the better the injection wells are contacting and sweeping the reservoir. It also serves as a guide to how much water needs to be injected to achieve optimal recovery.

17.7. Oil Recovery and the Impact of Wettability

We can now return to our previous analysis of relative permeability and use these curves to compute recovery for a linear displacement. It is important to emphasise that in most field cases we rarely inject more than one pore volume of water. Hence, the fact that very low oil saturations can be achieved by oil layer drainage when hundreds or thousands of pore volumes are injected is not economically relevant, even if it can be seen in laboratory experiments. Of much greater import is the shock front saturation; this generally determines the local efficiency of waterflooding, with a high shock front being most favourable.

In general, strongly water-wet samples give a Buckley–Leverett displacement that is all shock, with the maximum recovery reached at breakthrough. This is efficient, but the high residual saturation makes the process less than ideal. Strongly oil-wet systems also give poor recovery — often even poorer recovery — since the oil breaks through very early and there is only substantial production after breakthrough, where water is also produced. This is economically unfavourable. The best recovery is for wettability states in between: mixed-wet or weakly water-wet rocks. One of the new research areas at the moment is in how to control wettability through adjusting the chemistry of the injected brine. This is the idea behind low-salinity waterflooding that aims to adjust the rock wettability to a favourable near water-wet state.

I will now illustrate the remarks above with some experimental data on sandstones. Note, however, that for the carbonates discussed earlier, the most favourable conditions for waterflooding were more mixed-to-oil-wet than shown here.

The example I choose uses the computations of relative permeability for mixed-wet Berea sandstone made by Valvatne and Blunt (2004). In this work, a fixed fraction of the pores contacted by oil after primary drainage became oil-wet. Different initial saturations of water (i.e. saturations after primary drainage) were considered; if the initial water saturation is high, the system appears to be more water-wet, as the pores that remain full of water remain water-wet. As the initial saturation decreases, the system becomes more oil-wet.

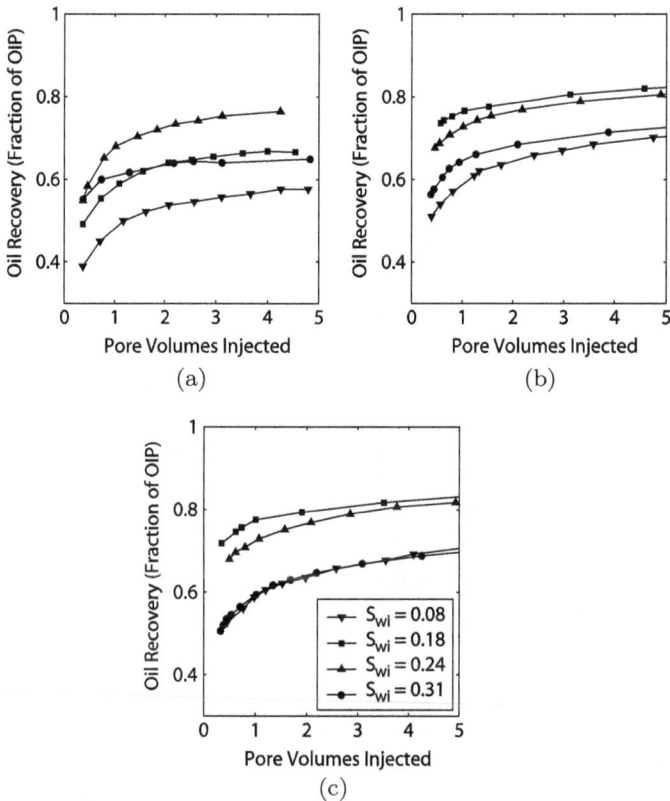

Figure 17.16. Measured (left) and predicted (middle and right) oil recoveries as a function of prove volumes injected in a mixed-wet Berea sandstone. The different initial water saturations are indicated. Note that the most favourable recoveries occur for intermediate saturations. The two modelling predictions use slightly different assignments of wettability. From Valuvatne and Blunt (2004).

This is the trend with height that is observed in the transition zone of an oil reservoir.

The graphs in Fig. 17.16 show measured recovery profiles for waterflooding from Jadhunandan and Morrow (1995) — with different initial water saturations — compared to different predictions using pore-scale modelling. For this sandstone sample, we see unfavourable recovery for the most water-wet case (because of the high residual saturation) and for the most oil-wet case (because of early water breakthrough and the slow drainage of oil layers). The ideal cases have an intermediate initial water saturation and are overall weakly water-wet to mixed-wet in character, meaning that the Amott–Harvey wettability index is close to zero.

This behaviour, as alluded to previously, is different from the trend we expect in carbonates, where the most favourable recoveries are seen when most of the pores are oil-wet; the difference is a result of the connectivity of the pore space, the nature of the pore and throat size distributions and the local variations in wettability (contact angle). Carbonates display a very wide range of connectivity and pore sizes, and can have very low recoveries in the water-wet limit; however, a definitive characterisation requires detailed analysis of the sample of interest coupled with an accurate assessment of wettability. At present, we do not have a way to assign contact angle on a pore-by-pore basis unambiguously, and so, at present, we have to rely on macroscopic measurements of wettability (such as Amott index) to tune the contact angle distribution in our pore-scale models.

One last comment: the pore volumes produced is NOT the recovery factor shown in the graphs above. The recovery factor, R_F, is a ratio of the volume of oil produced to the total of oil initially in place. It is easy to relate these two quantities as follows:

$$R_F = N_{PD} \frac{B_{oi}}{B_o(1 - S_{wc})}. \tag{17.26}$$

Chapter 18

Analytic Solutions for Spontaneous Imbibition

18.1. Counter-current Imbibition

We will now present a solution for spontaneous imbibition, where displacement is controlled entirely by capillary forces. This is a useful complement to the Buckley–Leverett solution. It is valuable experimentally as a way to determine, or at least constrain, capillary pressure and relative permeability. It is also useful for the analysis of recovery in fractured reservoirs, as discussed previously.

The formulation here is quite new in the literature. While the solutions were first proposed by McWhorter and Sunada (1990), it was not until the work of Schmid *et al.* (2011), and Schmid and Geiger (2012) that it was appreciated that this was indeed a closed-form solution generally applicable for spontaneous imbibition.

Before wading into the mathematical details, let us review the physical situation. It is illustrated in Fig. 18.1, where bubbles of non-wetting phase escape from a core when the wetting phase imbibes from all sides. It is similar to the bubbles of air seen around a cube of sugar dropped into a drink — if we ignore the dissolution. CT images of imbibition into a Ketton limestone core are shown in Fig. 18.2.

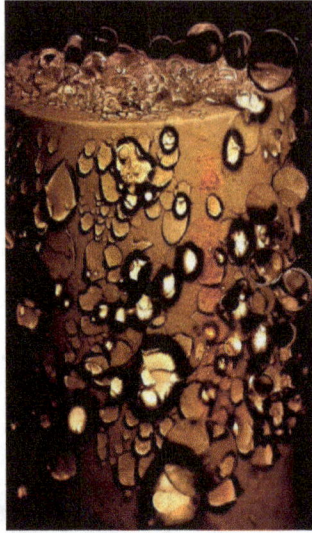

Figure 18.1. Photograph illustrating imbibition in a rock core. The bubbles are the displaced non-wetting phase. Picture from drahellkat.deviantart.com.

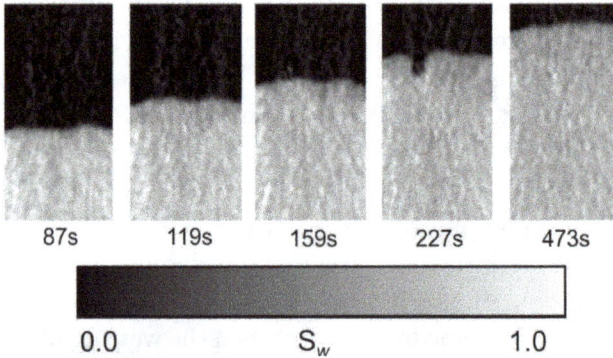

Figure 18.2. Experimental measurements of imbibition. Here the arrangement is similar to that proposed mathematically — except that we ignore gravity; there is a reservoir of fluid at one end of the core and water imbibes. The saturation profiles are measured using medical CT scanning. This is cocurrent imbibition of water into a dry (air-saturated) Ketton carbonate core. In the text we consider a slightly simpler case where the total velocity is zero, which means that the non-wetting phase must escape through the inlet. This is counter-current imbibition. Both co- and counter-current imbibition can be analysed analytically in one dimension. Figure courtesy of Nayef Alyafei, Imperial College.

For clarity, we start with the conservation equations, Eqs. (16.3) and (16.2) which are written here in terms of the water Darcy velocity (rather than the total velocity, since this will be zero):

$$\phi \frac{\partial S_w}{\partial t} + \frac{\partial q_w}{\partial x} = 0, \tag{18.1}$$

$$q_w = \frac{\lambda_w}{\lambda_t} \left\{ q_t + K\lambda_o \left(\frac{\partial P_c}{\partial x} + (\rho_w - \rho_w)g_x \right) \right\}. \tag{18.2}$$

For spontaneous imbibition, we ignore gravitational forces (assume that they are either small compared to capillary forces at the core (cm scale), or that the displacement is horizontal) and the total velocity. Setting the total velocity q_t to zero means that no fluid is injected and the flow is counter-current; the movement of water into the porous medium is matched exactly by the volume of oil (or gas) that leaves. Then we write

$$q_w = \frac{K\lambda_w\lambda_o}{\lambda_t} \frac{\partial P_c}{\partial x}. \tag{18.3}$$

The conservation equation, Eq. (18.1), becomes

$$\phi \frac{\partial S_w}{\partial t} + \frac{\partial}{\partial x} \left(\frac{K\lambda_w\lambda_o}{\lambda_t} \frac{dP_c}{dS_w} \frac{\partial S_w}{\partial x} \right) = 0. \tag{18.4}$$

Assuming a constant porosity we can write Eq. (18.4) as a non-linear diffusion equation,

$$\frac{\partial S_w}{\partial t} = \frac{\partial}{\partial x} \left(D(S_w) \frac{\partial S_w}{\partial x} \right), \tag{18.5}$$

where the non-linear capillary diffusion coefficient is

$$D(S_w) = -\frac{K\lambda_w\lambda_o}{\phi\lambda_t} \frac{dP_c}{dS_w}. \tag{18.6}$$

Note the negative sign. D is positive, so we assert that the gradient of the capillary pressure as a function of water saturation is always negative. You should by now appreciate that this is indeed correct.

The boundary conditions are a porous medium containing initially irreducible water $(S = S_{wc})$ and a non-wetting phase (which we call oil for convenience here). At the inlet, $x = 0$, we maintain a capillary pressure of zero. In a strongly water-wet system, the saturation will be $1 - S_{or}$; in any other case this will simply be the saturation at which the capillary pressure is zero, which we define as S^*.

We will now try to find a solution as follows. We write

$$\omega = \frac{x}{\sqrt{t}}, \tag{18.7}$$

where we assume that the solution can be stated as $S_w(\omega)$ only. I also will state that we can write

$$\omega = \frac{dF}{dS_w}, \tag{18.8}$$

for some capillary fractional flow $F(S_w)$. We assume that F has a maximum value $F^* = F(S^*)$ and is zero for the irreducible water saturation: $F(S_{wc}) = 0$.

Note the approach here. I have assumed a certain functional form of the solutions, based — please note — on both the previous solution for diffusion and the Buckley–Leverett analysis. However, this is an educated guess and we have to test if we can solve the governing partial differential equations and the boundary conditions. There is no manner to know — before we start — if this approach is correct. Despite what you may be told in mathematics classes, solving partial differential equations is more inspiration (guessing the correct approach) than application (of all the methods you are given in class).

Then we define the following derivatives:

$$\frac{\partial S_w}{\partial t} = -\frac{\omega}{2t} \frac{dS_w}{d\omega}, \tag{18.9}$$

$$\frac{\partial S_w}{\partial x} = \frac{1}{\sqrt{t}} \frac{dS_w}{d\omega}. \tag{18.10}$$

Then Eq. (18.5) becomes an ordinary differential equation:

$$\omega \frac{dS_w}{d\omega} + 2\frac{d}{d\omega}\left(D\frac{dS_w}{d\omega}\right) = 0. \tag{18.11}$$

We integrate once:

$$\int \omega \, dS_w = -2D\frac{dS_w}{d\omega}, \tag{18.12}$$

where the integration constant is zero since we define $F(S_{wc}) = 0$ and also $D(S_{wc}) = 0$. Then substitute in F from Eq. (18.8) to find

$$F\frac{d^2 F}{dS_w^2} = -2D. \tag{18.13}$$

Equation (18.13) is the key equation to define F and hence construct a solution.

In a formal mathematical sense, the solution can be expressed in closed form simply by integrating Eq. (18.13) twice — this is the solution presented in Schmid *et al.* (2011). The problem, which appears — at first sight — somewhat off-putting, is that the solution is expressed in terms of implicit integrals — i.e. the integral to find F involves F itself. Indeed, it is presented in terms of two implicit integrals, since we have to define $F(1 - S_{or})$, noting that F is a dimensional quantity and so we cannot set it to 1.

We have already defined one boundary condition on F: $F(S_{wc}) = 0$. Since we have a second-order equation for F, we require two conditions. The second, applicable in most cases, is that for $S_w = 1 - S_{or}, dF/dS_w = 0$ (the saturation front does not move at the inlet, which again makes physical sense).

However, before proceeding, let us first study the implications for the amount of water that enters the porous medium. The water Darcy velocity can be found from Eqs. (18.3) and (18.6):

$$q_w = -\phi D\frac{\partial S_w}{\partial x}, \tag{18.14}$$

which from Eq. (18.10) is

$$q_w = -\frac{\phi D}{\sqrt{t}} \frac{dS_w}{d\omega},$$

(18.15)

and substituting in Eq. (18.13) yields

$$q_w = \frac{\phi F F''}{2\sqrt{t}} \frac{dS_w}{d\omega} = \frac{\phi F}{2\sqrt{t}}.$$

(18.16)

Note the identity $F^* = d\omega/dS_w$.

The inlet flux is then related to the value of F^*. The total amount of water that enters the system, Q_w, is

$$Q_w = \int_0^t \frac{\phi F^*}{2\sqrt{t}} dt = \phi F^* \sqrt{t}.$$

(18.17)

The amount imbibed scales as the square root of time. Note that this expression differs from the solutions presented previously, where we allowed the flux injected to reach a maximum in a block of finite size; here we consider only the early-time behaviour before the imbibing front has reached a boundary.

We will find a solution using a simple numerical approach. For given relative permeabilities and capillary pressures, we compute D. We then solve Eq. (18.13) numerically in a spreadsheet starting from S^* and then decreasing saturation in small increments. We guess F^* and impose $F'(S^*) = 0$. We then iterate to find the value of F^* such that when $S_w = S_{wc}$, $F = 0$. I will not go through the details, but it is readily computed using a backwards difference scheme. I show some example results here. Notice that using a Buckley–Leverett analogy, this is a simple case, since there are no shocks; the saturation profile is smeared out and we have, technically, an all-rarefaction solution.

I show the capillary pressure and relative permeabilities in the saturation range for which there is spontaneous imbibition Figs. 18.3 and 18.4, respectively. I then show the dimensionless fractional flow that varies between 0 and 1, Fig. 18.5; this is defined as $f = F/F^*$. Similarly I can define dimensionless wavespeeds $\omega_D = \omega/F^*$, Fig. 18.6. In the example, I show $F^* = 2.57 \times 10^{-4} \text{m}/\sqrt{\text{s}}$. I then show

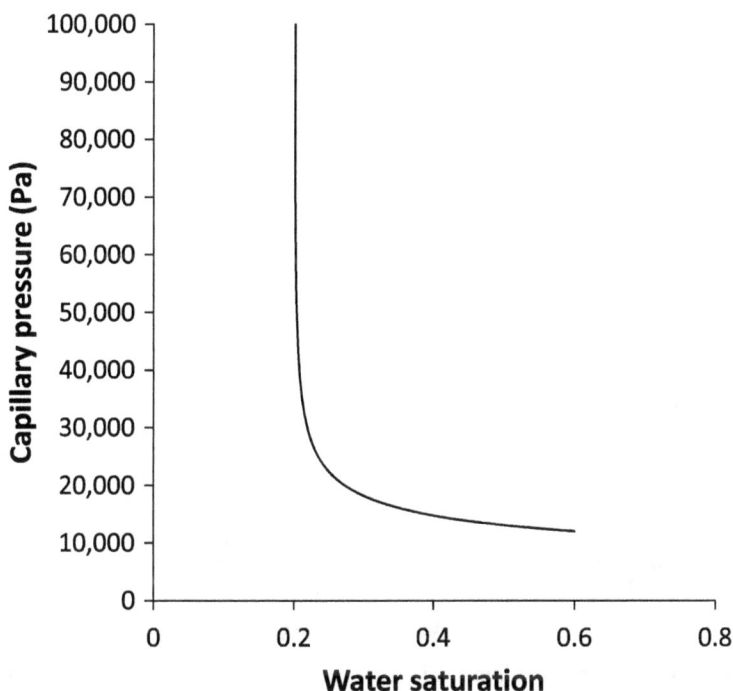

Figure 18.3. The capillary pressure used in the example calculation for imbibition. This resembles a primary drainage curve, but in fact — as we see below — this is a mixed-wet case. I assume that for saturations beyond $S^* = 0.6$ the capillary pressure is negative.

the saturation as a function of ω, Fig. 18.7; this is the full solution, with the distance moved by the saturation scaling not linearly with time (as for water injection in the Buckley–Leverett analysis) but as the square root of time.

If, instead, we have experimental measurements, then we can use the measured saturation profiles, obtained from *in situ* scanning (which give us F') to find D using Eq. (18.13). This is now the topic of ongoing research. My hope is that combining pore-scale modelling from images, macroscopic corefloods using Buckley–Leverett theory, this imbibition solution and steady-state measurements of relative permeability, we can readily and very reliably determine capillary pressure and relative permeability. At present, the subject woefully lacks good experimental data; the methods described in this volume

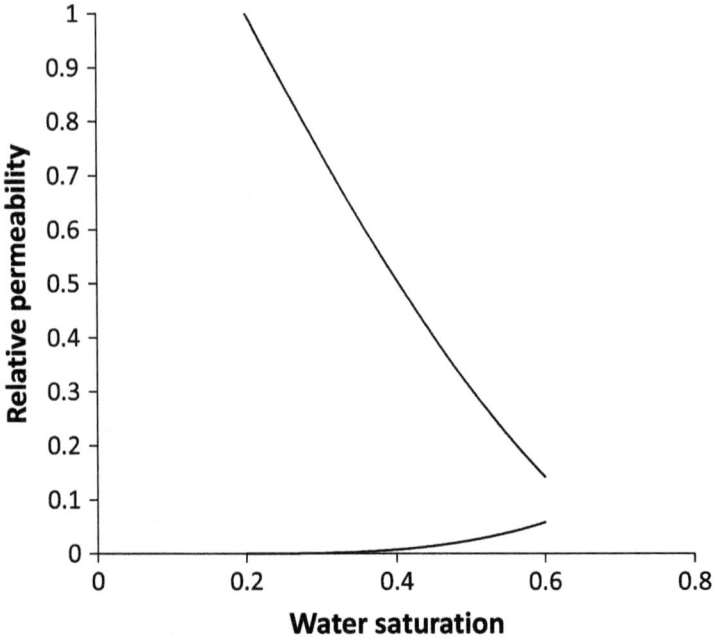

Figure 18.4. The relative permeabilities used in the example calculation for imbibition. The values are truncated when the capillary pressure is zero.

offer a new opportunity to produce results that are reliable and accurate.

18.2. Extensions to Analytic Theory and Reservoir Simulation

The method of characteristics can be extended to study a whole range of 1D displacements, including tracer flooding (combined with water injection), floods involving the injection of both a miscible solvent (miscible gas) and water, polymer flooding, as well as three-phase flow.

A complete discussion of this is beyond the scope of this volume, but uses the same ideas — derive an appropriate conservation equation and identify wave and shock speeds.

Needless to say, however, this is a powerful and relatively simple way to assess recovery and displacement in porous media and serves

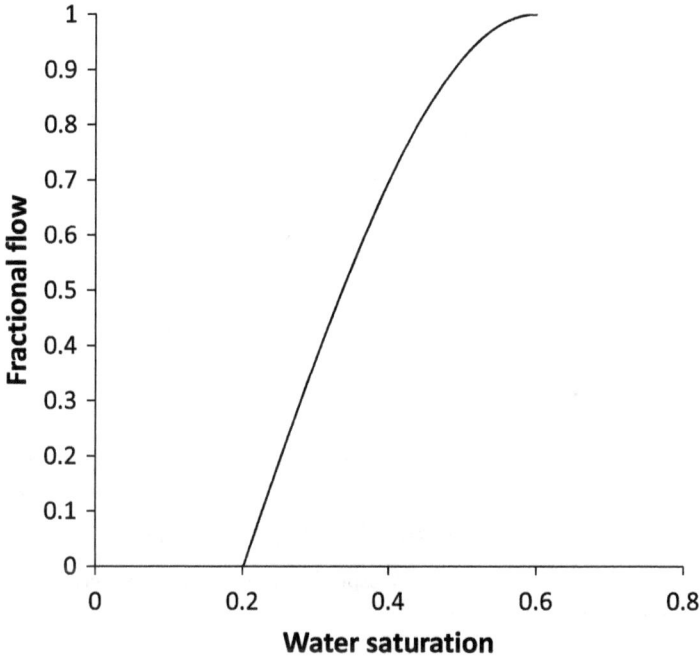

Figure 18.5. The dimensionless capillary fractional flow F. From the Buckley–Leverett analogy, we can see that the solution is a dispersed profile with no shock front. In this example I use the capillary pressure and relative permeabilities shown in Figs. 18.3 and 18.4 respectively, a permeability of 300 mD, a porosity of 0.2, and equal oil and water viscosities of 1 mPa·s.

as a useful complement to more sophisticated numerical approaches. In any event, numerical models require relative permeability curves as input and it is important to be able to understand their impact on the flow behaviour.

In the end, to describe flow in heterogeneous reservoirs with many wells and complex constraints on pressure and rates, it is necessary to perform a numerical analysis, solving the flow and transport equations presented in these notes in three dimensions. This is a rich and fascinating topic in its own right. However, it is important to retain a physical insight into the displacement and recovery processes and how core-scale analysis and measurements relate to field-scale recovery.

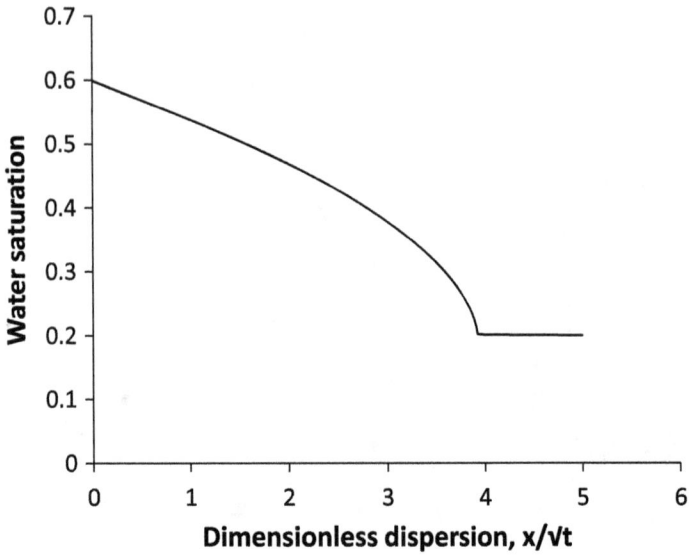

Figure 18.6. The dimensionless wavespeed found from the derivative of the fractional flow shown in Fig. 18.5.

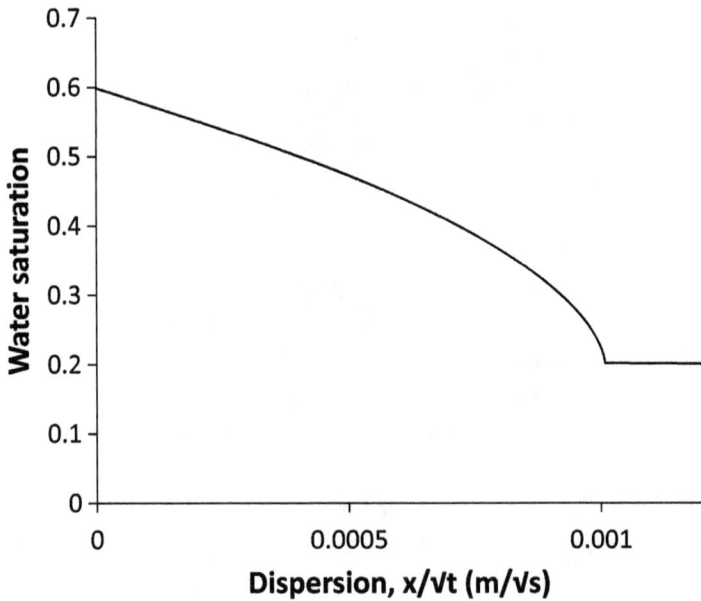

Figure 18.7. The same curve as before, but now multiplied by F^* to give the dimensional wavespeed.

Chapter 19

Bibliography and Further Reading

Here are the references of the papers mentioned in the notes as well as other papers that are useful for reference and further reading.

Many of these papers can be accessed through the Society of Petroleum Engineers database, if they are not available through Web of Science or Google Scholar — www.onepetro.org.

19.1. Relevant Research Papers and Other References

Akbarabadi, M. and Piri, M. (2013). Relative permeability hysteresis and capillary trapping characteristics of supercritical CO_2/brine systems: an experimental study at reservoir conditions, *Adv. Water Resour.*, **52**, 190–206.

Anderson, W.G. (1986). Wettability literature survey part 1: Rock/oil/brine interactions and the effects of core handling on wettability, *J. Petrol. Technol.*, **38**(10), 1125–1144.

Anderson, W.G. (1987). Wettability literature survey-part 6: The effects of wettability on waterflooding, *J. Petrol. Technol.*, **39**(12), 1605–1622.

Arns, C.H., Knackstedt, M.A., Pinczewski, V. *et al.* (2001). Accurate estimation of transport properties from microtomographic images, *Geophys. Res. Lett.*, **17**, 3361–3364.

Arns, C.H., Bauget, F., Limaye, A. *et al.* (2005). Pore-scale characterization of carbonates using X-ray microtomography, *SPE J.*, **10**(4), 475–484.

Arns, J.Y., Sheppard, A.P., Arns, C.H. *et al.* (2007). Pore-level validation of representative pore networks obtained from micro-CT images, *Proceedings of the Annual Symposium of the society of Core Analysis*, 10–12 September, Calgary, Canada.

Aronofsky, J.S., Masse, L. and Natanson, S.G. (1958). A model for the mechanism of oil recovery from the porous matrix due to water invasion in fractured reservoirs, *Petrol. Trans. AIME*, **213**, 17–19.

Arps, J.J. (1956). Estimation of primary oil reserves, *Petrol. Trans. AIME*, **207**, 182–191.

Bakke, S. and Øren, P.-E. (1997). 3-D pore-scale modelling of sandstones and flow simulations in the pore networks, *SPE J.*, **2**, 136–149.

Bartell, F.E. and Osterhoff, H.J. (1927). Determination of the wettability of a solid by a liquid. *Ind. Eng. Chem.*, **19**, 1277.

Bear, J. (1972). *Dynamics of Fluids in Porous Media*, Dover Science Publications Inc., New York.

BP statistical review of world energy (2016). http://www.bp.com/en/ global/corporate/energy-economics/statistical-review-of-world-energy. html

Buckley, S.E. and Leverett, M.C. (1942). Mechanisms of fluid displacement in sands, *Trans. AIME*, **146**, 107116.

Carslaw, H.S. and Jaeger, J.C. (1946). *Conduction of Heat in Solids, 2nd Edition*, Oxford Science Publications, Clarendon Press, Oxford.

Chatzis, I. and Morrow, N. (1984). Correlation of capillary number relationships for sandstone, *SPE J.*, **24**(5), 555–562.

Craig, Jr. F.F. (1971). *The Reservoir Engineering Aspects of Waterflooding*, Society of Petroleum Engineers, ISBN 0-89520-202-6.

Dake, L.P. (1991). *Fundamentals of Reservoir Engineering*, Elsevier, ISBN 0-444-41830-X.

De Gennes, P.-G., Brochard-Wyart, F. and Quéré, D. (2002). *Capillary and Wetting Phenomena: Drops, Bubbles, Pearls, Waves*, Springer.

dehaanservices.ca, Accessed 1st April (2013).

DiCarlo, D.A., Sahni, A. and Blunt, M.J. (2000) The effect of wettability on three-phase relative permeability, *Transport in Porous Med.*, **39**, 347–366.

Dullien, F.A.L. (1992). *Porous Media: Fluid Transport and Pore Structure*, 2nd Edition, Academic Press, San Diego.

Dunsmuir, J.H., Ferguson, S.R., D'Amico, K.L. *et al.* (1991). X-ray microtomography. A new tool for the characterization of porous media, *Proceedings of the 1991 SPE Annual Technical Conference and Exhibition*, 6–9 October, Dallas, TX.

Fatt, I. (1956). The network model of porous media I. Capillary pressure characteristics, *Trans. AIME*, **207**, 144–159.

Flannery, B.P., Deckman, H.W., Roberge, W.G. *et al.* (1987). Three-dimensional X-ray microtomography, *Science*, **237**, 1439–1444.

Gelhar, L.W. (1993). *Stochastic Subsurface Hydrology*, Prentice-Hall, Upper Saddle River, NJ.

Jadhunandan, P.P. and Morrow, N.R. (1995). Effect of wettability on waterflood recovery for crude oil/brine/rock systems, *SPE Reservoir Eng.*, **10**(1), 40–46.

Jadhunandan, P.P and Morrow, N.R. (1991). Spontaneous imbibition of water by crude oil/brine/rock systems, *In Situ*, **15**(4), 319–345.

Jerauld, G.R. and Salter, S.J. (1990). Effect of pore-structure on hysteresis in relative permeability and capillary pressure: Pore-level modeling, *Transport Porous Med.*, **5**, 103–151.

Krevor, S., Pini, R., Zuo, L. *et al.* (2012). Relative permeability and trapping of CO_2 and water in sandstone rocks at reservoir conditions, *Water Resour. Res.*, **48**(2), W02532.

Krevor, S.C.M., Pini, R., Li, B. *et al.* (2011). Capillary heterogeneity trapping of CO_2 in a sandstone rock at reservoir conditions, *Geophys. Res. Lett.*, **38**, L15401.

Killins, C.R., Nielsen, R.F. and Calhoun, J.C. (1953). Capillary Desaturation and Imbibition in Porous Rocks, *Producers Monthly*, **18**(2), 30–39.

Lake, L. W. (1989).*Enhanced Oil Recovery*, Prentice Hall, Englewood Cliffs.

Land, C. (1968). Calculation of imbibition relative permeability for two- and three-phase flow from rock properties, *SPE Journal*, **24**, 149–156.

Lenormand, R., Zarcone, C. and Sarr, A. (1983). Mechanisms of the displacement of one fluid by another in a network of capillary ducts, *J. Fluid Mech.*, **135**, 337–353.

Lenormand, R. and Zarcone, C. (1984). Role of roughness and edges during imbibition in square capillaries, *Proceedings of the 59th Annual Technical Conference and Exhibition of the Society of Petroleum Engineers of AIME*, 16–19 September, Houston, TX.

Lenormand, R. (1985). Invasion Percolation in an etched network: measurement of a fractal dimension, *Phys. Rev. Lett.*, **54**(20), 2226–2231.

McCain, W.D. (1990)*The Properties of Petroleum Fluids*, 2nd Edition, PenWell Books.

McWhorter, D.B. and Sunada, D.K. (1990). Exact integral solutions for two-phase flow, *Water Resour. Res.*, **26**(3), 399–413.

Meissner, J.P., Wang, F.H.L., Kralik, J.G. *et al.* (2009). State of the art special core analysis program design and results for effective reservoir management, Dukhan field, Qatar, *Proceedings of the International Petroleum Technology Conference*, 7–9 December, Doha, Qatar.

Morrow, N.R. (1975). Effects of surface roughness on contact angle with special reference to petroleum recovery, *J. Can. Pet. Technol.*, **14**, 42–53.

Morrow, N.R. and Mason, G. (2001). Recovery of oil by spontaneous imbibition, *Curr. Opin. Colloid Interface Sci.*, **6**(4), 321–337.

Muskat, M. (1949). *Physical Principles of Oil Production*, McGraw Hill, New York.

Muskat, M. and Meres, M.W. (1936). The flow of heterogeneous fluids through porous media, *J. Appl. Phys.*, **7**, 346–344. doi: 10.1063/1.1745403.

Oak, M.J., Baker, L.E. and Thomas, D.C. (1990). Three-phase relative permeability of Berea sandstone, *J. Petrol. Technol.*, **42**(8), 1054–1061.

Øren, P.-E., Bakke, S. and Arntzen, O.J. (1998). Extending predictive capabilities to network models, *SPE J.*, **3**, 324–336.

Øren, P.-E. and Bakke, S. (2002). Process based reconstruction of sandstones and prediction of transport properties, *Transport Porous Med.*, **46**(2–3), 311–343.

Øren, P.-E. and Bakke, S. (2003). Reconstruction of Berea sandstone and pore-scale modelling of wettability effects, *J. Petrol. Sci. Eng.*, **39**, 177–199.

Okasha, T.M., Funk, J.J. and Rashidi, H.N. (2007). Fifty years of wettability measurements in the Arab-D carbonate reservoir, *Proceedings of the SPE Middle East Oil and Gas Show and Conference*, 11–14 March Manama, Kingdom of Bahrain.

Patzek, T.W. (2001). Verification of a complete pore network simulator of drainage and imbibition, *SPE J.*, **6**, 144–156.

Sahimi, M. (1995). *Flow and Transport in Porous Media and Fractured Rock*, Wiley-VCH Verlag GmbH, Weinheim, Germany.

Sahni, A., Burger, J. and Blunt, M. J. (1998). Measurement of three phase relative permeability during gravity drainage using CT scanning. SPE 39655, proceedings of the SPE/DOE Improved Oil Recovery Symposium, Tulsa, OK, April.

Salathiel, R.A. (1973). Oil recovery by surface film drainage in mixed-wettability rocks, *SPE J.*, **25**(10), 1216–1224.

Schilthuis, R.J. (1936). Active oil and reservoir energy *Petrol. Trans. AIME*, **118**, 33–52, Society of Petroleum Engineers. doi:10.2118/936033-G.

Schmid, K.S., Geiger, S. and Sorbie, K.S. (2011). Semianalytical solutions for cocurrent and countercurrent imbibition and dispersion of solutes in immiscible two-phase flow, *Water Resour. Res.*, **47**, W02550, doi:10.1029/2010WR009686.

Schmid, K.S. and Geiger, S. (2012). Universal scaling of spontaneous imbibition for water-wet systems, *Water Resour. Res.*, **48**, W03507.

Terzaghi, K. and Peck, R.B. (1996). Soil mechanics in engineering practice, 1948. *Publications of the Disaster Prevention Research Institute*, 448.

Wardlaw, N.C. and Taylor, R.P. (1976). Mercury capillary pressure curves and the interpretation of pore structure and capillary behaviour in reservoir rocks, *B. Can. Petrol. Geol.*, **24**(2), 225–262.

Wikipedia, accessed 6th January (2015): http://en.wikipedia.org/wiki/List_of_oil_fields.

Wilkinson, D. and Willemsen, J.F. (1983). Invasion percolation: a new form of percolation theory, *J. Phys. A*, **16**, 3365–3376.

Zhou, X., Morrow, N.R. and Ma, S. (2000). Interrelationship of wettability, initial water saturation, aging time, and oil recovery by spontaneous imbibition and waterflooding, *SPE J.*, **5**(2), 199–207.

Zimmerman, R. W. (1991). *Compressibility of Sandstones*, Elsevier Science Publishers, New York, NY, USA, ISBN 0444-88325-8, (1991).

19.2. Papers from Imperial College

These are presented in date order. Most of these papers can be downloaded from our website: http://www.imperial.ac.uk/earth-science/research/research-groups/perm/research/pore-scale-modelling/

Blunt, M.J. (1998). Physically based network modeling of multiphase flow in intermediate-wet media, *J. Petrol. Sci. Eng.*, **20**, 117–125.

Valvatne, P.H. and Blunt, M.J. (2004). Predictive pore-scale modeling of two-phase flow in mixed wet media, *Water Resour. Res.*, **40**, W07406.

Okabe, H. and Blunt, M.J. (2004). Prediction of permeability for porous media reconstructed using multiple-point statistics, *Phys. Rev. E*, **70**, 066135.

Al-Gharbi, M.S. and Blunt, M.J. (2005). Dynamic network modeling of two-phase drainage in porous media, *Phys. Rev. E*, **71**, 016308.

Piri, M. and Blunt, M.J. (2005a). Three-dimensional mixed-wet random pore-scale network modeling of two- and three-phase flow in porous media. I. Model description, *Phys. Rev. E*, **71**, 026301.

Piri, M. and Blunt, M.J. (2005b). Three-dimensional mixed-wet random pore-scale network modeling of two- and three-phase flow in porous media. II. Results, *Phys. Rev. E*, **71**, 026302.

Valvatne, P.H., Piri, M., Lopez, X. *et al.* (2005). Predictive pore-scale modeling of single and multi-phase flow, *Transport Porous Med.*, **58**, 23–41.

Tavassoli, Z., Zimmerman, R.W. and Blunt, M.J. (2005). Analytic analysis for oil recovery during counter-current imbibition in strongly water-wet systems, *Transport Porous Med.*, **58**, 173–189.

Jackson, M.D., Valvatne, P.H. and Blunt, M.J. (2005). Prediction of wettability variation within an oil/water transition zone and its impact on production, *SPE J.*, **10**(2), 184–195.

Tavassoli, Z., Zimmerman, R.W. and Blunt, M.J. (2005). Analysis of counter-current imbibition with gravity in weakly water-wet systems, *J. Petrol. Sci. Eng.*, **48**, 94–104.

Behbahani, H. and Blunt, M.J. (2005). Analysis of imbibition in mixed-wet rocks using pore-scale modeling, *SPE J.*, **10**(4), 466–474.

Bijeljic, B. and Blunt, M.J. (2006). Pore-scale modeling and continuous time random walk analysis of dispersion in porous media, *Water Resour. Res.*, **42**, W01202.

Juanes, R., Spiteri, E.J., Orr, Jr., F.M. *et al.* (2006). Impact of relative permeability hysteresis on geological CO_2 storage, *Water Resour. Res.*, **42**, W12418.

Suicmez, V.S., Piri, M. and Blunt, M.J. (2007). Pore-scale simulation of water alternate gas injection, *Transport Porous Med.*, **66**, 259–286.

Al-Kharusi, A.S. and Blunt, M.J. (2007). Network extraction from sandstone and carbonate pore space images, *J. Petrol. Sci. Eng.*, **56**, 219–231.

Bijeljic, B. and Blunt, M.J. (2007). Pore-scale modeling of transverse dispersion in porous media, *Water Resour. Res.*, **43**, W12S11.

Okabe, H. and Blunt, M.J. (2007). Pore space reconstruction of vuggy carbonates using microtomography and multiple-point statistics, *Water Resour. Res.*, **43**, W12S02.

van Dijke, M.I.J., Piri, M., Helland, J.O. *et al.* (2007). Criteria for three-fluid configurations including layers in a pore with nonuniform wettability, *Water Resour. Res.*, **43**, W12S05.

Suicmez, V.S., Piri, M. and Blunt, M.J. (2008). Effects of wettability and pore-level displacement on hydrocarbon trapping, *Adv. Water Resour.*, **31**, 503–512.

Spiteri, E.J., Juanes, R., Blunt, M.J. *et al.* (2008). A new model of trapping and relative permeability hysteresis for all wettability characteristics, *SPE J.*, **13**(3), 277–288.

Al-Kharusi, A.S. and Blunt, M.J. (2008). Multi-phase flow predictions from carbonate pore space images using extracted network models, *Water Resour. Res.*, **44**, W06S01.

Al-Sayari, S.S. (2009). The influence of wettability and carbon dioxide injection on hydrocarbon recovery, PhD thesis, Imperial College London.

Talabi, O., Al-Sayari, S.S., Iglauer, S. *et al.* (2009). Pore-scale simulation of NMR response, *J. Petrol. Sci. Eng.*, **67**(3–4), 168–178.

Dong, H. and Blunt, M.J. Pore-network extraction from micro-computerized-tomography images, *Phys. Rev. E*, **80**, 036307.

Al Mansoori, S.K., Iglauer, S., Pentland, C.H. *et al.* (2009). Three-phase measurements of oil and gas trapping in sand packs, *Adv. Water Resour.*, **32**, 1535–1542.

Al Mansoori, S.K., Itsekiri, E., Iglauer, S. *et al.* (2010). Measurements of non-wetting phase trapping applied to carbon dioxide storage, *Int. J. Greenh. Gas Con.*, **4**, 283–288.

Idowu, N.A. and Blunt, M.J. (2010). Pore-scale modelling of rate effects in waterflooding, *Transport Porous Med.*, **83**, 151–169.

Zhao, X., Blunt, M.J. and Yao, J. (2010). Pore-scale modeling: Effects of wettability on waterflood oil recovery, *J. Petrol. Sci. Eng.*, **71** 169–178.

Pentland, C.H., Itsekiri, E., Al-Mansoori, S. *et al.* (2010). Measurement of non-wetting phase trapping in sandpacks, *SPE J.*, **15**, 274–281.

Iglauer, S., Favretto, S., Spinelli, G. *et al.* (2010). X-ray tomography measurements of power-law cluster size distributions for the nonwetting phase in sandstones, *Phys. Rev. E*, **82**, 056315.

Pentland, C.H., El-Maghraby, R., Iglauer, S. *et al.* (2011). Measurements of the capillary trapping of super-critical carbon dioxide in Berea sandstone, *Geophys. Res. Lett.*, **38**, L06401.

Bijeljic, B., Mostaghimi, P. and Blunt, M.J. (2011). The signature of non-Fickian solute transport in complex heterogeneous porous media, *Phys. Rev. Lett.*, **107**, 204502.

Iglauer, S., Paluszny, A., Pentland, C.H. *et al.* (2011a). Residual CO_2 imaged with X-ray micro-tomography, *Geophys. Res. Lett.*, **38**, L21403.

Iglauer, S., Wülling, W., Pentland, C.H. *et al.* (2011b). Capillary trapping capacity of rocks and sandpacks, *SPE J.*, **16**(4), 778–783.

Iglauer, S., Fernø, M.A., Shearing, P. *et al.* (2012). Comparison of residual oil cluster size distribution, morphology and saturation in oil-wet and water-wet sandstone, *J. Coll. Interf. Sci.*, **375**, 187–192.

Raeini, A.Q., Blunt, M.J. and Bijeljic, B. *et al.* (2012). Modelling two-phase flow in porous media at the pore scale using the volume-of-fluid method, *J. Comput. Phys.*, **231**(17), 5653–5668.

Mostaghimi, P., Bijeljic, B. and Blunt, M.J. (2012). Simulation of flow and dispersion on pore-space images, *SPE Journal*, **17**, 1131–1141.

Gharbi, O. and Blunt, M.J. (2012). The impact of wettability and connectivity on relative permeability in carbonates: A pore network modeling analysis, *Water Resour. Res.*, **48**, W12513.

El-Maghraby, R.M. (2013). Measurements of CO_2 trapping in carbonate and sandstone rocks, PhD Thesis, Department of Earth Science and Engineering, Imperial College London.

Iglauer, S., Paluszny, A. and Blunt, M.J. (2013). Simultaneous oil recovery and residual gas storage: A pore-level analysis using in situ X-ray micro-tomography, *Fuel*, **103**, 905–914.

Bijeljic, B., Raeini, A., Mostaghimi, P. *et al.* (2013a). Predictions of non-Fickian solute transport in different classes of porous media using direct simulation on pore-scale images, *Phys. Rev. E*, **87**, 013011.

Blunt, M.J., Bijeljic, B., Dong, H. *et al.* (2013). Pore-scale imaging and Modelling. *Adv. Water Resour.*, **51**, 197–216.

Bijeljic, B., Mostaghimi, P. and Blunt, M.J. (2013b). Insights into non-Fickian solute transport in carbonates, *Water Resour. Res.*, **49**, 2714–2728.

Andrew, M., Bijeljic, B. and Blunt, M.J. (2013). Pore-scale imaging of geological carbon dioxide storage under in situ conditions, *Geophys. Res. Lett.*, **40**, 3915–3918.

Tanino, Y. and Blunt, M.J. (2013). Laboratory investigation of capillary trapping under mixed-wet conditions, *Water Resour. Res.*, **49**(7), 4311–4319.

Raeini, A.Q., Bijeljic, B. and Blunt, M.J. (2014). Numerical modelling of sub-pore scale events in two-phase flow through porous media, *Transport Porous Med.*, **101**, 191–213.

Andrew, M., Bijeljic, B. and Blunt, M.J. (2014). Pore-scale imaging of trapped supercritical carbon dioxide in sandstones and carbonates, *Int. J. Greenh. Gas Con.*, **22**, 1–14.

Amaechi, B., Iglauer, S., Pentland, C.H. *et al.* (2014). An experimental study of three-phase trapping in sand packs, *Transport Porous Med.*, **103**(3), 421–436.

Chapter 20

Homework Problems

These problems are useful to understand much of the material in these notes.

1. Define in a single sentence the following terms: (i) capillary pressure; (ii) drainage; (iii) imbibition; (iv) forced water injection.
2. Draw a typical oil/water capillary pressure curve for a water-wet sandstone with a porosity of 0.2 and a permeability of 100 mD. Draw primary drainage, imbibition and secondary drainage curves. Also indicate on the graph typical values for the residual oil saturation and connate water saturation and typical values for the capillary pressure. These need only be estimates, but please explain clearly why you chose these values. Also explain why the primary and secondary drainage curves are different and why the imbibition curve is lower than the secondary drainage curve.
3. You are planning a surfactant flood. You perform some core flood experiments to find the effect of capillary number on the residual oil saturation. You find a correlation of the form:

$$S_{or} = \text{Max} \left[0, 0.4 - \frac{\sqrt{N_{\text{cap}}}}{0.08} \right].$$

You now perform a reservoir-scale flood. The injection and production wells are 100 m apart and have a pressure drop of

10 atm between them. The rock permeability is 200 mD. With surfactant the oil/water interfacial tension is 0.01 mN/m. What is the predicted residual oil saturation? What pressure difference between the wells is necessary to remove all the oil?

4. In this question we consider the relative permeability of a bundle of capillary tubes of different radii. A fraction $f(r)dr$ tubes have a radius between r and $r + dr$. By definition,

$$\int_0^\infty f(r)dr = 1.$$

All the tubes are aligned horizontally and have the same pressure drop across them. They are all of the same length. The flow (volume per unit time) in each tube is given by

$$Q = \frac{\pi r^4}{8\mu l}\Delta P.$$

The total volume of each tube is $\pi r^2 L$. The tubes are water-wet and are initially filled with water. Oil is now injected into the tubes. What size tubes does the oil preferentially occupy? The last tube to be filled with oil has a radius R. Derive an expression for the oil saturation and the oil relative permeability.

5. You are planning a tracer test in a portion of a reservoir that is entirely full of water. The tracer dissolves in water and flows with the water. You also know that the tracer absorbs to the reservoir rock. The mass of tracer that absorbs per unit volume of the reservoir, ρ^a is given by the following equation:

$$\rho^a = ac,$$

where a is a dimensionless constant and c is the concentration of tracer in the water, measured in units of mass of tracer per unit volume of water.

Starting with the expression

$$\frac{\partial}{\partial t}(\textit{mass per unit volume of the reservoir})$$

$$+ \frac{\partial}{\partial x}(\textit{Mass flux per unit area}) = 0$$

derive a conservation equation for tracer in one dimension, where q is the water volumetric flux per unit area and ϕ is the rock porosity. You may assume that q and ϕ are both constant.

With what speed does the tracer flow through the rock? Find this speed when $q = 10^{-6} \text{ m} \cdot \text{s}^{-1}$, $\phi = 0.2$ and $a = 3$.

6. Derive a conservation equation in one direction (the x-direction) for a radioactive tracer moving through a fully water-saturated porous medium. Write the equation in terms of the tracer concentration (mass per unit volume of fluid), c. As well as moving with the fluid, the tracer also diffuses. Fick's law of diffusion states that the diffusive flux is

$$F_D = -D\frac{\partial c}{\partial x},$$

where D is the diffusion coefficient. Since the tracer is radioactive its concentration in a static fluid with no concentration gradient will decrease with time as $c(t) = c(t = 0)e^{-at}$ which is the same as saying that with no flow $\frac{\partial c}{\partial t} = -ac$.

7. **Unit conversions**

A committee is set up to increase North Sea oil production by 1,000,000 barrels per day. The committee members disagree on what units to use. For all the sections below express 1,000,000 barrels per day in the unit systems described.

(a) Milton Keynes, an economist, wants to use millions of \$ per year. The oil price is \$105 per barrel.
(b) Napoleon Laval, a Frenchman, wants to use SI units.
(c) Abdus Goldstein, a theoretical physicist, wants to use units in which $h/2\pi = 1$, $c = 1$ and $G = 1$. $h/2\pi = 1.055 \times 10^{-36}\text{J} \cdot \text{s}$, $c = 3.0 \times 10^8 \text{m} \cdot \text{s}^{-1}$, $G = 6.7 \times 10^{-11}\text{m}^3 \cdot \text{kg}^{-1} \cdot \text{s}^{-2}$.
(d) Jerry R. Beltbuckle III, an oil industry representative, prefers to use acre-feet per month. What is the target in (i) February 1999, (ii) December?

8. **Oil formation volume factor.** A production well in an undersaturated reservoir has a bottom-hole water-cut (f_w) of 0.2. If $B_w = 1.01$ and $B_o = 1.3$, then what is the water–oil ratio into the stock tank?

9. Derive a conservation equation for radial two-phase flow in a cylindrical geometry appropriate for near-wellbore flow of a water injector. Show that in the Buckley–Leverett rarefaction, the flow speed is given by

$$v = \frac{\partial f}{\partial S},$$

where f is the fractional flow, S is the water saturation and

$$v = \frac{\pi r^2 h \phi}{Qt},$$

where r is the radial distance from the well, h is the perforated interval, t is the time, ϕ is the porosity and Q is the flow rate at the well.

10. For each of the examples below provide a graph of: (i) S_w as a function of $v_D = x_D/t_D$; and (ii) pore volumes of oil produced N_{pD} as a function of t_D.

   In all cases $S_{wi} = S_{wc} = 0.2$.

   There is no gravity and the relative permeabilities are given by

$$k_{rw} = k_{rw}^{\max} \frac{(S_w - S_{wc})^a}{(1 - S_{or} - S_{wc})^a},$$

$$k_{ro} = k_{ro}^{\max} \frac{(1 - S_{or} - S_w)^b}{(1 - S_{or} - S_{wc})^b},$$

$$M = \frac{\mu_o}{\mu_w} \frac{k_{rw}^{\max}}{k_{ro}^{\text{Max}}}.$$

   (a) A strongly water-wet rock with $a = 3$, $b = 1$, $k_{rw}^{\max} = 0.18$, $k_{ro}^{\max} = 0.9$, $S_{wc} = 0.2$, $S_{or} = 0.4$ and $\mu_w = \mu_o(M = 0.2)$.
   (b) An oil-wet rock with $a = 1, b = 3$, $k_{rw}^{mx} = 0.9$, $k_{ro}^{\max} = 0.18$, $S_{wc} = 0.2$, $S_{or} = 0.1$ and $\mu_w = \mu_o(M = 5)$.
   (c) Repeat part (a) for $M = 5$ and $M = 50$.
   (d) Repeat part (b) for $M = 0.2$ and $M = 50$.

   Comment on your results. Is it better to waterflood a water-wet or an oil-wet reservoir? What is the effect of mobility ratio on recovery?

Hints: Yes, this is a tedious exercise, but after it you will really know how to perform a Buckley–Leverett analysis. Either plot out the fractional flow and find the shock height graphically by hand, or do everything analytically/numerically. It is not possible to find a closed-form expression for the shock saturation for this type of relative permeability, but you could write a small computer program to find everything you need automatically. Once you have worked it out for one case, the others should be easy.

Chapter 21

Previous Exam Papers

21.1. Reservoir Engineering Examinations

PE 300(ii) RESERVOIR ENGINEERING I (RESERVOIR MECHANICS AND SECONDARY RECOVERY)

May 1999
Answer any THREE questions.

1. You have the following production data for a dry gas field: *(50 marks total)*

Pressure (MPa)	Z	$G_p(10^7\,\text{m}^3)$
30	0.75	0
29	0.76	1.16
28	0.77	2.25
27	0.78	3.26
26	0.79	4.20
25	0.80	5.07

Total aquifer compressibility is 2×10^{-9} Pa^{-1}.
Reservoir temperature is 310 K.
$B_w = 1$.
$P_{\text{atm}} = 0.101$ MPa, $T_{\text{atm}} = 288.7$ K, and $Z_{\text{atm}} = 1$.

(i) Assuming a simple aquifer model, estimate the aquifer size and the original gas in place. You may neglect the compressibility of the connate water and rock in the gas field itself. *(25 marks)*

(ii) Gas sales are planned for this field until the reservoir pressure drops to 2 MPa. Make an APPROXIMATE estimate of the Z-factor at 2 MPa. Think carefully about the limit of Z for low pressure. *(5 marks)*

(iii) Estimate the gas produced at 2 MPa and the recovery factor. *(6 marks)*

(iv) Comment on your answer to part (iii). Is your answer physically reasonable and if not why not? What will happen that will prevent the reservoir pressure reaching 2 MPa? *(8 marks)*

(v) What further information would you need to estimate the likely recovery factor for the field and the pressure of likely abandonment? *(6 marks)*

2. You have the following data for an oil reservoir: *(50 marks total)*

N_p (10^7 stb)	G_p (10^9 scf)	R_s (scf/stb)	B_o (rb/stb)	P (atm)	B_g (rb/scf)
0	0	600	1.653	250	0.00123
0.59	4.25	550	1.604	230	0.00137
1.38	11.5	504	1.568	210	0.00156
2.11	18.4	470	1.524	190	0.00178
2.61	23.6	450	1.498	170	0.00195

(i) Is this reservoir being produced above or below the bubble point? Explain your reasoning. *(5 marks)*

(ii) Use the material balance equation to determine the type of reservoir drive, the amount of initial oil in place and the size of the gas cap (if any). You may assume that there is no water influx and that the compressibility of the formation is negligible. *(30 marks)*

(iii) What is the recovery factor at a pressure of 170 atm? Is the recovery factor high or low compared to typical values for primary production? Explain physically what might be happening in the reservoir to explain the recovery factor that has been obtained. *(8 marks)*

(iv) You are considering improved oil recovery for this field. What would you consider injecting into the field and why? *(7 marks)*

3. A core has the following properties: *(50 marks total)*

$$k_{rw} = 0.2(S_w - 0.2)^2$$
$$k_{ro} = 0.8(S_0 - 0.4)^2$$

$\mu_o = 0.0025$ Pa·s
$\mu_w = 0.001$ Pa·s
$\phi = 0.3$
$L = 30$ cm
$Q = 0.1$ cm^3/minute
$A = 1$ cm^2

(i) For the Buckley–Leverett analysis what assumptions are made about the displacement? *(6 marks)*

(ii) Plot the fractional flow curve and from this draw a plot of water saturation against dimensionless velocity for this system. *(20 marks)*

(iii) Plot a graph of pore volumes recovered against pore volumes injected. *(10 marks)*

(iv) Find the time to breakthrough of water for this core. *(6 marks)*

(v) After injecting water for 10 hours, what is the recovery factor and what volume of oil has been recovered? *(8 marks)*

4. Partitioning tracers. *(50 marks total)*

Tracers that dissolve in both oil and water are often used to determine residual oil saturation after waterflooding. This then can be used to evaluate the reservoir's potential for enhanced oil recovery.

In this question you are asked to derive a conservation equation for such a tracer (a partitioning tracer) and then use it to determine the residual oil saturation for an example problem.

The tracer has a concentration c in water. The units of c are mass of tracer per unit volume of water. The tracer does not absorb to the rock, but it does dissolve in oil. There is a saturation S_w of water and a residual oil saturation $1 - S_w = S_{or}$. The oil is not flowing. Both the oil and water saturations are constant throughout the tracer test.

If the concentration of tracer in the water is c, then the concentration in oil is ac.

(i) Starting from the equation

$$\frac{\partial}{\partial t}\left(\begin{array}{c} mass\ per\ unit \\ volume\ of\ the\ reservoir \end{array}\right) + \frac{\partial}{\partial x}\left(\begin{array}{c} Mass\ flux \\ per\ unit\ area \end{array}\right) = 0,$$

derive a conservation equation for the tracer. *(25 marks)*

(ii) Find the speed with which the tracer travels through the porous medium. *(10 marks)*

(iii) In a tracer test, the distance between the injection and production wells is 100 m. The reservoir has undergone water-flooding until there is no further oil production. The tracer is injected with water with a Darcy velocity of 1 m/day. The reservoir porosity is 0.25. $a = 5$. The tracer breaks through at the production well after 50 days. What is the residual oil saturation? *(15 marks)*

5. Pressure measurements in an oil field. *(50 marks total)*

During the exploration phase for a new field, three wells are drilled. The following information is collected:

Well 1. Produced: water. Depth: 1350 m. Pressure: 12.50 MPa.
Well 2. Produced: oil. Depth: 1250 m. Pressure: 11.61 MPa.
Well 3. Produced: gas. Depth: 1056 m. Pressure: 10.48 MPa.
Depth is depth from the water table.

All pressures are measured relative to atmospheric pressure — i.e. atmospheric pressure is defined to be zero.

Acceleration due to gravity, $g = 9.81 \, \mathrm{m \cdot s^{-2}}$.

Density of water $= 1{,}000 \, \mathrm{kg \cdot m^{-3}}$.

Density of oil $= 850 \, \mathrm{kg \cdot m^{-3}}$.

Density of gas $= 340 \, \mathrm{kg \cdot m^{-3}}$.

(i) Is the reservoir normally pressured, over-pressured or under-pressured? *(5 marks)*

(ii) Explain why the pressure of gas and oil initially in a reservoir is normally higher than surrounding water at the same depth. Draw a diagram if you think it will help your explanation. *(10 marks)*

(iii) Find the depths of the gas/oil and oil/water contacts. *(25 marks)*

(iv) What is the bubble point pressure of the oil? *(10 marks)*

UNIVERSITY OF LONDON

PE 300(ii) RESERVOIR ENGINEERING I (RESERVOIR MECHANICS AND SECONDARY RECOVERY)

May 2000

Answer any THREE questions.

1. You have the following production data for a dry gas field: *(50 marks total)*

Pressure (MPa)	Z	$G_p(10^8 \text{ m}^3)$
25	0.85	0
24	0.86	6.09
23	0.87	11.8
22	0.88	17.1
21	0.89	22.1

Total aquifer compressibility is $2.5 \times 10^{-9} \text{ Pa}^{-1}$.

Reservoir temperature is 330 K.

$B_w = 1$.

$P_{\text{atm}} = 0.101 \text{ MPa}$, $T_{\text{atm}} = 288.7 \text{ K}$, and $Z_{\text{atm}} = 1$.

(i) Assuming a simple aquifer model, estimate the aquifer size and the original gas in place. You may neglect the compressibility of the connate water and rock in the gas field itself. *(25 marks)*

(ii) What is the initial *reservoir* volume of gas? *(4 marks)*

(iii) During water influx, the residual gas saturation is 0.35. The initial water saturation is 0.25. What reservoir volume of water influx would be necessary to sweep the reservoir to residual gas saturation? *(10 marks)*

(iv) Using your estimate of the aquifer size, find the pressure drop necessary to give the water influx in part (iii). What is the reservoir pressure at this point? At this pressure water will sweep the entire field. You would expect no further recovery of gas below this pressure. *(6 marks)*

(v) You might want to estimate the recovery factor at the pressure in part (iv). What other information would you need to estimate this recovery factor? *(5 marks)*

2. You have the following data for an oil reservoir: *(50 marks total)*

N_p (10^6stb)	R_p (scf/stb)	R_s (scf/stb)	B_o (rb/stb)	P (atm)	B_g (rb/scf)
0	0	500	1.514	380	0.00234
6.38	550	450	1.498	330	0.00256
10.35	520	410	1.445	310	0.00278
14.35	490	380	1.416	290	0.00289

(i) Use the material balance equation to determine the type of reservoir drive, the amount of initial oil in place and the size of the gas cap (if any). You may assume that there is no water influx and that the compressibility of the formation is negligible. *(25 marks)*

(ii) Explain carefully the trend in R_p observed for this field. Is the field above or below the bubble point at 290 atm? As part of your answer explain what is meant by the critical gas saturation. Why is production normally stopped once the critical gas saturation is reached in the reservoir? *(10 marks)*

(iii) What is the gas saturation when the reservoir pressure is 290 atm? The connate water saturation is 0.25. *(5 marks)*

(iv) The critical gas saturation is 0.2. Estimate the recovery factor at this saturation. What parameter do you need to estimate?

Make a sensible estimate of this parameter, based on the other values that have been measured. *(10 marks)*

3. A reservoir has the following relative permeabilities: *(50 marks total)*

$$k_{rw} = 0.3(S_w - 0.2)^2,$$
$$k_{ro} = 0.8(S_0 - 0.3)^2,$$

$\mu_o = 0.004$ Pa·s
$\mu_w = 0.001$ Pa·s
$\phi = 0.2$

(i) Plot the fractional flow curve and from this draw a plot of water saturation against dimensionless velocity for this system. Indicate clearly the shock front saturation and dimensionless speed. *(20 marks)*

(ii) Plot a graph of pore volumes recovered against pore volumes injected. *(10 marks)*

(iii) Between wells the average total velocity is approximately 0.1 m/day. If the injection and production wells are 500 m apart, then how long will it be before water breaks through? *(6 marks)*

(iv) What is the reservoir water fractional flow at water breakthrough? If $B_o = 1.4$ and $B_w = 0.9$, then what is the surface fractional flow at water breakthrough? *(8 marks) (6 marks)*

(v) After injecting water for 2 years, what pore volumes of oil are recovered? *(6 marks)*

4. Tracer flow with reaction. *(50 marks total)*
Two compounds, 1 and 2, are flowing in a porous medium that contains only water. There is no adsorption. The two compounds react to form a third compound. R_1 is the rate at which compound 1 reacts (measured in units of mass per unit volume of water). R_2 is the rate at which compound 2 reacts. All of the mass of 1 and 2 that react go to form compound 3.

(i) Starting from the equation

$$\frac{\partial}{\partial t}\left(\begin{array}{c} mass\ per\ unit \\ volume\ of\ the\ reservoir \end{array}\right) + \frac{\partial}{\partial x}\left(\begin{array}{c} Mass\ flux \\ per\ unit\ area \end{array}\right)$$
$$= source\ or\ loss,$$

or from explicitly considering mass conservation in a small element of the porous medium, derive a 1D conservation equation for the concentration of compounds 1, 2 and 3 (c_1, c_2 and c_3, measured in units of mass per unit volume of water. *(30 marks)*

(ii) It is suggested that the behaviour of reacting tracers could be studied in 3D heterogeneous aquifers using streamline-based simulation. Do you think that using streamlines would be appropriate for this type of problem? *(5 marks)*

(iii) Write down the conservation equations for compounds 1, 2 and 3 along a streamline, using the time-of-flight coordinate τ. *(15 marks)*

5. Streamline-based simulation and history matching. *(50 marks total)*

In recent years there has been considerable interest in the use of streamlines as a history matching tool. In this question you will derive an expression to find the permeability along a streamline that will match production data for a tracer flood.

A tracer test is performed in a heterogeneous reservoir. Tracer is injected and recovered at a producer. There is a fixed pressure drop between the wells. You may assume single-phase flow with no adsorption. The time for tracer breakthrough, t_{meas}, is measured. An estimated permeability distribution is used in a streamline-based simulation of the tracer test. The predicted tracer breakthrough time is calculated as t_{pred}.

(i) Is streamline-based simulation an appropriate tool for simulating a field-scale tracer displacement? What assumptions does the simulation make? *(6 marks)*

(ii) Define the time-of-flight, τ. *(4 marks)*

(iii) The streamline with the smallest time-of-flight between injector and producer is found from the simulation. Why is this the predicted breakthrough time t_{pred}? The length of this streamline is L and the average permeability along the streamline is K_{pred}. Using Darcy's law, find an expression for t_{pred} in terms of L, K_{pred} and the pressure drop DP between the wells. *(15 marks)*

(iv) The true breakthrough time is t_{meas}. Using the expression from (iii), find the permeability K_{meas} that would give the correct breakthrough time. What assumptions have you made? *(15 marks)*

(v) In a tracer test the breakthrough time $t_{meas} = 250$ days. From a streamline simulation $L = 600$ m. $DP = 10^6$ Pa, $\phi = 0.15$ and $\mu = 10^{-3}$ Pa·s. From this data find K_{meas}. *(10 marks)*

UNIVERSITY OF LONDON

MSc. EXAMINATION 2001

For internal students of the Imperial College of Science, Technology and Medicine.

Taken by students of the T.H. Huxley School (Engineering).

This paper is also taken for the relevant examination for the Associateship of the Royal School of Mines.

PE 300(ii) RESERVOIR ENGINEERING I (RESERVOIR MECHANICS AND SECONDARY RECOVERY)

Friday 7 May 2001: 10.00–13.00

Answer any THREE questions.

1. You have the following production data for a large, dry gas field that is produced by a strong natural water drive: *(50 marks total)*

Pressure (MPa)	G_p (10^8 scf)
41	0
40	0.571
39	1.123
38	1.658
37	2.175

The initial water saturation is 0.2.

Experiments have determined the following empirical expression for B_g as a function of pressure (B_g is measured in units of rb/scf and P is measured in MPa):

$B_g = 0.03/P^{1.2}$

(i) Assuming a simple aquifer model, estimate the original gas in place and the value of the aquifer size times the aquifer compressibility. You may neglect the compressibility of the connate water and rock in the gas field itself. (*25 marks*)

(ii) From geological information it is estimated that there will be significant water breakthrough once 85% of the reservoir volume of the field is swept by water, leaving behind a residual gas saturation of 0.3. At this point the field will be abandoned. Find the reservoir pressure and recovery factor. Explain your working carefully. (*25 marks*)

2. You have the following data for an oil reservoir: (*50 marks total*)

N_p (10^8 stb)	R_p (scf/stb)	R_s (scf/stb)	B_o (rb/stb)	P (atm)	B_g (rb/scf)
0	0	800	1.634	280	0.00345
2.33	900	700	1.603	260	0.00387
3.61	950	600	1.584	240	0.00412
4.59	970	500	1.554	220	0.00435

(i) Use the material balance equation to determine the type of reservoir drive, the amount of initial oil in place and the size of the gas cap (if any). You may assume that there is no water influx and that the compressibility of the formation is negligible. (*25 marks*)

(ii) What is the average gas saturation in the reservoir initially occupied by oil when the reservoir pressure is 220 atm? The connate water saturation is 0.25. Comment on the likely consequences of having this amount of gas in the reservoir. What could be done to prevent the production of excessive amounts of gas? (*13 marks*)

(iii) Re-injection of produced gas into the top of the reservoir is being considered for this field. Comment on the advantages and disadvantages of such a strategy with reference to your answer to part (ii). (*12 marks*)

3. A reservoir has the following relative permeabilities: *(50 marks total)*

$$k_{rw} = 0.4(S_w - 0.2)^2,$$
$$k_{ro} = 0.6(S_0 - 0.35)^2,$$

$\mu_o = 0.003$ Pa·s

$\mu_w = 0.001$ Pa·s

$\phi = 0.2.$

(i) Plot the fractional flow curve and from this draw a plot of water saturation against dimensionless velocity for this system. Indicate clearly the shock front saturation and dimensionless speed. *(20 marks)*

(ii) Plot a graph of pore volumes recovered against pore volumes injected. *(8 marks)*

(iii) The injection and production wells are 300 m apart. The average cross-sectional area of the reservoir is 1,600 m². Water is injected at a rate of 120 m³/day (measured at surface conditions). At the initial reservoir pressure, $B_o = 1.3$ and $B_w = 0.96$. After 1,000 days, $B_o = 1.4$ and $B_w = 0.96$. What is the recovery factor (based on *surface* volumes)? *(12 marks)*

(iv) At late times, the reservoir pressure is allowed to fall. Assuming that the Buckley–Leverett analysis is still correct, find the recovery factor after the injection of 200,000 m³ water (at surface conditions) when $B_o = 1.15$ and $B_w = 0.96$. Comment on your answer compared to part (iii). How can any apparent inconsistency be resolved? *(10 marks)*

4. Flow with equilibrium reaction. *(50 marks total)*
Compound 1 flows in a water-saturated porous medium that contains only water. Compound 1 reacts to form compound 2. One mole of compound 1 reacts to form 1 mole of compound 2. There is no adsorption of either compound. Both compounds are dissolved in water.

(i) Starting from the equation

$$\frac{\partial}{\partial t}\left(\begin{array}{c} mass\ per\ unit \\ volume\ of\ the\ reservoir \end{array}\right) + \frac{\partial}{\partial x}\left(\begin{array}{c} Mass\ flux \\ per\ unit\ area \end{array}\right)$$
$$= source\ or\ loss,$$

or from explicitly considering mass conservation in a small element of the porous medium, derive 1D conservation equations for the concentration of compounds 1 and 2 (c_1 and c_2 respectively, measured in units of mass per unit volume of water). *(30 marks)*

(ii) The two compounds are in chemical equilibrium, which means that $c_1/c_2 = a$, where a is a constant. Eliminate c_2 from the equations, to derive an equation for the transport of compound 1. *(10 marks)*

(iii) Find an expression for the speed with which compound 1 travels. Explain physically why this is different from the speed of a non-reacting tracer. *(10 marks)*

5. Streamline-based simulation. *(50 marks total)*
   The following is a quotation from "Full-Field Modeling Using Streamline-Based Simulation: 4 Case Studies," by R.O. Baker, F. Kuppe, S. Chugh, R. Bora, S. Stojanovic, and R. Batycky, SPE 66405, in the proceedings of the SPE Reservoir Simulation Symposium held in Houston, Texas, 11–14 February 2001.
   "For waterfloods, streamline simulation is a practical tool. Streamline simulation has many advantages, compared to conventional simulation, in terms of the:

   1. flow visualisation,
   2. ability to model larger models and/or more wells,
   3. computational speed enhancements,
   4. ability to decouple the various history matching stages,
   5. generation of well allocation factors,
   6. quantification of drainage volumes, and
   7. easy identification of flow, or drainage, patterns.

   Streamline simulation is not the panacea or catch-all tool for modeling reservoirs. The prerequisite of a relatively consistent voidage replacement, over the production life of the pool, must be adhered to. In some field cases, capillary cross-flow or a depletion drive mechanism may be dominant in which case FD simulation would be the preferred option."

Write a brief essay that amplifies and explains the remarks above. Discuss each of the seven advantages listed above and comment on the cases where streamline-based simulation does not work so well. The essay must be written *in your own words*.

UNIVERSITY OF LONDON

MSc. EXAMINATION 2002

For internal students of the Imperial College of Science, Technology and Medicine.

   Taken by students of the Department of Earth Science and Engineering.

   This paper is also taken for the relevant examination for the Associateship of the Royal School of Mines.

PE 300(ii) RESERVOIR ENGINEERING I (RESERVOIR MECHANICS AND SECONDARY RECOVERY)

Answer any THREE questions.

1. You have the following production data for a dry gas field that is produced by a strong natural water drive: *(50 marks total)*

Pressure (MPa)	G_p (10^{10} scf)	B_g (rb/scf)
30	0	0.000560
29	4.52	0.000575
28	9.08	0.000595
27	13.55	0.000620
26	17.85	0.000650

The initial water saturation is 0.25.

(i) Define the terms dry gas, wet gas and gas condensate. *(7 marks)*

(ii) Assuming a simple aquifer model, estimate the original gas in place and the value of the aquifer size times the aquifer compressibility. You may neglect the compressibility of the connate water and rock in the gas field itself. *(25 marks)*

(iii) At a pressure of 24 MPa the wells are watered out and there is no further gas production. If we assume that at this point the entire reservoir has been swept by gas, then estimate the residual gas saturation. *(18 marks)*

2. You have the following data for an oil reservoir: *(50 marks total)*

N_p (10^6stb)	R_p (scf/stb)	R_s (scf/stb)	B_o (rb/stb)	P (MPa)	B_g (rb/scf)
0	0	500	1.453	40.0	0.000425
40.3	800	450	1.432	39.8	0.000456
69.6	900	400	1.412	39.6	0.000480
101.9	1000	350	1.395	39.4	0.000508

(i) Use the material balance equation to determine the type of reservoir drive, the amount of initial oil in place and the size of the gas cap (if any). You may assume that there is no water influx and that the compressibility of the formation is negligible. (Hint: Find Nm and then *estimate* N and m separately). *(25 marks)*

(ii) Comment on your results and on the recovery so far — what do they indicate about the reservoir drive? Provide a clear and reasoned discussion about what the material balance equation has told you about the reservoir. *(12 marks)*

(iii) Discuss the reservoir management of this field. What problems are you likely to encounter as the pressure falls further? What other recovery strategies might you consider? What other information about the field would you need to know before making a final decision on reservoir management? *(13 marks)*

3. A reservoir has the following relative permeabilities: *(50 marks total)*

$$krw = 0.3(S_w - 0.25)^2,$$
$$kro = 0.8(S_0 - 0.30)^2,$$

$\mu_o = 0.002$ Pa·s

$\mu_w = 0.001$ Pa·s

$\phi = 0.15$.

(i) Plot the fractional flow curve and from this draw a plot of water saturation against dimensionless velocity for this system. Indicate clearly the shock front saturation and dimensionless speed. *(20 marks)*

(ii) Plot a graph of pore volumes recovered against pore volumes injected. *(10 marks)*

(iii) The injection and production wells are 200 m apart. The average cross-sectional area of the reservoir is 2,000 m². Water is injected at a rate of 200 m³/day (measured at surface conditions). At the reservoir pressure, $B_o = 1.5$ and $B_w = 0.98$. Re-plot the graph in part (ii) as oil produced (measured as surface volume in m³) against time (in days). *(12 marks)*

(iv) Estimate the time at which the oil production rate falls below 50 m³/day (measured at surface conditions). *(8 marks)*

4. Dual porosity modelling of fracture flow. *(50 marks total)*

The conventional approach to model flow in fractured reservoirs is the dual porosity approach. The reservoir is assumed to be composed of a connected network of high permeability fractures in communication with a porous matrix of much lower permeability. All the flow is assumed to take place through the fracture network. Fluid is transported from matrix to fracture and vice versa by capillary pressure, but it assumed that there is no flow in the matrix.

(i) Starting from the equation

$$\frac{\partial}{\partial t}\left(\begin{array}{c} mass\ per\ unit \\ volume\ of\ the\ reservoir \end{array}\right) + \frac{\partial}{\partial x}\left(\begin{array}{c} Mass\ flux \\ per\ unit\ area \end{array}\right)$$
$$= source\ or\ loss,$$

or from explicitly considering mass conservation in a small element of the porous medium, derive a 1D conservation equation for the water saturation in the fracture network. The transfer of water from the fracture to the matrix can be viewed as a sink term and is represented by an empirical transfer function T. T has the units of 1/time. Also derive a conservation equation for water in the matrix, remembering that the mass flux due to matrix flow is zero. *(40 marks)*

(ii) Normally T is written as a shape factor, σ multiplied by a pressure difference between fracture and matrix (essentially the capillary pressure, P_{cap}) multiplied by the effective matrix permeability, K_{matrix}, divided by the water viscosity, μ_w. That is: $T = \sigma \frac{K_{matrix} P_{cap}}{\mu_w}$. What are the units of the shape factor? *(10 marks)*

5. FAQs in streamline-based simulation. *(50 marks total)*

(i) You perform a streamline-based simulation on the Maureen field using a reservoir model with 10,000 grid blocks. The run time is 35 s. The same model run using grid-based simulation takes 20 s. You then produce a more detailed model with 40,000 grid blocks. The run times are now 185 s for streamlines and 380 s for grid-based simulation. Assuming that run times scale as a power of the number of grid blocks, estimate the run times for a model with 100,000 and 1,000,000 grid blocks. Comment on your results. *(15 marks)*

(ii) Here are some frequently asked questions about streamline-based simulation. Answer each one briefly and clearly. *(35 marks)*

(a) How many streamlines do I need to have in a simulation?

(b) How is gravity handled using streamline-based simulation?

(c) What are the criteria used for deciding the number of timesteps in a simulation?

(d) What situations are ideal for streamlines?

(e) What cases can't streamlines handle very well?

UNIVERSITY OF LONDON MSc.
EXAMINATION 2003

For internal students of the Imperial College of Science, Technology and Medicine.

Taken by students of the Department of Earth Science and Engineering.

This paper is also taken for the relevant examination for the Associateship of the Royal School of Mines.

PE 300(ii) RESERVOIR ENGINEERING I
(RESERVOIR MECHANICS AND SECONDARY RECOVERY)

Answer any THREE questions.

1. *(50 marks total)*

   (i) You have the following production data for a dry gas field that is produced by a natural water drive:

Pressure (MPa)	G_p (10^6 scf)	B_g (rb/scf)
25.0	0	0.00167
24.5	45.8	0.00170
24.0	94.8	0.00174
23.5	143.4	0.00180
23.0	190.4	0.00186
22.5	234.0	0.00192

   The initial water saturation is 0.25.
   The material balance equation is

   $$G_p = G\left[1 - \frac{B_{gi}}{B_g}\right] + \frac{W_e}{B_g}.$$

   Assuming a simple aquifer model, estimate the original gas in place and the value of the aquifer size times the aquifer compressibility. You may neglect the compressibility of the connate water and rock in the gas field itself. *(30 marks)*

(ii) The residual gas saturation is 0.3. What is the water influx at the current reservoir pressure of 22.5 MPa? What fraction of the reservoir volume has been swept by water? What is expected to happen when the pressure is dropped further? *(20 marks)*

2. *(50 marks total)*

The following pressure measurements are made for an oil field:

Well 1. Water. Depth $= 2{,}100$ m. Pressure $= 18.50$ MPa. Water density $= 1030\ \text{kg}\cdot\text{m}^{-3}$.

Well 2. Oil. Depth $= 2{,}000$ m. Pressure $= 17.75$ MPa. Oil density $= 750\ \text{kg}\cdot\text{m}^{-3}$.

Well 3. Gas. Depth $= 1{,}950$ m. Pressure $= 17.50$ MPa. Gas density $= 380\ \text{kg}\cdot\text{m}^{-3}$. All depths are relative to sea level. All pressures are relative to atmospheric pressure (that is, assume atmospheric pressure $= 0$). The acceleration due to gravity $= 9.81\ \text{m}\cdot\text{s}^{-2}$.

(i) Is the reservoir normally pressured, over-pressured or under-pressured? *(5 marks)*

(ii) Find the depths of the oil/water and gas/oil contacts. Hence find the depth of the oil column. *(30 marks)*

(iii) The areal extent of the reservoir is $1.6 \times 10^6\ \text{m}^2$ and the average porosity is 0.16. The oil formation volume factor is 1.7. Find the oil volume in the reservoir, measured at surface conditions if $S_{wc} = 0.2$. *(15 marks)*

3. You have the following data for an oil reservoir: *(50 marks total)*

N_p (10^8stb)	R_p (scf/stb)	R_s (scf/stb)	B_o (rb/stb)	P (MPa)	B_g (rb/scf)
0	0	600	1.514	30.0	0.000576
0.563	900	550	1.502	29.0	0.000598
0.979	1500	500	1.496	28.0	0.000623
1.676	2300	450	1.480	27.0	0.000684

(i) Use the material balance equation,

$$N_P(B_o + (R_p - R_s)B_g)$$
$$= NB_{oi} \left[\frac{(B_o - B_{oi}) + (R_{si} - R_s)B_g}{B_{oi}} + m\left(\frac{B_g}{B_{gi}} - 1\right) \\ +(1+m)\left(\frac{c_wS_{wc} + c_f}{1 - S_{wc}}\right)|\Delta P| \right]$$
$$+(W_e - W_pB_w)$$

to determine the type of reservoir drive, the amount of initial oil in place and the size of the gas cap (if any). You may assume that there is no water influx and that the compressibility of the formation is negligible. *(25 marks)*

(ii) What options are there for dealing with the produced gas in this field. What options would you recommend? *(8 marks)*

(iii) Discuss what you would do to arrest the pressure decline in this field and boost recovery. *(8 marks)*

(iv) Explain physically why oil recovery can be very high in regions of the reservoir swept by gas cap expansion. *(7 marks)*

4. A core sample has the following relative permeabilities: *(50 marks total)*

$$k_{rw} = 0.25(S_w - 0.2)^2,$$
$$k_{ro} = 0.9(S_0 - 0.3)^2,$$

$\mu_o = 0.003$ Pa·s

$\mu_w = 0.0001$ Pa·s

$\phi = 0.25$.

(i) Plot the fractional flow curve and from this draw a plot of water saturation against dimensionless velocity for this system. Indicate clearly the shock front saturation and dimensionless speed. *(20 marks)*

(ii) Plot a graph of pore volumes recovered against pore volumes injected. *(10 marks)*

(iii) The core has a cross-sectional area of 5 cm² and a length of 15 cm. In a single-phase water flow test the flow rate was 1 cm^3/s for an imposed pressure drop across the core of 0.1 Mpa. What is the permeability of the core? *(8 marks)*

(iv) With the same fixed injection rate of 1 cm^3/s use the results of part (ii) to find the volume of oil recovered after 20 s of injection. *(12 marks) (6 marks)*

5. Streamline-based simulation. *(50 marks total)*

(i) Define the time-of-flight, τ. *(5 marks)*

(ii) Starting from the volume conservation equation for water, derive the following equation for transport along a streamline:

$$\frac{\partial S_w}{\partial t} + \frac{\partial f_w}{\partial \tau} = 0 \quad (20\ marks)$$

(iii) Write a brief critique of streamline-based simulation that addresses the following points: *(25 marks)*

(a) Very briefly, what is streamline-based simulation and how does it differ from conventional grid-based simulation?

(b) What types of simulation is streamline-based simulation best suited for?

(c) What cases are better handled by conventional grid-based approaches?

(d) Why can streamline-based simulation be faster and have less numerical dispersion than grid-based codes?

(e) What practical applications of streamline-based simulation have been pursued in the oil industry?

UNIVERSITY OF LONDON

MSc. EXAMINATION 2004

For internal students of the Imperial College of Science, Technology and Medicine.

Taken by students of the Department of Earth Science and Engineering.

This paper is also taken for the relevant examination for the Associateship of the Royal School of Mines.

PE 300(ii) RESERVOIR ENGINEERING I (RESERVOIR MECHANICS AND SECONDARY RECOVERY)

Answer any THREE questions.

1. Material balance for a gas field. *(50 marks total)*

   (i) Draw a schematic phase diagram as a function of temperature and pressure for a hydrocarbon mixture. Indicate clearly the critical point and the two-phase region. Show the regions of the phase diagram that represent an oil, a gas condensate, a dry gas and a wet gas. Provide a few words of explanation if necessary. *(17 marks)*

   (ii) You have the following data for a large dry gas reservoir:

Pressure (MPa)	G_p (10^8 scf)	B_g (rb/scf)
30.0	0	0.00201
29.5	970	0.00206
29.0	2066	0.00213
28.5	3244	0.00222
28.0	4530	0.00234

The initial water saturation is 0.3.

The material balance equation is

$$G_p = G\left[1 - \frac{B_{gi}}{B_g}\right] + \frac{W_e}{B_g}.$$

Assuming a simple aquifer model, estimate the original gas in place and the value of the aquifer size times the aquifer compressibility. You may neglect the compressibility of the connate water and rock in the gas field itself. *(20 marks)*

(iii) The residual gas saturation is 0.25. What will be the pressure at which water has swept the entire reservoir? What will be the recovery factor? Hint: to do this you will need to *estimate* one quantity — find a sensible estimate of this from an extrapolation of the data given above. *(13 marks)*

2. Pressure distribution. *(50 marks total)*

The following pressure measurements are made for an oil field:

Well 1. Water. Depth = 2,700 m. Pressure = 30.1 MPa. Water density = 1050 kg·m^{-3}.

Well 2. Oil. Depth = 2,300 m. Pressure = 26.4 MPa. Oil density = 650 kg·m^{-3}.

Well 3. Gas. Depth = 2,100 m. Pressure = 25.5 MPa. Gas density = 350 kg·m^{-3}.

All depths are relative to sea level. All pressures are relative to atmospheric pressure (i.e. assume atmospheric pressure = 0). The acceleration due to gravity = 9.81 m·s^{-2}.

(i) Is the reservoir normally pressured, over-pressured or under-pressured? *(4 marks)*

(ii) Find the depths of the oil/water and gas/oil contacts and the depth of the oil column. *(26 marks)*

(iii) The areal extent of the reservoir is 6.8×10^6 m^2 and the average porosity is 0.15. The initial water saturation is 0.2. The oil formation volume factor is 1.3. Find the oil volume in the reservoir, measured at surface conditions. *(10 marks)*

(iv) Explain the difference between the depth of the oil column obtained from the pressure data and obtained from logs. Is your estimate of the oil volume in part (iii) likely to be an overestimate or an underestimate. *(10 marks)*

3. You have the following data for an oil reservoir: *(50 marks total)*

N_p (10^7stb)	R_p (scf/stb)	R_s (scf/stb)	B_o (rb/stb)	P (MPa)	B_g (rb/scf)
0	—	800	1.321	32.0	0.000341
0.996	800	800	1.356	31.0	0.000389
2.122	800	800	1.423	30.0	0.000432
3.566	800	800	1.587	29.0	0.000501

(i) Is the field being produced above or below the bubble point? Explain your reasoning. *(3 marks)*

(ii) Use the material balance equation

$$N_P(B_o + (R_p - R_s)B_g)$$
$$= NB_{oi} \left[\frac{(B_o - B_{oi}) + (R_{si} - R_s)B_g}{B_{oi}} + m\left(\frac{B_g}{B_{gi}} - 1\right) \\ +(1+m)\left(\frac{c_w S_{wc} + c_f}{1 - S_{wc}}\right)|\Delta P| \right]$$
$$+(W_e - W_p B_w)$$

to determine the initial oil in place and the size of the aquifer times the compressibility. Assume that there is no gas cap and that the compressibility of the formation is negligible. *(20 marks)*

(iii) What is the recovery factor so far? Is this a good, bad or average recovery factor for this process? *(7 marks)*

(iv) What options would you consider for the further development of this field? What extra information would you need? *(10 marks)*

(v) What problems are you likely to encounter later in the field life if you simply to continue the pressure decline? *(10 marks)*

4. You have the following relative permeabilities: *(50 marks total)*

$$k_{rw} = 0.4(S_w - 0.25)^2,$$
$$k_{ro} = 0.8(S_o - 0.3)^2,$$

$\mu_o = 0.002$ Pa·s

$\mu_w = 0.0005$ Pa·s

$\phi = 0.18.$

(i) Plot the fractional flow curve and from this draw a plot of water saturation against dimensionless velocity for this system. Indicate clearly the shock front saturation and dimensionless speed. *(15 marks)*

(ii) Plot a graph of pore volumes recovered against pore volumes injected. *(10 marks)*

(iii) Explain the difference between pore volumes recovered and recovery factor. *(5 marks)*

(iv) You use these relative permeabilities to estimate recovery at the field scale. The initial oil in place in the reservoir is 350 MMstb. You plan to inject a total of 50,000 stb water per day. Plot oil recovery in stb against time in days. $B_w = 1.1$ and $B_o = 1.55$. *(10 marks)*

(v) What is the recovery factor after 10,000 days? *(4 marks)*

(vi) You perform a 3D reservoir simulation of waterflooding and predict a recovery factor of 0.3 after 10,000 days. Comment on how your answer compares with part (v). *(6 marks)*

5. Sorbing tracer. *(50 marks total)*

Rate-limited sorption can be modelled by assuming that the tracer sorbs at a rate k_f (forward reaction) and desorbs at a rate k_b (backward reaction). The forward reaction rate is proportional to the tracer concentration while the backward reaction rate is proportional to the sorbed tracer concentration. Mathematically this is expressed (with no flow):

$$\phi \frac{\partial C}{\partial t} = k_b C_s - k_f C,$$

where C is the tracer concentration (mass per unit water volume) and C_s is the sorbed tracer concentration (mass per unit volume of the porous medium).

(i) Derive conservation equations in one dimension for the tracer concentration and the sorbed tracer concentration. *(20 marks)*

(ii) In chemical equilibrium, when the overall reaction rate is zero, what is the relationship between C and C_s? *(10 marks)*

(iii) In chemical equilibrium (imagine that the reaction rates k are very large) how fast does the tracer flow compared to the Darcy velocity? What is the relationship between k_b, k_f and the retardation factor? *(20 marks)*

UNIVERSITY OF LONDON

MSc. EXAMINATION 2005

For internal students of the Imperial College of Science, Technology and Medicine.

Taken by students of the Department of Earth Science and Engineering.

This paper is also taken for the relevant examination for the Associateship of the Royal School of Mines.

PE 300(ii) RESERVOIR ENGINEERING I (RESERVOIR MECHANICS AND SECONDARY RECOVERY)

Answer any THREE questions.

1. Material balance for a gas field. *(50 marks total)*

    (i) Explain briefly but clearly what the following terms mean: wet gas; gas condensate; solution gas; bubble point; and dew point. *(10 marks)*

    (ii) In an oil field that also produces gas, measured at reservoir conditions, the oil fractional flow is 0.5. If $B_o = 1.4$ rm^3/sm^3 and $B_g = 0.0056$ rm^3/sm^3, then what are the oil and gas fractional flows at surface conditions? *(10 marks)*

    (iii) You have the following data for a dry gas reservoir:

Pressure (MPa)	G_p (10^8 scf)	B_g (rb/scf)
35.0	0	0.00345
34.5	87.4	0.00355
34.0	178	0.00367
33.5	266	0.00380
33.0	402	0.00411

The initial water saturation is 0.4.

Assuming a simple aquifer model, estimate the original gas in place and the value of the aquifer size times the aquifer

compressibility. You may neglect the compressibility of the connate water and rock in the gas field itself. *(17 marks)*
The material balance equation is

$$G_p = G \left[1 - \frac{B_{gi}}{B_g} \right] + \frac{W_e}{B_g}.$$

(iv) The residual gas saturation is 0.3. What will be the pressure at which water has swept the entire reservoir? What will be the recovery factor? To do this you will need to *estimate* one quantity — find a sensible estimate of this based on the data given above. *(8 marks)*

(v) What will happen once water has swept the entire reservoir? *(5 marks)*

2. You have the following data for an oil reservoir: *(50 marks total)*

N_p (10^6stb)	R_p (scf/stb)	R_s (scf/stb)	B_o (rb/stb)	P (MPa)	B_g (rb/scf)
0	—	800	1.405	45	0.000134
3.00	1000	700	1.397	44	0.000167
6.58	2000	600	1.386	43	0.000205
10.9	5000	450	1.368	42	0.000267

(i) Is the field being produced above or below the bubble point? Explain your reasoning. *(4 marks)*

(ii) The material balance equation is

$$N_P(B_o + (R_p - R_s)B_g)$$
$$= N B_{oi} \left[\frac{(B_o - B_{oi}) + (R_{si} - R_s)B_g}{B_{oi}} + m \left(\frac{B_g}{B_{gi}} - 1 \right) \atop +(1+m) \left(\frac{c_w S_{wc} + c_f}{1 - S_{wc}} \right) |\Delta P| \right]$$
$$+ (W_e - W_p B_w)$$

What is the principal reservoir drive mechanism? There has been no water production and you do not consider it likely that there is a strong aquifer drive. Estimate the initial oil in place. You may assume that the compressibility of the formation is negligible. *(20 marks)*

(iii) What is the recovery factor so far? Is this a good, bad or average recovery factor for this process? *(6 marks)*

(iv) What is the average gas saturation in the field at the current reservoir pressure, 42 MPa? The connate water saturation is 0.3. Comment on this value. Why is there significant gas production? *(10 marks)*

(v) What options would you consider for the further development of this field? What problems are you likely to encounter? What extra information would you need? *(10 marks)*

3. You have the following relative permeabilities. *(50 marks total)*

$$k_{rw} = 0.15(S_w - 0.3)^2,$$
$$k_{ro} = 0.8(S_o - 0.3)^2,$$

$\mu_o = 0.004$ Pa·s
$\mu_w = 0.001$ Pa·s
$\phi = 0.12$.

(i) Plot the fractional flow curve and from this draw a plot of water saturation against dimensionless velocity for this system. Indicate clearly the shock front saturation and dimensionless speed. *(15 marks)*

(ii) Plot a graph of pore volumes recovered against pore volumes injected. *(10 marks)*

(iii) The field has an estimated STOIIP of 400 MMstb. What is the pore volume of the reservoir? $B_o = 1.4$. *(7 marks)*

(iv) Water is injected into several wells at a total rate of 300,000 stb/day. Plot oil recovery in stb against time in days. $B_w = 1.05$. *(10 marks)*

(v) What is the recovery at 1,000 days? What is the recovery factor at 1,000 days? What is the number of pore volumes produced? Why are the recovery factor and number of pore volumes produced different? *(8 marks)*

4. Streamline-based simulation. *(50 marks total)*

   (i) Briefly explain what streamline-based simulation is, what cases it works well for and what cases it does not work well for. *(15 marks)*

  (ii) Explain why streamline-based simulation is useful in history matching. Explain how the reservoir properties in regions affecting each well can be modified and how these regions may be defined using streamlines. *(15 marks)*

 (iii) Imagine that the measured breakthrough at a well occurs after 200 days. The simulation model predicts breakthrough at 400 days. The wells are running with a fixed pressure drop between injector and producer. By what factor do you need to modify the permeability in order to match breakthrough? What assumptions are you making? *(20 marks)*

5. Carbon dioxide injection into an aquifer. *(50 marks total)*
One way to limit carbon dioxide emissions to the atmosphere is to collect them from the exhaust stream of fossil-fuel-burning power stations and then inject them into saline aquifers. The carbon dioxide moves in its own phase and also dissolves into water.

   (i) Derive a conservation equation for the mass of carbon dioxide. Remember that it can be both in the aqueous phase with a concentration C (units mass per unit water volume) and in its own phase with a saturation S and density. *(35 marks)*

  (ii) On physical grounds do you think that dissolution will speed up, slow down or have no effect on the speed of the injected carbon dioxide compared to an identical case with no dissolution? Explain your reasoning. *(15 marks)*

UNIVERSITY OF LONDON

MSc. EXAMINATION 2006

For internal students of the Imperial College of Science, Technology and Medicine.

Taken by students of the Department of Earth Science and Engineering.

This paper is also taken for the relevant examination for the Associateship of the Royal School of Mines.

PE 300(ii) RESERVOIR ENGINEERING I (RESERVOIR MECHANICS AND SECONDARY RECOVERY)

Answer any THREE questions.

1. Material balance for a gas field. *(50 marks total)*

    (i) Explain briefly but clearly what the following terms mean: dry gas; wet gas; gas condensate; solution gas/oil ratio and oil formation volume factor. *(10 marks)*

    (ii) Discuss how the solution gas/oil ratio can vary from infinity to zero; what sorts of oil and gas fields do different values represent and what does this mean in terms of hydrocarbon composition? *(10 marks)*

    (iii) You have the following data for a dry gas reservoir:

Pressure (MPa)	G_p (10^{12} scf)	B_g (rb/scf)
35	0	0.00245
34	0.079	0.00255
33	0.158	0.00270
32	0.228	0.00285
31	0.302	0.00316

The initial water saturation is 0.4.

Assuming a simple aquifer model, estimate the original gas in place and the value of the aquifer size times the aquifer

compressibility. You may neglect the compressibility of the connate water and rock in the gas field itself. *(17 marks)*

(iv) The residual gas saturation is 0.25. What will be the pressure at which water has swept the entire reservoir? What will be the recovery factor? To do this you will need to *estimate* one quantity — find a sensible estimate of this based on the data given above. *(8 marks)*

(v) Water has just broken through in the reservoir. What future problems are anticipated as the pressure is dropped further? *(5 marks)*

2. Pressure regimes. *(50 marks total)*

The following pressure measurements are made for an oil field with a small gas cap:

Well 1. Water. Depth = 1,950 m. Pressure = 22.65 MPa. Water density = $1040 \, \text{kg} \cdot \text{m}^{-3}$.

Well 2. Oil. Depth = 1,900 m. Pressure = 22.25 MPa. Oil density = $650 \, \text{kg} \cdot \text{m}^{-3}$.

Well 3. Gas. Depth = 1,850 m. Pressure = 22.05 MPa. Gas density = $300 \, \text{kg} \cdot \text{m}^{-3}$.

All depths are relative to sea level. All pressures are relative to atmospheric pressure (i.e. assume atmospheric pressure = 0). The acceleration due to gravity = $9.81 \, \text{m} \cdot \text{s}^{-2}$.

(i) Draw schematic pressure-temperature phase diagrams for an oil and its associated gas in the gas cap. Mark the temperature and pressure conditions on the diagram. Why do they have to be located where they are? *(10 marks)*

(ii) Is the reservoir normally pressured, over-pressured or under-pressured? *(4 marks)*

(iii) Find the depths of the oil/water and gas/oil contacts and the depth of the oil column. *(26 marks)*

(iv) The areal extent of the reservoir is $50 \times 10^6 \, \text{m}^2$, the average porosity is 0.13 and the net-to-gross ratio is 0.8. The oil formation volume factor is 1.41 rm^3/sm^3. The initial water saturation is 0.3. Find the oil volume in the reservoir, measured at surface conditions. *(10 marks)*

3. You have the following data for an oil reservoir: *(50 marks total)*

N_p (10^6stb)	R_p (scf/stb)	R_s (scf/stb)	B_o (rb/stb)	P (MPa)	B_g (rb/scf)
0	—	700	1.612	40	0.000224
33.6	800	600	1.601	39	0.000289
55.5	1,800	475	1.587	38	0.000345
67.3	3,000	300	1.541	37	0.000407

(i) Discuss the types of process for which the material balance equation provides valuable information and the cases where material balance is unlikely to be useful. *(5 marks)*

(ii) Is the field being produced above or below the bubble point? Explain your reasoning. *(4 marks)*

(iii) The material balance equation is

$$N_P(B_o + (R_p - R_s)B_g)$$
$$= NB_{oi} \left[\frac{(B_o - B_{oi}) + (R_{si} - R_s)B_g}{B_{oi}} + m\left(\frac{B_g}{B_{gi}} - 1\right) \right. $$
$$\left. + (1 + m)\left(\frac{c_w S_{wc} + c_f}{1 - S_{wc}}\right)|\Delta P| \right]$$
$$+ (W_e - W_p B_w).$$

What is the principal reservoir drive mechanism? There has been no water production and you do not consider it likely that there is a strong aquifer drive. Estimate the initial oil in place and the size of the gas cap, if any. You may assume that the compressibility of the formation is negligible. *(25 marks)*

(iv) What is the recovery factor so far? Is this a good, bad or average recovery factor for this process? *(6 marks)*

(v) What options would you consider for the further development of this field? What problems are you likely to encounter? What extra information would you need? *(10 marks)*

4. You have the following relative permeabilities: *(50 marks total)*

$$k_{rw} = 0.4(S_w - 0.2)^2,$$
$$k_{ro} = 0.8(S_o - 0.25)^2,$$

$\mu_o = 0.0032$ Pa·s,

$\mu_w = 0.0008$ Pa·s,

$\phi = 0.12$.

(i) Plot the fractional flow curve and from this draw a plot of water saturation against dimensionless velocity for this system. Indicate clearly the shock front saturation and dimensionless speed. *(17 marks)*

(ii) Plot a graph of pore volumes recovered against pore volumes injected. *(10 marks)*

(iii) The field has an estimated STOIIP of 200 MMstb. What is the pore volume of the reservoir? $B_o = 1.35$. *(5 marks)*

(iv) Water is injected into several wells at a total rate of 50,000 stb/day. Plot oil recovery in stb against time in days. $B_w = 1.02$. *(10 marks)*

(v) How much oil has been recovered at 10,000 days? What is the recovery factor at 10,000 days? Is this likely to be an overestimate or underestimate of the real recovery at this time? Explain your answer. *(8 marks)*

5. Tracer injection. *(50 marks total)*
Partitioning tracers are used to determine residual oil saturation in reservoirs that have been waterflooded, to assess the target for gas injection. The tracer dissolves in both water and oil. If the concentration in the water is C, then the concentration in the oil is aC.

(i) Derive a conservation equation for the concentration C in water (units mass per unit water volume). *(20 marks)*

(ii) At what speed does the tracer move? What is the speed of a conservative tracer that does not dissolve in the oil? *(15 marks)*

(iii) A conservative and a sorbing tracer are both injected into a water injection well. The conservative tracer breaks through at the producer after 200 days and the sorbing tracer after 500 days. If $a = 2$, then what is the residual oil saturation? *(15 marks)*

UNIVERSITY OF LONDON

MSc. EXAMINATION 2007

For internal students of the Imperial College of Science, Technology and Medicine.

Taken by students of the Department of Earth Science and Engineering.

This paper is also taken for the relevant examination for the Associateship of the Royal School of Mines.

RESERVOIR ENGINEERING I (RESERVOIR MECHANICS AND SECONDARY RECOVERY)

Answer any THREE questions.

1. Material balance for a gas field. *(50 marks total)*

    (i) Explain briefly but clearly what the following terms mean: dry gas; wet gas; and gas condensate. *(6 marks)*

    (ii) Explain carefully how the management of a gas field would differ from that of a gas condensate or oil field. *(5 marks)*

    (iii) Why would you consider water injection for pressure maintenance in an oil field, but be wary of doing this in a gas field? What are the problems associated with water injection in a gas field? *(7 marks)*

    (iv) You have the following data for a dry gas reservoir:

Pressure (MPa)	G_p (10^{10} scf)	B_g (rb/scf)
28	0	0.00578
27	1.06	0.00621
26	2.00	0.00667
25	2.88	0.00725

The initial water saturation is 0.3.

Assuming a simple aquifer model, estimate the original gas in place and the value of the aquifer size times the aquifer

compressibility. You may neglect the compressibility of the connate water and rock in the gas field itself. *(20 marks)*

(v) The residual gas saturation is 0.3. What will be the pressure at which water has swept the entire reservoir? What will be the recovery factor? To do this you will need to *estimate* one quantity — find a sensible estimate of this based on the data given above. *(8 marks)*

(vi) Water has just broken through in the reservoir. What future problems are anticipated as the pressure is dropped further? *(4 marks)*

2. Reservoir pressure regimes. *(50 marks total)*

The following pressure measurements are made for an oil field with a small gas cap:

Well 1. Water. Depth $= 1,200$ m. Pressure $= 12.1$ MPa. Water density $= 1030$ kg·m^{-3}.

Well 2. Oil. Depth $= 1,100$ m. Pressure $= 11.2$ MPa. Oil density $= 750$ kg·m^{-3}.

Well 3. Gas. Depth $= 900$ m. Pressure $= 10.5$ MPa. Gas density $= 300$ kg·m^{-3}.

All depths are relative to sea level. All pressures are relative to atmospheric pressure (i.e. assume atmospheric pressure $= 0$). The acceleration due to gravity $= 9.81$ m·s^{-2}.

(i) What is drilling mud and what is it used for? What precautions need to be taken when drilling a well into a hydrocarbon-bearing formation? *(6 marks)*

(ii) Is the reservoir normally pressured, over-pressured or under-pressured? *(4 marks)*

(iii) Find the depths of the oil/water and gas/oil contacts and the depth of the oil column. *(23 marks)*

(iv) The areal extent of the reservoir is 1.7 km by 3.6 km. The average porosity is 0.15 and the net-to-gross ratio is 0.76. The oil formation volume factor is 1.37 rm^3/sm^3. The initial water saturation is 0.35. Find the oil volume in the reservoir, measured at surface conditions. *(4 marks)*

(v) Log analysis finds an oil/water contact that is 5 m above the contact estimated in part (iii). Why is this? Which estimate would give the better estimate of initial oil volume? Use this information to make an order-of-magnitude estimate of the reservoir permeability. Explain your working. *(13 marks)*

3. You have the following data for an oil reservoir: *(50 marks total)*

N_p (10^6stb)	R_p (scf/stb)	R_s (scf/stb)	B_o (rb/stb)	P (MPa)	B_g (rb/scf)
0	—	300	1.367	39	0.000456
24.3	800	230	1.337	38	0.000518
37.6	1,800	160	1.301	37	0.000587
43.6	3,000	90	1.263	36	0.000665

(i) Explain why it is worthwhile to perform a material balance analysis before embarking on a reservoir simulation study. *(5 marks)*

(ii) Is the field being produced above or below the bubble point? Explain your reasoning. *(4 marks)*

(iii) The material balance equation is

$$N_P(B_o + (R_p - R_s)B_g)$$

$$= N B_{oi} \left[\frac{(B_o - B_{oi}) + (R_{si} - R_s)B_g}{B_{oi}} + m\left(\frac{B_g}{B_{gi}} - 1\right) \right.$$
$$\left. + (1+m)\left(\frac{c_w S_{wc} + c_f}{1 - S_{wc}}\right)|\Delta P| \right]$$
$$+ (W_e - W_p B_w).$$

What is the principal reservoir drive mechanism? You may assume that there is no aquifer drive. Estimate the initial oil in place and the relative size of the gas cap, if any. The compressibility of the formation is negligible. *(25 marks)*

(iv) What is the recovery factor so far? Is this a good, bad or average recovery factor for this process? *(4 marks)*

(v) Comment on the accuracy of your result. Why is it not possible to determine N accurately? *(5 marks)*

(vi) What options would you consider for the further development of this field? What problems are you likely to encounter? What extra information would you need? *(7 marks)*

4. You have the following relative permeabilities: *(50 marks total)*

$$k_{rw} = 0.2(S_w - 0.3)^2,$$

$$k_{ro} = 0.8(S_0 - 0.3)^2,$$

$\mu_o = 0.002$ Pa·s,

$\mu_w = 0.001$ Pa·s,

$\phi = 0.15$.

(i) Plot the fractional flow curve and from this draw a plot of water saturation against dimensionless velocity for this system. Indicate clearly the shock front saturation and dimensionless speed. *(17 marks)*

(ii) Plot a graph of pore volumes recovered against pore volumes injected. *(10 marks)*

(iii) Explain the difference between recovery factor and pore volumes recovered. *(2 marks)*

(iv) The field has an estimated STOIIP of 300 MMstb. What is the pore volume of the reservoir? $B_o = 1.45$. *(3 marks)*

(v) Water is injected into several wells at a total rate of 300,000 stb/day. What real time (in days) corresponds to one pore volume injected? Plot oil recovery in stb against time in days. $B_w = 1.04$. *(10 marks)*

(vi) How much oil has been recovered at 2,000 days? What is the recovery factor at 5,000 days? Is this likely to be an overestimate or underestimate of the real recovery at this time? Explain your answer. *(8 marks)*

5. Carbon dioxide injection. *(50 marks total)*
Carbon dioxide is injected into an aquifer for long-term storage in order to reduce greenhouse gas emissions and mitigate climate change. The carbon dioxide will dissolve in water as well as flow

in its own (supercritical) phase. Assuming equilibrium, wherever carbon dioxide is present in its own phase, the concentration in water is C_s (units mass per unit volume of water). You may assume incompressible flow where the carbon dioxide density ρ_c and water density ρ_w are constant.

(i) Consider 1D flow into and out of a small box. Consider conservation of mass of carbon dioxide for a shock in saturation where carbon dioxide in its own phase is moving with a saturation S_c and fractional flow f_c and contacting water initially with no carbon dioxide either in its own phase or dissolved in water. Derive an equation for the shock speed. Write the equation in its simplest form in terms of the fractional flow of carbon dioxide in its own phase f_c. *(30 marks)*

(ii) What is the speed of the shock divided by the speed of a shock with no dissolution in water? Explain this result physically. *(13 marks)*

(iii) Find the ratio of the speed of non-dissolving gas to carbon dioxide if $\rho_c = 700\,\text{kg·m}^{-3}$, $C_s = 100\,\text{kg·m}^{-3}$, $S_c = 0.5$ and $f_c = 0.7$. *(7 marks)*

MSc. EXAMINATION 2008

For internal students of the Imperial College of Science, Technology and Medicine.

Taken by students of the Department of Earth Science and Engineering.

This paper is also taken for the relevant examination for the Associateship of the Royal School of Mines.

RESERVOIR ENGINEERING I (RESERVOIR MECHANICS AND SECONDARY RECOVERY)

Answer any THREE questions.

1. Material balance for a gas field. *(50 marks total)*

    (i) Define the following terms: dry gas; wet gas; gas condensate; dew point; and bubble point. *(10 marks)*

    (ii) You have the following production data for a dry gas field that is produced by a natural water drive:

Pressure (MPa)	G_p (10^{10} scf)	B_g (rb/scf)
25.0	0	0.00167
24.0	45.8	0.00170
23.0	94.8	0.00174
22.0	143.4	0.00180
21.0	190.4	0.00186
20.0	234.0	0.00192

The initial water saturation is 0.25.
The material balance equation is

$$G_p = G \left[1 - \frac{B_{gi}}{B_g} \right] + \frac{W_e}{B_g}.$$

Assuming a simple aquifer model, estimate the original gas in place and the value of the aquifer size times the aquifer

compressibility. You may neglect the compressibility of the connate water and rock in the gas field itself. *(24 marks)*

(iii) The residual gas saturation is 0.3. What is the water influx at the current reservoir pressure of 20 MPa? What fraction of the reservoir volume has been swept by water? What do expect to happen when the pressure is dropped further? *(16 marks)*

2. Reservoir pressure regimes. *(50 marks total)*

(i) The following pressure measurements are made for an oil field:

Well 1. Water. Depth $= 3{,}000$ m. Pressure $= 42.0$ MPa. Water density $= 1050$ kg·m^{-3}.

Well 2. Oil. Depth $= 2{,}800$ m. Pressure $= 40.0$ MPa. Oil density $= 700$ kg·m^{-3}.

Well 3. Gas. Depth $= 2{,}600$ m. Pressure $= 39.0$ MPa. Gas density $= 350$ kg·m^{-3}.

All depths are relative to sea level. All pressures are relative to atmospheric pressure (i.e. assume atmospheric pressure $= 0$). The acceleration due to gravity $= 9.81$ m·s^{-2}. From the expression $\frac{\partial P}{\partial z} = \rho g$ write down expressions for the water, oil and gas pressures as a function of depth. *(12 marks)*

(ii) What is drilling mud and what functions does it perform? *(6 marks)*

(iii) Is the reservoir normally pressured, over-pressured or under-pressured? *(4 marks)*

(iv) Find the depths of the oil/water and gas/oil contacts and the depth of the oil column. *(12 marks)*

(v) There is an error in the measurement of the gas pressure and the gauge reads 39.1 MPa. What is your estimate of the depth of oil column now? Comment on your result. *(10 marks)*

(vi) The areal extent of the reservoir is 3.5×10^6 m^2 and the average porosity is 0.15. The oil formation volume factor is 1.7. The average water saturation in the oil zone is 0.2. Find the oil volume in the reservoir (using the original estimate of

the depth of the oil column), measured at surface conditions.
(6 marks)

3. You have the following data for an oil reservoir: *(50 marks total)*

N_p (10^7stb)	R_p (scf/stb)	R_s (scf/stb)	B_o (rb/stb)	P (MPa)	B_g (rb/scf)
0	—	900	1.331	25.0	0.000785
0.922	900	900	1.356	24.0	0.000818
1.810	900	900	1.381	23.0	0.000854
2.667	900	900	1.406	22.0	0.000892
3.840	1000	800	1.356	21.0	0.000935

(i) Was the field initially above or below the bubble point? Has this changed during production? Explain your reasoning. *(7 marks)*

(ii) Use the material balance equation,

$$N_P(B_o + (R_p - R_s)B_g)$$
$$= NB_{oi} \left[\frac{(B_o - B_{oi}) + (R_{si} - R_s)B_g}{B_{oi}} + m\left(\frac{B_g}{B_{gi}} - 1\right) \right. $$
$$\left. + (1+m)\left(\frac{c_w S_{wc} + c_f}{1 - S_{wc}}\right)|\Delta P| \right],$$
$$+ (W_e - W_p B_w)$$

to determine the initial oil in place. Assume that there is no aquifer and no gas cap. *(16 marks)*

(iii) What is the recovery factor so far? Is this a good, bad or average recovery factor for this process? *(7 marks)*

(iv) Is this the best way to operate this field? Explain your reasons. *(12 marks)*

(v) You have been asked to recommend the best recovery process for a recently discovered oil field which is known to be below the bubble point. What would you recommend if it were in the North Sea? Would this change if the field were located in the Sahara Desert? Explain your reasons. *(8 marks)*

4. You have the following relative permeabilities: *(50 marks total)*

$$k_{rw} = 0.6(S_w - 0.2)^3,$$
$$k_{ro} = 0.8(S_0 - 0.25)^4,$$

$\mu_o = 0.002$ Pa·s,

$\mu_w = 0.001$ Pa·s,

$\phi = 0.15.$

(i) Plot the fractional flow curve and from this draw a plot of water saturation against dimensionless velocity for this system. Indicate clearly the shock front saturation and dimensionless speed. *(15 marks)*

(ii) Plot a graph of pore volumes recovered against pore volumes injected. *(10 marks)*

(iii) Explain the difference between pore volumes recovered and recovery factor. *(5 marks)*

(iv) You use these relative permeabilities to estimate recovery at the field scale. The initial oil in place in the reservoir is 450 MMstb. You plan to inject a total of 90,000 stb water per day. Plot oil recovery in stb against time in days. $B_w = 1.05$ and $B_o = 1.45$. *(10 marks)*

(v) What is the recovery factor after 4,000 days? *(4 marks)*

(vi) You perform a 3D reservoir simulation of waterflooding and predict a recovery factor of 0.3 after 4,000 days. Comment on your answer compared with part (v). *(6 marks)*

5. Polymer flooding. *(50 marks total)*
Polymers are injected with water into reservoirs to increase the viscosity of the aqueous phase; this leads to a more favourable mobility contrast between injected and displaced (oil) phases, giving better sweep efficiency and improved recovery. The polymer may also sorb to the solid surface.

Assume that the concentration of polymer in the aqueous phase is C_p (units mass per unit volume of water). Then the mass of polymer sorbed per unit rock volume is aC_p.

(i) Derive a conservation equation for polymer concentration. *(25 marks)*

(ii) What is the speed with which the polymer travels through the porous medium? *(15 marks)*

(iii) Find the polymer flow speed for injection for a polymer front moving through a residual oil saturation of 0.3 where $a = 4$. The Darcy velocity is $10^{-7}\,\mathrm{m\cdot s^{-1}}$ and the porosity is 0.2. *(10 marks)*

MSc. EXAMINATION 2009

For internal students of the Imperial College of Science, Technology and Medicine.

Taken by students of the Department of Earth Science and Engineering.

This paper is also taken for the relevant examination for the Associateship of the Royal School of Mines.

RESERVOIR ENGINEERING I (RESERVOIR MECHANICS AND SECONDARY RECOVERY)

Answer any THREE questions.

1. Material balance for a gas field. *(50 marks total)*

   (i) Explain what is meant by: dry gas; a gas condensate field; drilling mud; waterflooding; and pressure decline. *(10 marks)*
   (ii) You have the following production data for a dry gas field that is produced by a natural water drive:

Pressure (MPa)	G_p (10^6 scf)	B_g (rb/scf)
32.00	0	0.00089
31.75	358	0.00095
31.50	684	0.00102
31.25	968	0.00110

   The initial water saturation is 0.31.
   The material balance equation is

   $$G_p = G \left[1 - \frac{B_{gi}}{B_g} \right] + \frac{W_e}{B_g}.$$

   Assuming a simple aquifer model, estimate the original gas in place and the value of the aquifer size times the aquifer

compressibility. You may neglect the compressibility of the connate water and rock in the gas field itself. *(20 marks)*

(iii) At a pressure of 31.25 MPa there is significant water production. Estimate the average gas saturation in the field. Comment on this value. Is there likely to be significant future production? Why is there significant water production? What further development options would you consider in this field? *(20 marks)*

2. *(50 marks total)*

The following pressure measurements are made for a huge oil field:

Well 1. Water. Depth = 3,500 m. Pressure = 27.80 MPa. Water density = $1060\,kg{\cdot}m^{-3}$.

Well 2. Oil. Depth = 3,350 m. Pressure = 26.52 MPa. Oil density = $800\,kg{\cdot}m^{-3}$.

Well 3. Gas. Depth = 3,200 m. Pressure = 25.57 MPa. Gas density = $350\,kg{\cdot}m^{-3}$.

All depths are relative to sea level. All pressures are relative to atmospheric pressure (i.e. assume atmospheric pressure = 0). The acceleration due to gravity = $9.81\,m{\cdot}s^{-2}$.

(i) Explain what precautions need to be made when drilling a well through a hydrocarbon-bearing formation. Why is a drilling mud used? *(4 marks)*

(ii) Is the reservoir normally pressured, over-pressured or under-pressured? *(4 marks)*

(iii) Find the depths of the oil/water and gas/oil contacts and the depth of the oil column. *(24 marks)*

(iv) The reservoir structure can be approximated as a spherical dome of radius 5,000 m with the top of the field at a depth of 3,050 m. The average porosity is 0.21, the net-to-gross is 0.85 and the oil formation volume factor is 1.65. The average water saturation in the oil zone is 0.28. *Estimate* the oil volume in the reservoir, measured at surface conditions. *(18 marks)*

3. You have the following data for a large oil reservoir: *(50 marks total)*

N_p (10^6 stb)	R_p (scf/stb)	R_s (scf/stb)	B_o (rb/stb)	P (MPa)	B_g (rb/scf)
0	—	700	1.320	43	0.000583
58.5	900	650	1.310	42	0.000603
113.0	1,200	630	1.295	41	0.000635
161.1	1,500	625	1.287	40	0.000678

(i) Based on the data you have, is the field being operated above or below the bubble point? What are the possible recovery mechanisms? *(7 marks)*

(ii) Use the material balance equation,

$$N_P(B_o + (R_p - R_s)B_g)$$
$$= NB_{oi} \left[\frac{(B_o - B_{oi}) + (R_{si} - R_s)B_g}{B_{oi}} + m\left(\frac{B_g}{B_{gi}} - 1\right) + (1+m)\left(\frac{c_w S_{wc} + c_f}{1 - S_{wc}}\right)|\Delta P| \right]$$
$$+ (W_e - W_p B_w)$$

to determine the initial oil in place. You may assume that there is no aquifer and ignore the compressibility of the formation and water. *(16 marks)*

(iii) What is the recovery factor so far? Is this a good, bad or average recovery factor for this process? *(7 marks)*

(iv) Discuss options for future development of this field. What factors would you consider to decide between water injection, produced gas injection, gas sales, or import of gas from a neighbouring field. *(12 marks)*

(v) It is suggested that this field be used to store carbon dioxide collected from a nearby power station. Ignoring costs, how would you decide if this were sensible option? With the data you have so far, is this likely to be feasible? *(8 marks)*

4. You have the following relative permeabilities: *(50 marks total)*

$$k_{rw} = 0.2(S_w - 0.3)^2,$$
$$k_{ro} = 0.9(S_0 - 0.3)^2,$$

$\mu_o = 0.003$ Pa·s,

$\mu_w = 0.005$ Pa·s,

$\phi = 0.15$.

(i) Plot the fractional flow curve and from this draw a plot of water saturation against dimensionless velocity for this system. Indicate clearly the shock front saturation and dimensionless speed. *(15 marks)*

(ii) Plot a graph of pore volumes recovered against pore volumes injected. *(10 marks)*

(iii) Is this likely to be a water-wet, intermediate-wet, mixed-wet or oil-wet rock? Explain your answer carefully. *(5 marks)*

(iv) You use these relative permeabilities to estimate recovery at the field scale. The initial oil in place in the reservoir is 300 MMstb. You plan to inject a total of 20,000 stb water per day for 2,000 days. What is the volume of oil produced measured in stb and the recovery factor? $B_w = 1.03$ and $B_o = 1.3$. *(10 marks)*

(v) Discuss how you would use these results in combination with a reservoir simulation study to predict recovery and design an optimal injection scheme. *(10 marks)*

5. Low-salinity waterflooding. *(50 marks total)*
Recently, some oil companies have considered the injection of low-salinity water since it leads to lower residual oil saturations than injecting normal, high salinity brine.

(i) Derive a conservation equation for the concentration C of salt. There is flow of both oil and water in the reservoir and the salt acts as a tracer in the water phase only. The salt does not sorb to the rock surface or react. *(15 marks)*

(ii) Imagine that water with a salt concentration C_i is injected to displace oil and connate water with a concentration C_c. The residual oil saturation for high salinity flooding is S_{orc} and for low-salinity brine is S_{ori}. Assume piston-like displacement of oil by water (all shock conditions). Draw the sequence of fluid fronts. Quantify the speed with which the fronts move where you can. Explain your results as carefully as possible and provide a physical explanation for the behaviour. *(25 marks)*

(iii) Find the time for breakthrough of low-salinity brine if the total velocity is 1 m/s, the porosity is 0.2, the production well is 100 m from the injection well, the connate water saturation is 0.3, S_{ori} is 0.1 and S_{orc} is 0.3. *(10 marks)*

MSc. EXAMINATION 2010

For internal students of the Imperial College of Science, Technology and Medicine.

Taken by students of the Department of Earth Science and Engineering.

This paper is also taken for the relevant examination for the Associateship of the Royal School of Mines.

RESERVOIR ENGINEERING I (RESERVOIR MECHANICS AND SECONDARY RECOVERY)

Answer any THREE questions.

1. Material balance for a gas field. *(50 marks total)*

   (i) Explain what is meant by: wet gas; a gas condensate field; primary recovery; secondary recovery; recovery factor. *(10 marks)*

   (ii) You have the following production data for a very small dry gas field that is produced by a natural water drive:

Pressure (MPa)	G_p (10^6 scf)	B_g (rb/scf)
31.0	0	0.00123
30.6	25.0	0.00137
30.2	47.4	0.00159
29.8	66.4	0.00203

   The initial water saturation is 0.28 and the residual gas saturation, from laboratory measurements, is 0.32.
   The material balance equation is

   $$G_p = G \left[1 - \frac{B_{gi}}{B_g} \right] + \frac{W_e}{B_g}.$$

   Assuming a simple aquifer model, make an approximate estimate of the original gas in place and the value of the aquifer size times the aquifer compressibility. You may neglect the compressibility of the connate water and rock in the gas field itself.

Is there evidence of significant water influx? *(20 marks)*

(iii) What is the recovery factor now? Estimate the maximum possible recovery factor for this field based on the laboratory measurement of residual gas saturation. Comment on your answer. What further development options would you consider in this field? *(20 marks)*

2. *(50 marks total)*

The following pressure measurements are made for a large oil field:

Well 1. Water. Depth = 2,850 m. Pressure = 16.41 MPa. Water density = $1040 \, \text{kg·m}^{-3}$.

Well 2. Oil. Depth = 2,720 m. Pressure = 15.33 MPa. Oil density = $820 \, \text{kg·m}^{-3}$.

Well 3. Gas. Depth = 2,690 m. Pressure = 15.12 MPa. Gas density = $410 \, \text{kg·m}^{-3}$.

All depths are relative to sea level. All pressures are relative to atmospheric pressure (i.e. assume atmospheric pressure = 0). The acceleration due to gravity = $9.81 \, \text{m·s}^{-2}$.

(i) Explain physically why gas and oil pressures in a reservoir are typically higher than the water pressure in the surrounding aquifer. *(8 marks)*

(ii) Is the reservoir normally pressured, over-pressured or under-pressured? *(4 marks)*

(iii) Find the depths of the oil/water and gas/oil contacts and the depth of the oil column. *(22 marks)*

(iv) The horizontal cross-section of the reservoir through the oil zone is approximately an ellipse with a maximum diameter of 3,450 m and a minimum diameter of 1,680 m. The average porosity is 0.24, the net-to-gross is 0.79 and the oil formation volume factor is 1.43. The average water saturation in the oil zone is 0.34. *Estimate* the oil volume in the reservoir, measured at surface conditions. *(16 marks)*

3. You have the following data for an oil reservoir: *(50 marks total)*

N_p (10^6 stb)	R_p (scf/stb)	R_s (scf/stb)	B_o (rb/stb)	P (MPa)	B_g (rb/scf)
0	—	200	1.210	23	0.000983
11.2	200	150	1.200	22	0.001034
24.6	350	125	1.194	21	0.001145
37.8	550	100	1.188	20	0.001302

(i) Discuss what the material balance equation can be used for. What types of reservoir and production process does it work well for and when does it not work? *(7 marks)*

(ii) Use the material balance equation,

$$N_P(B_o + (R_p - R_s)B_g)$$
$$= N B_{oi} \left[\frac{(B_o - B_{oi}) + (R_{si} - R_s)B_g}{B_{oi}} + m \left(\frac{B_g}{B_{gi}} - 1 \right) \right. $$
$$\left. + (1 + m) \left(\frac{c_w S_{wc} + c_f}{1 - S_{wc}} \right) |\Delta P| \right]$$
$$+ (W_e - W_p B_w)$$

to *estimate* the initial oil in place and the size of the gas cap (if any). You may assume that there is no aquifer and ignore the compressibility of the formation and water. *(16 marks)*

(iii) What is the principal recovery process? What is the recovery factor so far? Is this a good, bad or average recovery factor for this process? *(9 marks)*

(iv) Discuss options for future development of this field. What factors would you consider to decide between water injection, produced gas injection, gas sales, or import of gas from a neighbouring field? *(10 marks)*

(v) It is suggested that this field be used to store carbon dioxide collected from a nearby power station. Ignoring costs, how would you decide if this were sensible option? What other data would you collect? *(8 marks)*

4. You have the following relative permeabilities: *(50 marks total)*

$$k_{rw} = 0.3(S_w - 0.2)^3,$$
$$k_{ro} = 0.9(S_0 - 0.1)^4,$$

$\mu_o = 0.002$ Pa·s,

$\mu_w = 0.0005$ Pa·s,

$\phi = 0.2$.

(i) Plot the fractional flow curve and from this draw a plot of water saturation against dimensionless velocity for this system. Indicate clearly the shock front saturation and dimensionless speed. *(12 marks)*

(ii) Is this likely to be a water-wet, oil-wet or mixed-wet reservoir? Explain your answer carefully. What definitive test could you use to determine wettability? *(8 marks)*

(iii) Plot a graph of pore volumes recovered against pore volumes injected. *(10 marks)*

(iv) You use these relative permeabilities to estimate recovery at the field scale. The initial oil in place in the reservoir is 400 MMstb. You plan to inject a total of 10,000 stb water per day for 6,000 days. What is the volume of oil produced and the recovery factor? $B_w = 1.01$ and $B_o = 1.25$. *(10 marks)*

(v) Discuss how you would use these results in combination with a reservoir simulation study to predict recovery and design an optimal injection scheme. How would you estimate the sweep efficiency? *(10 marks)*

5. Design of carbon dioxide storage. *(50 marks total)*
Consider the injection of huge volumes of CO_2 into a saline aquifer. One problem is that injection causes pressure increases that may lead to fracturing of the rock, which may allow the CO_2 to escape to the surface.

(i) If the rock and brine in the aquifer have a compressibility of c, the volume of the aquifer is V and the rise in pressure is ΔP, then write down the volume of CO_2 that has been injected to cause this pressure increase. *(15 marks)*

(ii) What have you assumed in this analysis? Comment on the pressure at the injection well. *(10 marks)*

(iii) If $c = 10^{-9}$ Pa^{-1}, the aquifer is normally pressured at an average depth of 2,000 m and we consider storage in a large regional aquifer of thickness 1,000 m and extent 100 km by 100 km, then what mass of CO_2 can be stored if the pressure overall increases by no more than 10% and the average CO_2 density is 600 kg·m^{-3}? *(15 marks)*

(iv) Global emissions of CO_2 are approximately 30 Gt (1 Gt = 10^{12} kg) per year. Comment on your answer to part (iii). *(10 marks)*

MSc. EXAMINATION 2011

For internal students of the Imperial College of Science, Technology and Medicine.

Taken by students of the Department of Earth Science and Engineering.

This paper is also taken for the relevant examination for the Associateship of the Royal School of Mines.

RESERVOIR ENGINEERING I (RESERVOIR MECHANICS AND SECONDARY RECOVERY)

Answer any THREE questions.

1. Material balance for a gas field. *(50 marks total)*

   (i) Explain what is meant by: stb; black oil; dry gas; a gas condensate field; and a blowout. *(10 marks)*

   (ii) You have the following production data for a dry gas field that might be produced by a natural water drive:

Pressure (MPa)	G_p (10^6 scf)	B_g (rb/scf)
33.0	0	0.00089
32.6	290	0.00098
32.2	568	0.00113
31.8	814	0.00130

   The initial water saturation is 0.26 and the residual gas saturation, from laboratory measurements, is 0.27.
   The material balance equation is

   $$G_p = G \left[1 - \frac{B_{gi}}{B_g} \right] + \frac{W_e}{B_g}.$$

   Assuming a simple aquifer model, make an approximate estimate of the original gas in place and the value of the aquifer size times the aquifer compressibility. You may neglect the compressibility of the connate water and rock in the gas field itself. *(18 marks)*

(iii) What is the recovery factor now? What is the average gas saturation in the field? Estimate the maximum possible recovery factor for this field based on the laboratory measurement of residual gas saturation. *(12 marks)*

(iv) It is suggested that this field could be used for gas storage and that CO_2 or nitrogen could be injected near the gas/water contact as a "cushion gas". Comment on this proposal. What do you think a "cushion gas" is and what role does it play? *(10 marks)*

2. You are advised to draw a large, clear sketch to illustrate your work. You will be awarded marks for this sketch even if your calculations are in error. *(50 marks total)*

   (i) Write down the definition of isothermal compressibility. Write the definition in terms of density. *(4 marks)*

   (ii) An aquifer is used for CO_2 storage. An injection well and a monitoring well are drilled and pressures are taken at different depths. The aquifer brine has a density of $1{,}120\,kg{\cdot}m^{-3}$. Before any injection the following pressure measurements are taken. Find the depth of the monitoring well measurement. *(6 marks)*

   Well 1: injection. Depth $= 1{,}120$ m. Pressure $= 11.2\,MPa$.
   Well 2: monitoring. Depth $=??$. Pressure $= 12.1\,MPa$.

   (iii) After the injection of 1 million tonnes (10^9 kg) of CO_2 the following measurements are made. What is the depth of the CO_2-water contact? At reservoir conditions, the CO_2 has a density of $600\,kg{\cdot}m^{-3}$. Why has the pressure in the monitoring well increased? *(20 marks)*

   Well 1: injection of CO_2. Depth $= 1{,}120$ m. Pressure $= 13.1\,MPa$.
   Well 2: monitoring (brine). Depth $=$ answer to part (ii). Pressure $= 13.7\,MPa$.

   (iv) CO_2 has a constant compressibility of $10^{-8}\,Pa^{-1}$. Using the definition of compressibility, write an expression for density as a function of pressure. Derive an expression for pressure as a function of depth. Make a rough estimate of the error in depth associated with the assumption of incompressibility. (*20 marks*)

3. You have the following data for an oil reservoir: (*50 marks total*)

N_p (10^6stb)	R_p (scf/stb)	R_s (scf/stb)	B_o (rb/stb)	P (MPa)	B_g (rb/scf)
0	—	300	1.451	32	0.000750
23.4	400	250	1.432	31	0.000856
55.4	550	200	1.416	30	0.001019
78.2	900	150	1.400	29	0.001230

(i) Compare and contrast material balance and reservoir simulation. When would you use reservoir simulation, when material balance and when are they complementary? (*8 marks*)

(ii) Use the material balance equation,

$$N_P(B_o + (R_p - R_s)B_g)$$
$$= NB_{oi} \left[\frac{(B_o - B_{oi}) + (R_{si} - R_s)B_g}{B_{oi}} + m\left(\frac{B_g}{B_{gi}} - 1\right) \right.$$
$$\left. + (1+m)\left(\frac{c_w S_{wc} + c_f}{1 - S_{wc}}\right)|\Delta P| \right]$$
$$+ (W_e - W_p B_w)$$

to *estimate* the initial oil in place and the size of the gas cap (if any). You may assume that there is no aquifer and ignore the compressibility of the formation and water. (*18 marks*)

(iii) What is the principal recovery process? What is the recovery factor so far? Is this a good, bad or average recovery factor for this process? (*9 marks*)

(iv) Discuss in detail options for future development of this field. What extra data do you need to make a decision? Mention economic factors. (*15 marks*)

4. You have the following relative permeabilities: *(50 marks total)*

$$k_{rw} = 0.2(S_w - 0.3)^3,$$
$$k_{ro} = 0.6(S_0 - 0.25)^2,$$

$\mu_o = 0.0015$ Pa·s,

$\mu_w = 0.0005$ Pa·s,

$\phi = 0.25$.

(i) Define relative permeability from the multiphase Darcy equation. Discuss how relative permeability affects overall field-scale oil production. What features of the relative permeability curves have a key impact on recovery? *(18 marks)*

(ii) Plot the fractional flow curve and from this draw a plot of water saturation against dimensionless velocity for this system. Indicate clearly the shock front saturation and dimensionless speed. *(12 marks)*

(iii) Plot a graph of pore volumes recovered against pore volumes injected. *(10 marks)*

(iv) You use these relative permeabilities to estimate recovery at the field scale. The initial oil in place in the reservoir is 200 MMstb. You plan to inject a total of 45,000 stb water per day for 7,000 days. What is the volume of oil produced and the recovery factor? $B_w = 1.02$ and $B_o = 1.30$. *(10 marks)*

5. Streamline-based simulation. *(50 marks total)*

(i) Define the time-of-flight. Starting for the conservation equation for incompressible flow of water, derive the conservation equation along a streamline in terms of the time-of-flight. *(15 marks)*

(ii) Mention the steps in streamline-based simulation; how does it differ from grid-based simulation? *(15 marks)*

(iii) Discuss the strengths and weaknesses of streamline-based simulation. When should it be used and when is it less useful? *(20 marks)*

MSc. EXAMINATION 2012

For internal students of the Imperial College of Science, Technology and Medicine.

Taken by students of the Department of Earth Science and Engineering.

This paper is also taken for the relevant examination for the Associateship of the Royal School of Mines.

RESERVOIR ENGINEERING I (RESERVOIR MECHANICS AND SECONDARY RECOVERY)

Answer any THREE questions.

1. Material balance for a gas field. *(50 marks total)*

    (i) Draw phase diagrams that explain the differences between dry gas, wet gas, gas condensate and black oil fields. *(10 marks)*

    (ii) You have the following production data for a dry gas field that might be produced by a natural water drive:

G_p (10^6 scf)	Pressure (MPa)	B_g (rb/scf)
0	34	0.00234
72	33	0.00298
112	32	0.00345
129	31	0.00354
155	30	0.00399

The initial water saturation is 0.23 and the residual gas saturation, from laboratory measurements, is 0.41.

The material balance equation is

$$G_p = G \left[1 - \frac{B_{gi}}{B_g} \right] + \frac{W_e}{B_g}.$$

Assuming a simple aquifer model, make an approximate estimate of the original gas in place and the value of the aquifer size times the aquifer compressibility. You may neglect the compressibility of the connate water and rock in the gas field itself. *(18 marks)*

(iii) What is the recovery factor now? What is the average gas saturation in the field? Estimate the maximum possible recovery factor for this field based on the laboratory measurement of residual gas saturation. Comment on your values. *(12 marks)*

(iv) Does this field have a strong water drive? Discuss the advantages and disadvantages of a natural water drive in a gas field. *(10 marks)*

2. Pressure analysis. *(50 marks total)*

(i) Write down the ideal gas law. Use this to find an expression for the density of the gas in terms of the molecular mass M and pressure P. *(8 marks)*

(ii) Derive an expression for the pressure as a function of depth in a gas field containing an ideal gas at constant temperature. *(20 marks)*

(iii) A well is drilled into a large gas field with a tall gas column. The top of the gas is at a depth of 2,780 m and the bottom is at a depth of 3,650 m. The pressure at the top is 31.43 MPa. The molecular mass is $0.023\,kg\cdot mol^{-1}$. The absolute temperature is 350 K. The ideal gas constant is $8.314\ J\cdot K^{-1}\cdot mol^{-1}$ and $g = 9.81\ m\cdot s^{-2}$. Use the expression derived in part (ii) to find the pressure in the gas at the base the gas column. *(12 marks)*

(iv) A second well is injected into the field and a pressure of 32.48 MPa is measured at a depth of 3,450 m. Is this consistent with the pressure measured in the first well? If not, then what does this imply about the field? *(10 marks)*

3. Recovery mechanisms. *(50 marks total)*

(i) Write down the multiphase extension of Darcy's law and define relative permeability. Draw relative permeability curves that are typical of a water-wet, mixed-wet and oil-wet rock and explain carefully all the distinctive features of the curves. *(19 marks)*

(ii) Explain what is meant by recovery factor, sweep efficiency and local displacement efficiency. What controls the sweep efficiency of a waterflood? What controls the local displacement efficiency? *(12 marks)*

(iii) Two relative permeability curves are shown below. The first figure is for a sandstone and the second figure (overleaf) is a carbonate. What is the likely wettability of the two samples? Which sample indicates better waterflood recovery? Explain your answer carefully. *(19 marks)*

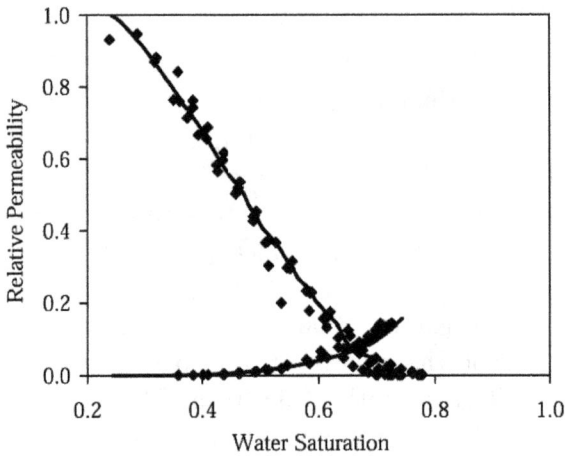

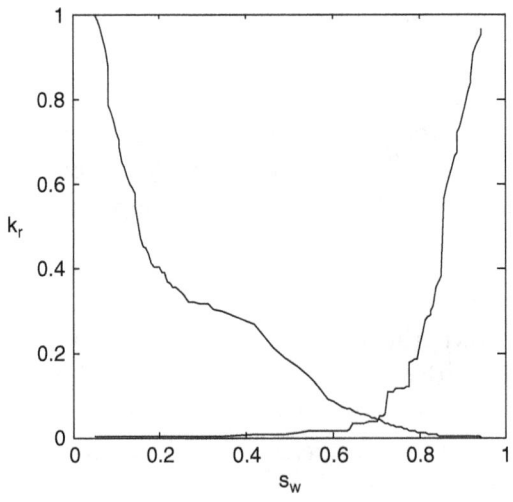

4. Analysis of viscous fingering. *(50 marks total)*

    (i) In miscible gas injection, the injected gas is less viscous than the oil it displaces. There is an instability — called viscous fingering — where there is a mixing zone where gas and oil are present, with the gas moving rapidly though the oil in thin channels. The average behaviour can be described by the following fractional flow of gas in oil:

$$f_g = \frac{c}{c + (1 - c)/M^{1-\omega}}, \qquad M = \frac{\mu_o}{\mu_g},$$

$\mu_o = 0.0020$ Pa·s; $\mu_g = 0.0002$ Pa·s; $\phi = 0.25$; $S_{wc} = 0$; $\omega = 2/3$.

c is the concentration of gas dissolved in the hydrocarbon (oil) phase. By definition the injected gas has $c = 1$.

Explain carefully what miscible gas injection means, and what phases and what components are present. Gas is injected into oil and connate water. For this question you can assume that the connate water saturation is zero. What is injected, what flows and what is recovered? *(8 marks)*

    (ii) Plot the fractional flow curve and from this draw a plot of gas concentration against dimensionless velocity for this system. *(22 marks)*

    (iii) What is the maximum dimensionless velocity with which the gas moves? What is the minimum dimensionless velocity? *(10 marks)*

    (iv) Plot a graph of pore volumes of oil recovered against pore volumes of gas injected. *(10 marks)*

5. Streamline-based simulation. *(50 marks total)*

    (i) Describe the steps in streamline-based simulation. Mention where the steps are the same or different from conventional grid-based simulation. *(14 marks)*

    (ii) Provide some detail on how streamlines are traced through the computational domain. Why is it important that this is performed semi-analytically? *(12 marks)*

    (iii) For the examples below, say what is the appropriate choice of analysis tool — material balance, conventional reservoir simulation or streamline-based simulation — and provide a brief justification. *(24 marks)*

(a) History matching a complex mature waterflood.

(b) Estimating reserves on a small field under primary production.

(c) Simulating production and water influx in a structurally complex gas condensate field.

(d) Ranking and screening many reservoir models as part of a proposed miscible gas injection project.

(e) Assessing the quality of an upscaled reservoir model for a waterflood simulation.

(f) History matching and simulating the development of a large field with an active gas cap and aquifer and a complex schedule of well rates.

MSc. EXAMINATION 2014

For internal students of the Imperial College of Science, Technology and Medicine.

Taken by students of the Department of Earth Science and Engineering.

This paper is also taken for the relevant examination for the Associateship of the Royal School of Mines.

RESERVOIR ENGINEERING I (RESERVOIR MECHANICS AND SECONDARY RECOVERY)

Answer any THREE questions.

1. Material balance for a gas field. *(50 marks total)*

   (i) Explain carefully the following terms: gas condensate field; shale gas; stock tank barrel; and oil formation volume factor. Draw diagrams where appropriate to illustrate your answers. *(10 marks)*

   (ii) You have the following production data for a dry gas field that might be produced by a natural water drive:

G_p (10^6 scf)	Pressure (MPa)	B_g (rb/scf)
0	32	0.000134
26	31	0.000146
63	30	0.000176
85	29	0.000201
105	28	0.000228

   The initial water saturation is 0.29 and the residual gas saturation, from laboratory measurements, is 0.31.

   The material balance equation is

   $$G_p = G \left[1 - \frac{B_{gi}}{B_g} \right] + \frac{W_e}{B_g}.$$

Assuming a simple aquifer model, make an approximate estimate of the original gas in place and the value of the aquifer size times the aquifer compressibility. You may neglect the compressibility of the connate water and rock in the gas field itself. *(18 marks)*

(iii) What is the recovery factor now? What is the average gas saturation in the field? What is the pressure at which water influx will have invaded the whole field? Estimate the maximum possible recovery factor for this field based on the laboratory measurement of residual gas saturation. *(12 marks)*

(iv) It is suggested that this gas field is used to store carbon dioxide. Comment on this suggestion. How could the injected carbon dioxide help or hinder production rates and ultimate recovery? *(10 marks)*

2. You measure the following from appraisal wells in an oil field: *(50 marks total)*

Depth (m)	Pressure (MPa)	Fluid and density ($kg \cdot m^{-3}$)
1,235	12.32	Gas, 250
1,356	12.67	Oil, 750
1,467	13.79	Water, 1,055

The acceleration due to gravity $= 9.81\,m \cdot s^{-2}$. Depths are measured from the surface.

(i) Explain physically why gas and oil pressures in a reservoir are typically higher than the water pressure in the surrounding aquifer. *(8 marks)*

(ii) Is the reservoir normally pressured, over-pressured or under-pressured? Explain your answer carefully. *(6 marks)*

(iii) Find the depths of the oil/water and gas/oil contacts and the depth of the oil column. *(18 marks)*

(iv) Define what is meant by primary and secondary production. *(4 marks)*

(v) Later, after a brief period of primary production, the pressures are 11.54 MPa, 11.79 MPa and 13.72 MPa for the gas, oil and water respectively. Explain carefully what this indicates concerning the mechanism for oil production in the field. What secondary production mechanisms would you recommend? *(14 marks)*

3. Recovery mechanisms and Buckley–Leverett analysis. *(50 marks total)*

(i) Write down the multiphase extension of Darcy's law, explain all the terms with units and use this to define relative permeability. *(5 marks)*

(ii) Plot the relative permeability curves shown below, as well as the corresponding fractional flow. What is the likely wettability of the rock? *(12 marks)*

$$k_{rw} = \frac{(S_w - 0.2)^6}{0.6^5},$$

$$k_{rw} = 0.8 \frac{(S_o - 0.2)^4}{0.6^4},$$

$\mu_o = 0.03$ Pa·s

$\mu_w = 0.001$ Pa·s.

(iii) Calculate the saturation as a function of dimensionless velocity and the pore volumes produced as a function of pore volumes injected. Plot your answers on a graph. *(15 marks)*

(iv) Low-salinity waterflooding is being considered for this field. This will alter the wettability of the rock towards being more water-wet. Schematically, using the approach followed in parts (ii) and (iii), explain how this will impact the fractional flow and local recovery efficiency. Would you recommend low-salinity waterflooding in this case? *(18 marks)*

4. Material balance for an oilfield. You are given the following data for an oilfield: *(50 marks total)*

N_p (MMstb)	G_p (MMscf)	P(MPa)	R_s (scf/stb)	B_0 (rb/stb)	B_g (rb/scf)
0	0	32	400	1.356	0.000187
1.41	480	30	400	1.361	0.000199
1.98	1568	28	370	1.355	0.000213
3.41	3016	26	345	1.349	0.000251
5.78	6890	24	295	1.335	0.000302

(i) Define, physically, the process of gas gravity drainage. During what recovery processes does it occur? Why can it provide a very high local displacement efficiency? *(10 marks)*

(ii) From the material balance equation below, find the size of the oilfield and the relative size of the gas cap (if any). You may assume that there is no active aquifer and can ignore the compressibility for the formation. *(18 marks)*

$$N_P(B_o + (R_p - R_s)B_g)$$
$$= NB_{oi} \left[\dfrac{\begin{array}{l}(B_o - B_{oi}) + (R_{si} - R_s)B_g\end{array}}{\begin{array}{l} B_{oi} \\ +(1+m)\left(\dfrac{c_w S_{wc} + c_f}{1 - S_{wc}}\right)|\Delta P| \end{array}} + m\left(\dfrac{B_g}{B_{gi}} - 1\right) \right]$$
$$+ (W_e - W_p B_w).$$

(iii) From comparing the expansion of oil to the expansion of gas (for the final set of data), quantify relatively how much recovery is contributed from gas expansion and how much from oil expansion. Comment on your answer. *(12 marks)*

(iv) What is the recovery factor now? Is this good as an ultimate recovery factor for this field? Comment on the relative pressure decline in the field. What further development options would you consider? *(10 marks)*

5. Streamline-based simulation. *(50 marks total)*

   (i) Describe the steps in streamline-based simulation. Define what is meant by time-of-flight and show how the conservation equation is transformed to an equation using the time-of-flight. Illustrate your answer with diagrams. *(20 marks)*

   (ii) Describe the oilfield applications where the use of streamline-based simulation is an appropriate tool and where it is inappropriate. *(10 marks)*

   (iii) For the examples below, provide the appropriate choice of analysis tool: material balance, conventional reservoir simulation or streamline-based simulation. Provide a brief justification of your answer. *(20 marks)*

   (a) Analysis of a shale gas field.

   (b) Prediction of carbon dioxide migration in a storage aquifer over thousands of years.

   (c) Coupling advection with geochemical reaction to understand the movement of pollutants in groundwater.

   (d) Modelling a laboratory displacement experiment in a core sample.

   (e) Elucidating the production mechanism for a large oilfield that has been producing for many decades without pressure maintenance.

21.2.   Flow In Porous Media Questions

MEng. EXAMINATION 2000

For internal students of the Imperial College of Science, Technology and Medicine.

Taken by students of the T.H. Huxley School (Engineering).

This paper is also taken for the relevant examination for the Associateship of the Royal School of Mines.

ERE 202. ENVIRONMENTAL AND RESERVOIR PHYSICS

Friday 7 May 2000: 10.00–13.00

Answer TWO questions from the section below.

1. Capillary pressure and the oil/water contact. *(20 marks total)*

    (i) Explain what is meant by the oil/water contact in a reservoir. *(2 marks)*

    (ii) Draw a sketch of the water saturation versus depth in an oil reservoir. Mark the location of the oil/water contact. *(3 marks)*

    (iii) In the appraisal stage of a reservoir, an estimate is made of the initial oil in place. Sometimes it is assumed that $S_o = 0$ below the oil/water contact and $S_o = 1 - S_{wc}$ above it. Is this likely to give a good estimate of the oil in place, overestimate the value or underestimate the value? Explain your answer with reference to your answer to part (ii). *(2 marks)*

    (iv) In order to determine the water saturation versus depth (part (ii)), a laboratory measurement of capillary pressure is performed on a rock sample. Should this be a primary drainage, imbibition or secondary drainage experiment? Explain your answer. *(2 marks)*

(v) Using the same fluids as in the reservoir the following measurement of capillary pressure is performed:

Water saturation	Capillary pressure (Pa)
1.0	0
1.0	15,000
0.5	18,000
0.3	24,000
0.3	50,000

Use this information to plot a graph of water saturation versus height above the oil/water contact in the reservoir (as in part (ii), but with real numbers!). The water (brine density) $= 1050\,\text{kg·m}^{-3}$, the oil density $= 750\,\text{kg·m}^{-3}$, the acceleration due to gravity $= 9.81\,\text{m·s}^{-2}$. The porosity of the rock sample $= 0.25$ and the permeability $= 500$ mD. The average porosity of the reservoir is 0.2 and the average permeability is 200 mD. *(11 marks)*

2. Planning water injection. *(20 marks total)*

(i) Vertical injection and production wells are 200 m apart in a reservoir with an average permeability of 200 mD. The oil column is 50 m deep, and the width of the reservoir is 200 m. The average porosity is 0.15. The oil and water viscosities $= 10^{-3}$ Pa·s. The pressure difference between the wells is 5 atm. Assuming simple linear flow, estimate the production rate of oil (in $\text{m}^3\text{·s}^{-1}$). *(6 marks)*

(ii) In part (i) is this a surface or reservoir rate? If $B_o = 1.5$, then find the surface production rate of oil. *(2 marks)*

(iii) Water is injected to maintain the reservoir pressure. If $B_w = 0.98$, then what is the surface injection rate of water? *(2 marks)*

(iv) Estimate *approximately* how long (in days) it will take for water to break through at the production well. *(4 marks)*

(v) To perform a full analysis of this problem would require the measurement of the oil/water relative permeabilities. List and briefly explain three differences between the relative permeabilities for an oil-wet rock and a water-wet rock. *(6 marks)*

MEng. EXAMINATION 2001

For internal students of the Imperial College of Science, Technology and Medicine.

Taken by students of the T.H. Huxley School (Engineering).

This paper is also taken for the relevant examination for the Associateship of the Royal School of Mines.

ERE 202. ENVIRONMENTAL AND RESERVOIR PHYSICS

Friday 7 May 2001: 10.00 – 13.00

Answer any TWO questions from the section below.

1. Calculating permeability. *(20 marks total)*

   (i) You perform an experiment on a core aligned as shown below. The cross-sectional area is 10 cm^2, the flow rate of water is 20 cm^3 per minute, the inlet pressure is 1.1 atm and the outlet pressure is 1 atm. The rock porosity is 0.25. The core has a length of 20 cm. The water density is 1,000 kg·m^{-3} and the acceleration due to gravity is 9.8 m·s^{-2}. The water viscosity is 10^{-3} Pa·s. You may assume that atmospheric pressure is 101 kPa. What is the permeability of the rock? *(15 marks)*

   (ii) At what speed is water moving through the rock? How long would it take for the injected water to be produced? *(5 marks)*

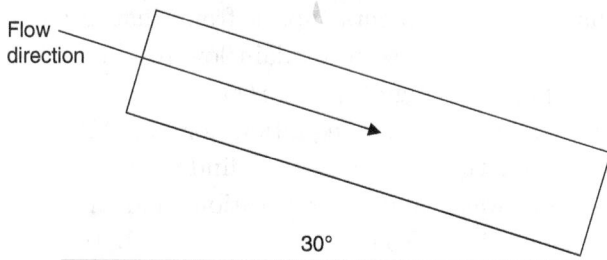

2. Wettability and capillary pressure. *(20 marks total)*

   (i) Describe the Amott wettability test and define the water and oil Amott wettability indices: A_w and A_o. *(4 marks)*

(ii) Draw schematic water injection and oil re-injection capillary pressure curves for the following situations. You need not indicate the magnitude of the capillary pressure, but clearly indicate the positive and negative portions of the capillary pressure curve, and representative values for the connate water and residual oil saturations. For each case indicate if the system would be described as "water-wet," "oil-wet," "mixed-wet," or "neutrally-wet." Explain your answers briefly. *(16 marks)*

(a) $A_w = 1$ and $A_o = 0$.
(b) $A_w = 0.1$ and $A_o = 0$.
(c) $A_w = 0.3$ and $A_o = 0.2$.
(d) $A_w = 0$ and $A_o = 0.6$.

3. Three-phase flow. *(20 marks total)*

(i) Consider the flow of oil, water and gas in a water-wet system. Two measurements of relative permeability are made. The first is a normal waterflood relative permeability with no gas present. The second is a measurement of gas and oil relative permeabilities for gas injection into oil and connate water. By considering the size of pores that the different phases might occupy, comment on whether you expect the residual oil saturation after waterflooding to be larger, smaller or the same size as the residual oil saturation after gas flooding. *(6 marks)*

(ii) What other arguments, apart from that used in part (i) above, could you use to explain low residual oil saturations in the presence of gas? *(4 marks)*

(iii) Derive a conservation equation for the 1D flow of water and gas for three-phase flow — find expressions relating the change in water and gas saturations and the water and gas Darcy velocities. You may assume that both the water and gas are incompressible. You may start with the equation below, if it helps. *(10 marks)*

$$\frac{\partial \ Mass \ per \ unit \ volume}{\partial t} + \frac{\partial \ Mass \ flux}{\partial x} = 0.$$

BSc. and MSci EXAMINATION 2004

For internal students of the Imperial College of Science, Technology and Medicine.

Taken by students of the Department of Earth Science and Engineering.

This paper is also taken for the relevant examination for the Associateship of the Royal School of Mines.

HYDROGEOLOGY EXAM

Answer any FIVE questions.

1. Hydrological Cycle. *(20 marks total)*

   (i) Explain the Hydrological Cycle and illustrate its various components with an appropriate diagram. *(10 marks)*

   (ii) A river catchment of 100 km^2 receives 800 mm of rainfall a year. Evaporation is 100 mm a year and the river run-off is estimated to be $4 \times 10^7 \, \mathrm{m}^3$ per year. Calculate how much water is expected to enter the ground per year. *(5 marks)*

   (iii) What assumptions have you made and what observations would you make to reduce these? *(5 marks)*

2. Measuring permeability. *(20 marks total)*

   (i) Give the definition and units of hydraulic conductivity and intrinsic permeability. *(5 marks)*

   (ii) Describe how permeability can be measured for a sand. *(5 marks)*

   (iii) Use the data below from a permeameter to compute the hydraulic conductivity of the sample of sand. The sample is cylindrical with a diameter of 20 mm. *(10 marks)*

Flow (mm^3)	Time (s)	Hydraulic gradient (mm/mm)
180	20	0.09
400	25	0.16

(Continued)

(*Continued*)		
Flow (mm^3)	Time (s)	Hydraulic gradient (mm/mm)
360	15	0.24
4200	60	0.7
800	10	0.8

3. Porosity and specific yield. *(20 marks total)*

   (i) Define total and effective porosity. Explain the units used to measure them. *(5 marks)*

   (ii) How may these quantities be measured on a sample of material taken to the surface? *(5 marks)*

   (iii) A saturated unconfined aquifer yields 0.02 m^3 of water per square metre of aquifer for a reduction in water table level of 0.1 m. What is the specific yield of the aquifer? *(5 marks)*

   (iv) What is the difference between the specific yield and effective porosity? *(5 marks)*

4. Coefficient of storage. *(20 marks total)*

   (i) Define coefficient of storage. Explain the units used to measure it. *(5 marks)*

   (ii) What happens physically when the pressure is dropped in a confined aquifer for which the coefficient of storage is defined? Why is the coefficient of storage much lower than the specific yield in an unconfined aquifer? *(10 marks)*

   (iii) A confined and fully saturated aquifer has a coefficient of storage of 0.002. How much water is released from an aquifer of area 1000 m^2 for a drop in head of 1 m? *(5 marks)*

5. Darcy's law. *(20 marks total)*

   (i) Define and explain Darcy's law. *(6 marks)*

   (ii) What is transmissivity and what are its units? *(4 marks)*

   (iii) A confined aquifer has a measured head gradient of 0.001 m/m, a depth of 15 m, a width of 150 m and an average intrinsic permeability of 10^{-12} m^2. What is the flow rate of water? Quote in units of m^3 water per day? You may assume that the water density times the acceleration due to gravity divided by the water viscosity is 9.81×10^6 m^{-1}·s^{-1}. *(10 marks)*

6. Capillary fringe. *(20 marks total)*

   (i) Draw a diagram showing the zones of sub-surface water. Define the water table and the capillary fringe. *(8 marks)*

   (ii) A well is being drilled. At a depth of 6 m below the ground surface moist sandy silt is encountered. At 10 m there is standing water in the hole. What do these results mean in terms of the zones defined in part (i) above? *(4 marks)*

   (iii) A pressure measurement is made at a depth of 12 m. The measured pressure is 30 kPa above atmospheric pressure. What can be said about the direction of groundwater flow? *(8 marks)*

7. Discharge and recharge of an aquifer. *(20 marks total)*

   (i) You have an unconfined aquifer. Define the irreducible water saturation. Explain how, by taking a sample of the aquifer sediment, the irreducible water saturation can be measured. *(6 marks)*

   (ii) The aquifer has an effective porosity of 0.25 and an irreducible water saturation of 0.2. The aquifer depth is 40 m and its area is 1.4 km^2. What is the water volume contained in the aquifer if it is completely saturated? What is the specific yield of the aquifer? How much water could be extracted from the aquifer if it were completely drained? *(10 marks)*

   (iii) The annual rainfall is 600 mm. Neglecting evaporation or run-off, how long would it take to recharge the aquifer completely? *(4 marks)*

8. Water quality. *(20 marks total)*

   (i) Discuss the issues to be addressed when assessing water quality in an unconfined aquifer. *(10 marks)*

(ii) What samples should be taken, how can you ensure that they are representative and what significance would you attach to different measurements? *(10 marks)*

END OF EXAM

BSc and MSci PRACTICE EXAMINATION 2005

For internal students of Imperial College London.

Taken by students of Geoscience.

This paper is also taken for the relevant examination for the Associateship of the Royal School of Mines.

HYDROGEOLOGY AND FLUID FLOW

Answer any FIVE questions.

1. Darcy's law. *(20 marks total)*

   An experiment is performed on a 30 cm-long cylindrical sand pack with a cross-sectional area of 3 cm^2. The pressure drop across the pack is 6,000 Pa and the flow rate is 0.1 cm^3/s. You may assume that the viscosity of water is 10^{-3} Pa·s. The pack is held at an angle 30° from horizontal so that the water is flowing uphill. The water density is 1,000 kg·m^{-3} and the acceleration due to gravity $g = 9.81$ m·s^{-2}.

   (i) Write down Darcy's law. Explain all the terms in the equation and give their units. *(4 marks)*

   (ii) What is the permeability of the sand pack? *(6 marks)*

   (iii) The porosity is 0.35. What are the Darcy velocity and interstitial velocities of the water? *(4 marks)*

   (iv) A sorbing tracer is injected that has a retardation coefficient of 5. How long will it take for the tracer to reach the end of the pack? *(6 marks)*

2. Three-phase relative permeability. *(20 marks total)*

   (i) Give two physical situations where the simultaneous flow of three fluid phases occurs. *(4 marks)*

   (ii) Define the spreading coefficient of oil. What does it mean physically? *(4 marks)*

   (iii) Explain why at low oil saturation the oil relative permeability in the presence of gas is proportional to the oil saturation squared. *(6 marks)*

   (iv) Why don't ducks get wet? *(6 marks)*

3. Dissolution. *(20 marks total)*

There has been a spill of a DNAPL. Below the water table, the spill is contained in a cross-sectional area of 20 m^2, a length (in the direction of the groundwater flow) of 5 m and with an average saturation of 0.01. The soil porosity is 0.4. The solubility of the DNAPL is 0.2 kg·m^{-3} and the groundwater flow speed is 10^{-7} m·s^{-1}. The DNAPL density is 1,200 kg·m^{-3}.

(i) What does DNAPL stand for? *(2 marks)*
(ii) What is the initial mass of DNAPL? *(4 marks)*
(iii) At what rate is DNAPL being dissolved? *(7 marks)*
(iv) Estimate how long it will take for all the DNAPL to dissolve. What approximations have you made? *(7 marks)*

4. Partitioning. *(20 marks total)*

There has been a spill of a hydrocarbon. The hydrocarbon density is 700 kg·m^{-3}, the solubility is 0.12 kg·m^{-3} and the saturated vapour density is 0.06 kg·m^{-3}. The hydrocarbon is spilled over a volume of soil 5 m by 40 m to a depth of 3 m in the unsaturated zone. The average water saturation in this region is 0.3 and the average oil saturation is 0.02. The retardation factor is 20. The porosity is 0.3.

(i) Estimate the maximum amount of hydrocarbon that is dissolved in water, in air, sorbed and in its own phase. What is the total mass of oil that has been spilled? *(14 marks)*
(ii) How might the spill be cleaned up? Mention the different options and discuss which ones are most likely to work. *(6 marks)*

5. Conservation equation — three-phase flow. *(20 marks total)*

(i) Starting from the equation

$$\frac{\partial}{\partial t}\left(\begin{array}{c}\text{mass per unit}\\\text{volume of the reservoir}\end{array}\right) + \frac{\partial}{\partial x}\left(\begin{array}{c}\text{Mass flux}\\\text{per unit area}\end{array}\right) = 0$$

derive a conservation equation for three-phase flow where oil, water and gas phases are all flowing. You can assume that the flow is incompressible. Hint: you need only consider a conservation equation for the water and gas phases. *(8 marks)*

(ii) Assume that the solution is a function of speed $v = x/t$ only. Find an expression for the possible wavespeed. *(8 marks)*

(iii) For given saturations, how many wavespeeds are possible? What might these correspond to physically? Will you always get a sensible solution? *(4 marks)*

6. Capillary pressure and the Leverett J-function. *(20 marks total)*

   (i) Write down the relation between capillary pressure and the Leverett J-function. *(4 marks)*

   (ii) A primary drainage mercury injection experiment is performed on a rock sample. A pressure of 10,000 Pa is measured. The rock permeability is 500 mD and the porosity is 0.35. The mercury/air interfacial tension is 140 mN/m. What is the estimated capillary pressure at the same saturation for an oil/water system with an interfacial tension of 30 mN/m, a permeability of 80 mD and a porosity of 0.18? *(8 marks)*

   (iii) Draw representative capillary pressure curves for a water-wet medium for primary drainage, imbibition (water displacement) and secondary drainage. Explain carefully the different features of the curves. *(8 marks)*

7. Aquifers. *(20 marks total)*

   (i) Define the following terms and give appropriate units: specific yield; coefficient of storage; and transmissivity. *(6 marks)*

   (ii) An unconfined aquifer has an area of 10,000 m^2 and a depth of 15 m. How much water can be released from the aquifer if the specific yield is 0.15. *(10'marks)*

   (iii) The porosity is 0.4. What is the connate or irreducible water saturation? *(4 marks)*

8. Relative permeability and the effect of flow rate. *(20 marks total)*

   (i) Write down the multiphase Darcy law and define relative permeability. *(5 marks)*

  (ii) Define the capillary number. What does it represent physically? *(5 marks)*

 (iii) Draw schematic figures that show the effect of capillary number on relative permeability and residual oil saturation. Explain the figures. *(10 marks)*

END OF EXAM

BSc and MSci EXAMINATION 2005

For internal students of Imperial College London.

Taken by students of Geoscience.

This paper is also taken for the relevant examination for the Associateship of the Royal School of Mines.

HYDROGEOLOGY AND FLUID FLOW

Answer any FIVE questions.

1. Darcy's law. *(20 marks total)*

   You perform an experiment on a 1 m-long cylindrical sand pack with a cross-sectional area of 2 cm^2. The pressure drop across the pack is 10,000 Pa and the flow rate is 0.072 cm^3/s. You may assume that the viscosity of water is 10^{-3} Pa·s. The pack is held horizontal.

   (i) Write down Darcy's law. Explain all the terms in the equation and give their units. *(5 marks)*

   (ii) What is the permeability of the sand pack? *(5 marks)*

   (iii) The porosity is 0.4. What are the Darcy velocity and interstitial velocities of the water? *(5 marks)*

   (iv) A conservative (non-sorbing) tracer is injected. How long will it take for the tracer to reach the end of the pack? *(5 marks)*

2. Sorption. *(20 marks total)*

   You perform a test in a long column of soil containing some organic material where you measure the speed at which toluene travels. The retardation factor of toluene is 25.

   (i) Explain physically why sorption causes a contaminant to move slower through a porous medium than a species that does not sorb. *(5 marks)*

   (ii) If the interstitial velocity is v and the retardation factor is R, then at what speed does the contaminant travel? *(2 marks)*

   (iii) The Darcy velocity of the water in the column is 1 mm/s and the porosity is 0.3. What is the speed of the toluene? *(5 marks)*

(iv) You perform two other experiments. In the first you use the same column but measure the speed of octane, which is much less soluble than toluene. In the second you measure the speed of toluene but for a clean soil containing no organic material. For these two cases would you expect the retardation factor to be higher, lower or the same? Explain your answer. *(8 marks)*

3. Diffusion and dispersion. *(20 marks total)*

(i) Explain physically what is meant by molecular diffusion and by dispersion. *(5 marks)*

(ii) If the diffusion coefficient is D, then approximately how far will a contaminant diffuse in a time t? How far will the contaminant move by advection in a time t if the flow speed is v? *(5 marks)*

(iii) $D = 10^{-9}\,\mathrm{m^2 \cdot s^{-1}}$ and $v = 10^{-5}\,\mathrm{m \cdot s^{-1}}$. What is the ratio of advective movement to diffusive movement for $t = 1$ s and for $t = 10^5$ s? Comment on your answers. *(10 marks)*

4. Partitioning. *(20 marks total)*

There has been a spill of a volatile hydrocarbon. Laboratory measurements indicate that the hydrocarbon density is $700\,\mathrm{kg \cdot m^{-3}}$, the solubility is $0.45\,\mathrm{kg \cdot m^{-3}}$ and the saturated vapour density is $0.25\,\mathrm{kg \cdot m^{-3}}$. The hydrocarbon is spilled over a volume of soil 10 m by 10 m to a depth of 5 m in the unsaturated zone. The average water saturation in this region is 0.2 and the average oil saturation is 0.01. The retardation factor is 11. The porosity is 0.4.

(i) Estimate the maximum amount of hydrocarbon that is dissolved in water, in air, sorbed and in its own phase. What is the total mass of oil that has been spilled? *(14 marks)*

(ii) How might the spill be cleaned up? Discuss briefly the different options and mention which ones are most likely to work. *(6 marks)*

5. Conservation equation — partitioning tracers. *(20 marks total)*

Tracers that dissolve in both oil and water are often used to determine how much oil is left behind (the residual oil saturation) after water is used to push out oil in a hydrocarbon reservoir.

In this question you are asked to derive a conservation equation for such a tracer (a partitioning tracer) and then use it to determine the residual oil saturation for an example problem.

The tracer has a concentration c in water. The units of c are mass of tracer per unit volume of water. The tracer does not absorb to the rock, but it does dissolve in oil. There is a saturation S_w of water and an oil saturation $1 - S_w = S_o$. The oil is not flowing. Both the oil and water saturations are constant throughout the tracer test.

If the concentration of tracer in the water is c, then the concentration in oil is ac.

(i) Starting from the equation

$$\frac{\partial}{\partial t} \left(\begin{array}{c} \textit{mass per unit} \\ \textit{volume of the reservoir} \end{array} \right) + \frac{\partial}{\partial x} \left(\begin{array}{c} \textit{Mass flux} \\ \textit{per unit area} \end{array} \right) = 0,$$

derive a conservation equation for the tracer. *(10 marks)*

(ii) Find the speed with which the tracer travels through the porous medium. *(5 marks)*

(iii) Two tracers are injected. One does not dissolve in oil ($a = 0$) and another has $a = 5$. The speed of the non-dissolving tracer (measured by how long it takes the tracer to go from an injection well to a producing well) is 4 times greater than for the dissolving tracer. What is the oil saturation? *(5 marks)*

6. Relative permeability. *(20 marks total)*

(i) Write down the multiphase Darcy law. Explain what relative permeability means and the assumptions that are made in its definition. *(5 marks)*

(ii) Explain physically how the wettability of soil and rock changes from water-wet to oil-wet, or mixed-wet, on contact with oil. *(5 marks)*

(iii) Draw representative relative permeability curves for a water-wet system and a mixed-wet system. Explain carefully the differences between the two curves. *(10 marks)*

7. Aquifer storage. *(20 marks total)*

   (i) Define coefficient of storage. Explain the units used to measure it. *(5 marks)*

  (ii) What happens physically when the pressure is dropped in a confined aquifer for which the coefficient of storage is defined? Why is the coefficient of storage much lower than the specific yield in an unconfined aquifer? *(10 marks)*

 (iii) A confined and fully saturated aquifer has a coefficient of storage of 0.001. How much water is released from an aquifer of area 2000 m^2 for a drop in head of 2 m? *(5 marks)*

8. Capillary fringe and pollution. *(20 marks total)*

   (i) Draw a diagram that carefully shows all the zones of subsurface water and marks clearly the capillary fringe. *(5 marks)*

  (ii) What do the terms LNAPL and DNAPL stand for? *(4 marks)*

 (iii) Explain what happens when an LNAPL and a DNAPL reach the capillary fringe. *(5 marks)*

 (iv) How does the behaviour of the two types of pollutant affect clean-up options? *(6 marks)*

END OF EXAM

BSc and MSci EXAMINATION 2007

For internal students of Imperial College London.

Taken by students of Geoscience.

This paper is also taken for the relevant examination for the Associateship of the Royal School of Mines.

HYDROGEOLOGY AND FLUID FLOW 2

Answer any FOUR questions.

1. Capillary pressure and Leverett J-function. *(25 marks total)*

   You measure the following primary drainage capillary pressure in the laboratory. The core has a permeability of 500 mD, a porosity of 0.25 and the interfacial tension is 50 mN/m.

Pressure (Pa)	Saturation
0	1
8,000	1
12,000	0.5
15,000	0.35
20,000	0.25
30,000	0.25

   (i) Write an equation that relates the capillary pressure to the Leverett J-function. Define all the terms and give appropriate units. *(5 marks)*

   (ii) In the field the average permeability is 100 mD, the porosity is 0.2 and the interfacial tension is 30 mN/m. Plot a graph of water saturation against height above the free water level in the reservoir. The oil density is $800 \, \text{kg·m}^{-3}$ and the brine density is $1100 \, \text{kg·m}^{-3}$. $g = 9.81 \, \text{m·s}^{-2}$. *(15 marks)*

   (iii) What approximations have you made in this analysis? *(5 marks)*

2. Relative permeability. *(25 marks total)*

   (i) Write down the multiphase Darcy equation, define all terms and give them suitable units. *(5 marks)*

   (ii) Draw a schematic of the waterflood relative permeabilities for oil and water for a water-wet sandstone. Label the graph and comment on the values given. *(7 marks)*

   (iii) Draw a schematic of the waterflood relative permeabilities for oil and water for a structurally similar sandstone to part (ii) but where the system is mixed-wet. Comment on the differences with a water-wet system. *(7 marks)*

   (iv) If oil has a density of $700\,\text{kg·m}^{-3}$ and brine a density of $1050\,\text{kg·m}^{-3}$, then what is the Darcy velocity of oil flowing vertically under gravity if the permeability is 100 mD, the oil viscosity is $2.5\,\text{mPa·s}$ and the relative permeability is 0.05? $g = 9.81\,\text{m·s}^{-2}$. *(6 marks)*

3. Young–Laplace equation and contact angles. *(25 marks total)*

   (i) Write down the Young–Laplace equation, define all the terms and give units. *(3 marks)*

   (ii) Derive the capillary pressure between two phases between parallel plates a distance d apart. The contact angle is θ. *(4 marks)*

   (iii) Find the capillary pressure between two fluids residing between two parallel glass plates a distance of $1\ \mu\text{m}$ apart with a contact angle of $40°$ and an interfacial tension of 40 mN/m. *(3 marks)*

   (iv) Derive the Young equation that relates interfacial tensions to contact angles on a flat surface for two fluid phases labelled 1 and 2 with a contact angle θ_{12} between them. The contact angle is measured through phase 2. *(3 marks)*

   (v) Write down the Young equations for each possible pair of fluids if there are three fluids 1, 2 and 3 in equilibrium. *(3 marks)*

   (vi) Use the answer to part (v) to derive an equation that relates the contact angles and interfacial tensions of the three phases. This is the Bartell–Osterhof equation. *(6 marks)*

(vii) If the interfacial tension between phases 1 and 2 is 30 mN/m and the contact angle is 50°, then the interfacial tension between phases 2 and 3 is 20 mN/m and the contact angle is 150° . If the interfacial tension between phases 1 and 3 is 45 mN/m, then what is the contact angle? *(3 marks)*

4. How do trees transpire? *(25 marks total)*

   Trees transpire — i.e. take up water — thanks to capillary action. There is an unresolved mystery concerning how tall trees can do this. A tree can be considered a porous medium — the water moves upwards though the tree in narrow vessels and then evaporates in the leaves.

   (i) Assume that the air pressure and the water pressure are the same at the water table. Neglecting the density of the air, write an equation for the water pressure as a function of height h above the water table. *(5 marks)*

   (ii) Is the answer to part (i) the correct equation for the water pressure insider the tree? Explain your answer. *(4 marks)*

   (iii) What is the capillary pressure between two phases in a cylindrical tube of radius r? *(3 marks)*

   (iv) Assuming that the contact angle between water and air in the tree is zero, find the radius of a vessel (assuming it is cylindrical) necessary to support water to a height of 50 m. The water density is $1000 \, \text{kg·m}^{-3}$ and the interfacial tension between water and air is 70 mN/m. $g = 9.81 \, \text{m·s}^{-2}$. *(7 marks)*

   (v) What is the water pressure at this height? Atmospheric pressure $= 10^5$ Pa. Comment on your answer. *(6 marks)*

5. Conservation equations for miscible WAG. *(25 marks total)*

   Water alternate gas (WAG) injection is often performed as an enhanced oil recovery technique in oil reservoirs. Here we assume that the gas injected is completely miscible with the oil. There are two phases: hydrocarbon and water. In the hydrocarbon phase gas (solvent) has a concentration c, measured in mass of solvent per unit volume of hydrocarbon phase.

(i) Derive conservation equations for 1D incompressible flow for the water saturation and for the solvent concentration. *(11 marks)*

(ii) Simplify the solvent concentration equation by writing out all the terms. *(5 marks)*

(iii) What is the speed of the solvent? *(4 marks)*

(iv) Find the speed of the solvent if only solvent is injected into the reservoir, the connate water saturation is 0.3, the total velocity is 1 m/day and the porosity is 0.2. *(5 marks)*

END OF EXAM

IMPERIAL COLLEGE LONDON

BSc and MSci EXAMINATION 2009

For internal students of Imperial College London.

Taken by students of Geoscience.

This paper is also taken for the relevant examination for the Associateship of the Royal School of Mines.

HYDROGEOLOGY AND FLUID FLOW 2

Answer any FOUR questions.

Wednesday 29th April 2009, 10:00–12:00

1. Capillary pressure and Leverett J-function. *(25 marks total)*

   You measure the following primary drainage capillary pressure in the laboratory using mercury. The core has a permeability of 600 mD, a porosity of 0.20, the interfacial tension is 487 mN/m and the contact angle is 140° .

Pressure (Pa)	Saturation
0	1
50,000	1
74,000	0.6
150,000	0.4
350,000	0.3
300,000	0.3

   (i) Write an equation that relates the capillary pressure to the Leverett J-function. Define all the terms and give appropriate units. *(5 marks)*

   (ii) In the field the average permeability is 200 mD, the porosity is 0.15 and the interfacial tension is 25 mN/m. Plot a graph of water saturation against height above the free water level in the reservoir. The oil density is $700 \, \text{kg·m}^{-3}$ and the brine density is $1050 \, \text{kg·m}^{-3}$. $g = 9.81 \, \text{m·s}^{-2}$. *(15 marks)*

(iii) What approximations have you made in this analysis and what did you have to assume? *(5 marks)*

2. Relative permeability. *(25 marks total)*

   (i) Write down the multiphase Darcy equation, define all terms and give them suitable units. *(5 marks)*

   (ii) Draw a schematic of the waterflood (water displacing oil) and gas flood (gas displacing oil and connate water) relative permeabilities for a water-wet sandstone. Label the graph and comment on the values given and any differences between the relative permeability functions. *(6 marks)*

   (iii) Discuss briefly how the three-phase oil relative permeability can be estimated when all three phases — oil, water and gas — are flowing. *(6 marks)*

   (iv) Estimate the oil production rate from a reservoir of cross-sectional area 1 km by 5 km. The oil drains under gas gravity drainage: oil has a density of $700\,\mathrm{kg\cdot m^{-3}}$ and gas a density of $300\,\mathrm{kg\cdot m^{-3}}$. The permeability is 50 mD and the oil viscosity is 1.5 mPa·s and the relative permeability is 0.001. $g = 9.81\,\mathrm{m\cdot s^{-2}}$. Comment on your result. Why is the oil relative permeability so low? *(8 marks)*

3. Wettability and contact angle. *(25 marks total)*

   (i) Define intrinsic, advancing and receding contact angles. *(3 marks)*

   (ii) Give all the reasons why the advancing contact angle in a porous medium is typically significantly higher than the receding contact angle. *(3 marks)*

   (iii) Draw a graph of primary drainage capillary pressure for a fine sand pack. Also show the waterflood capillary pressure if the sand pack becomes oil-wet after drainage. Explain why the waterflood capillary pressure is lower than the drainage capillary pressure. *(4 marks)*

   (iv) What is the capillary pressure for invasion through a tube of circular cross-section, but which has sides sloping at an angle α and a contact angle θ (see the figure below)? The interfacial tension is σ. *(10 marks)*

(v) Comment on your answer to part (iv). How can it be used to explain capillary pressure hysteresis (consider a porous medium with diverging and converging pores)? *(5 marks)*

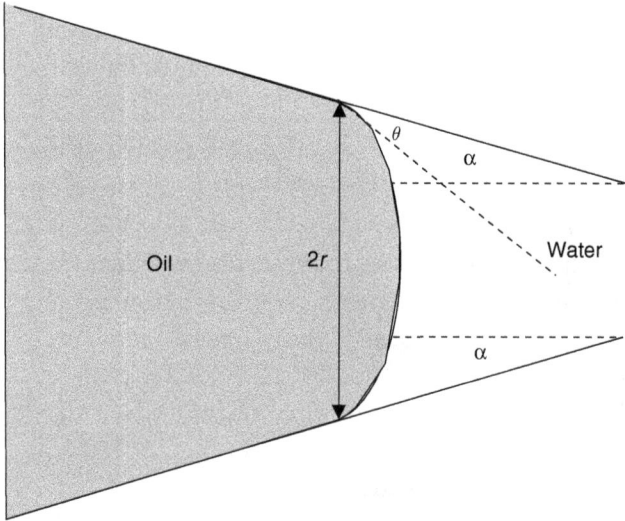

4. Gas storage. *(25 marks total)*

You are asked to consider using a depleted gas field to store CO_2 as part of a carbon-capture-and-storage project.

(i) The natural gas originally in the reservoir had a density of 300 kg·m^{-3}. The brine in the formation has a density of $1,100 \text{ kg·m}^{-3}$. The cap-rock has a porosity of 0.1 and a permeability of 0.01 mD. The gas/brine interfacial tension is 50 mN/m. Make an *approximate estimate* of the capillary pressure necessary for the gas to enter the cap-rock. Explain your calculation carefully. *(10 marks)*

(ii) Use the answer to part (i) to estimate the maximum height of a gas column that could be sustained under the cap rock. *(7 marks)*

(iii) If CO_2 were stored in the same formation, then what would the maximum height of the CO_2 be? The CO_2 density — at reservoir conditions — is 600 kg·m^{-3} and the interfacial tension is 20 mN/m. Comment on the result. Can CO_2

be safely stored in this field if it were to replace the gas?
(8 marks)

5. Conservation equations for a partitioning tracer. *(25 marks total)*
   After waterflooding, there is a non-flowing residual oil saturation
   S_{or}. A tracer is injected in the water that partitions (dissolves)
   in the oil. If the concentration in the water is C, then the
   concentration in oil is aC.

   (i) Derive a conservation equation for the tracer concentration
   for 1D incompressible flow. Explain all the terms carefully.
   (10 marks)

   (ii) With what speed does the tracer move? What is the speed of
   a conservative tracer that does not dissolve in oil? *(5 marks)*

   (iii) A conservative tracer and a partitioning tracer $(a = 2)$ are
   injected into a waterflooded oil field. The conservative tracer
   breaks through at a production well after 100 days, while
   the partitioning tracer breaks through at 150 days. Estimate
   the residual oil saturation. *(10 marks)*

END OF EXAM

IMPERIAL COLLEGE LONDON

BSc and MSci EXAMINATION 2011

For internal students of Imperial College London.

Taken by students of Geoscience.

This paper is also taken for the relevant examination for the Associateship of the Royal School of Mines.

HYDROGEOLOGY AND FLUID FLOW 2

Answer any FOUR questions.

1. Capillary pressure and Leverett *J*-function. *(25 marks total)* You measure the following waterflood capillary pressure in the laboratory. The core has a permeability of 10 mD, a porosity of 0.20, the interfacial tension is 50 mN/m. The core is carbonate taken from a fractured oil field.

Pressure (Pa)	Saturation
300,000	0.35
200,000	0.40
15,000	0.50
0	0.60
−15,000	0.65
−100,000	0.75
−300,000	0.80

(i) Write an equation that relates the capillary pressure to the Leverett *J*-function. Define all the terms and give appropriate units. *(5 marks)*

(ii) In the field the average permeability is 1 mD, the porosity is 0.15 and the interfacial tension is 25 mN/m. Plot a graph of capillary pressure as a function of water saturation for the field. *(10 marks)*

(iii) The reservoir is waterflooded. The matrix is surrounded by fractures, forming blocks approximately 6 m tall. The brine density is $1{,}150\,\text{kg·m}^{-3}$ and the oil density is $800\,\text{kg·m}^{-3}$. Estimate the final water (brine) saturation in the matrix. *(10 marks)*

2. Relative permeability. *(25 marks total)*

(i) Write down the multiphase Darcy equation, define all terms and give them suitable units. *(5 marks)*

(ii) Draw a schematic of the waterflood (water displacing oil) relative permeabilities for water-wet, oil-wet and mixed-wet rock, noting differences between them. *(7 marks)*

(iii) Comment on the implications for waterflood recovery in an oil field. For a light oil with a viscosity similar to water, which wettability type gives the most favourable recovery? *(6 marks)*

(iv) Polymer flooding is being proposed for an oil field. Polymer is injected with the water to increase the water viscosity. This method only works if it improves recovery beyond normal waterflooding. Why does polymer flooding work? For what wettability type(s) is this likely to be most favourable? Explain your answer carefully. *(7 marks)*

3. Pore-scale displacement. *(25 marks total)*

(i) Define snap-off and piston-like advance. What is meant by pore filling? Explain the processes that control the degree of non-wetting phase trapping. *(7 marks)*

(ii) Derive an equation for the entry pressure for piston-like advance through a cylindrical throat of inscribed radius r with contact angle θ. *(4 marks)*

(iii) Derive an equation for the threshold capillary pressure for filling by snap-off for a throat with an equilateral triangular cross-section of inscribed radius r and contact angle θ. What is the ratio of the threshold pressures for snap-off divided by piston-like advance? *(8 marks)*

(iv) Comment on your answer to part (iii). Use this result to explain how the residual non-wetting phase saturation varies with contact angle, for contact angles less than $90°$. *(6 marks)*

4. Gravity drainage and three-phase flow. *(25 marks total)*

   (i) Explain the concept of layer drainage in three-phase flow and explain carefully why the oil relative permeability is proportional to the square of the oil saturation in the layer drainage regime. *(7 marks)*

   (ii) In an oil field that is being produced by gravity drainage, the oil relative permeability is $k_{ro} = 0.1 \times S_o^2$. Write down the conservation equation for oil saturation, putting in the expression for the oil Darcy velocity for vertical flow under gravity. Use this to find the speed with which a saturation S_o travels. *(9 marks)*

   (iii) Estimate the time needed to drain the oil saturation to 20%. The reservoir has an oil column of height 50 m, the oil density is $850 \, \text{kg·m}^{-3}$, the gas density is $350 \, \text{kg·m}^{-3}$, the oil viscosity is 0.3 mPa·s and the vertical permeability is 50 mD. The porosity is 0.2. Comment on your answer. *(9 marks)*

5. Conservation equations for CO_2 storage in a fractured medium. *(25 marks total)*

   Huge fractured aquifers are possible storage locations for CO_2 collected from power stations and other industrial plants. The CO_2 flows through the fractures. CO_2 also dissolves in brine — this CO_2 saturated brine can enter the matrix.

   (i) The CO_2 in its own phase remains in the fractures. Explain physically what prevents the CO_2 in its own phase entering the matrix. *(5 marks)*

   (ii) By what physical mechanism does the CO_2 dissolved in brine move through the matrix? *(4 marks)*

   (iii) If the fractures are closely spaced, then all the water in the matrix in contact with a fracture containing CO_2 will have the same dissolved CO_2 concentration, equal to the solubility. If this solubility is C_s, write down a conservation equation for the flow of CO_2. You may assume that the only Darcy flow is in the fractures and can ignore water in the fractures themselves. To simplify the analysis, you can assume the Darcy flow of CO_2 is $S_c q_t$, where S_c is the saturation and q_t is the total (Darcy) velocity. ϕ_f is the porosity of the fractures and ϕ_m is the porosity of the matrix. Derive an expression for the speed of the CO_2 if the fracture saturation is 1. Draw a sketch to illustrate how the CO_2 moves that explains your answer. *(13 marks)*

(iv) Find the speed of the CO_2 in the fractures with dissolution if the total Darcy velocity is $10^{-7}\,\mathrm{m \cdot s^{-1}}$, $\phi_f = 0.05\%$, $\phi_m = 30\%$. $C_s = 40\,\mathrm{kg \cdot m^{-3}}$ and the density of CO_2 in its own phase is $600\,\mathrm{kg \cdot m^{-3}}$. *(3 marks)*

END OF EXAM

IMPERIAL COLLEGE LONDON

BSc and MSci EXAMINATION 2013

For internal students of Imperial College London.

Taken by students of Geoscience.

This paper is also taken for the relevant examination for the Associateship of the Royal School of Mines.

HYDROGEOLOGY AND FLUID FLOW 2

Answer ANY FOUR questions.

1. Pore-scale displacement. *(25 marks total)*

   (i) Define and explain the pore-scale filling processes that govern trapping in porous media when water displaces oil in a water-wet porous medium. Under what circumstances do you expect to see a significant amount of capillary trapping (a high residual non-wetting phase saturation). Draw pictures to help illustrate your explanation. *(7 marks)*

   (ii) Explain clearly what is meant by an oil layer in two-phase flow. Under what circumstances are oil layers observed? How do they affect the degree of trapping of oil? *(4 marks)*

   (iii) Derive an equation for the threshold capillary pressure for filling by snap-off for a throat with a square cross-section of inscribed radius r and contact angle θ. What is the largest contact angle possible for snap-off to occur (for the threshold capillary pressure to be positive)? *(10 marks)*

   (iv) Comment on your answer to part (iii). What happens to the amount of trapping for larger contact angles than the value found in part (iii)? *(4 marks)*

2. Carbon dioxide storage. *(25 marks total)*

   (i) Write down the multiphase Darcy equation. Define all the terms and provide units. *(5 marks)*

(ii) Write down the equation that describes the Darcy flow — under gravity only — of one fluid in the presence of another fluid. Write down the equation when the fluid that is moving is much more mobile than the fluid it displaces. *(7 marks)*

(iii) In carbon dioxide storage, find the Darcy flow rate under gravity of CO_2 if the density of CO_2 is $600\,\text{kg}\cdot\text{m}^{-3}$, the density of brine is $1{,}050\,\text{kg}\cdot\text{m}^{-3}$, and the viscosity of CO_2 is 2×10^{-5} Pa·s. The permeability is $2 \times 10^{-13}\,\text{m}^2$ and the relative permeability of CO_2 is 0.8. Assume negligible water mobility. In what direction does the CO_2 move? *(8 marks)*

(iv) CO_2 is injected at the bottom of a storage aquifer that is 200 m tall. Use the answer to part (iii) to estimate how long it takes for the CO_2 to rise to the top of the formation. You may assume that the CO_2 saturation is 1 and the porosity is 0.25. What prevents CO_2 from escaping to the surface? *(5 marks)*

3. Relative permeability. *(25 marks total)*

(i) In words explain what is meant by the concept of relative permeability. *(4 marks)*

(ii) Explain what is meant by "wettability". Why do we often encounter oil-wet surfaces in oil fields? *(5 marks)*

(iii) Define the terms water-wet, oil-wet and mixed-wet. *(4 marks)*

(iv) The relative permeability curve below is measured on a core sample from a giant oil field in the Middle East. What is the likely wettability of the sample? Explain your answer carefully. *(5 marks)*

(v) In this field, water is injected to displace oil. If the oil and water viscosities are similar, estimate — approximately — the water saturation at which more water than oil will be produced from the field. What fraction of the oil that was originally in the reservoir will be produced? *(7 marks)*

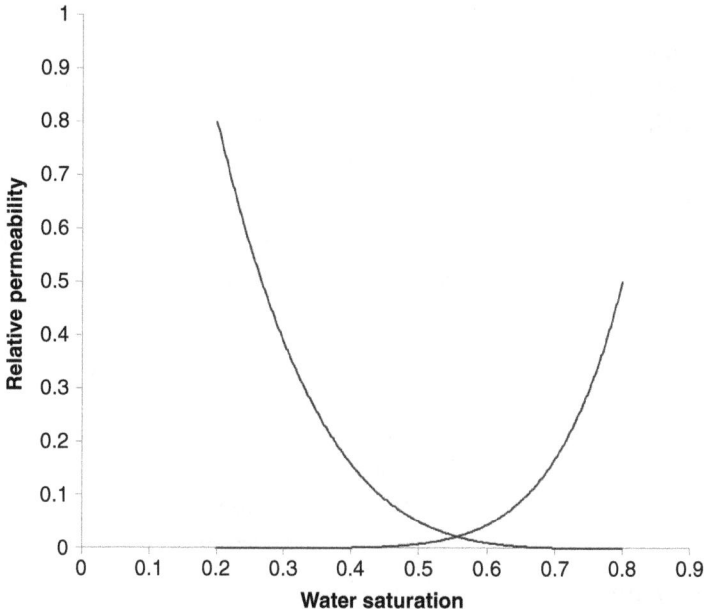

4. Capillary-controlled displacement. *(25 marks total)*

(i) Write down the Young–Laplace equation. Define all the terms with units. *(4 marks)*

(ii) In a rock sample, a typical pore radius is 1 μm and the interfacial tension is 25 mN/m. What — approximately — is a typical capillary pressure? *(4 marks)*

(iii) If I have another rock sample, where all the pores are twice the size as before, by what amount does the typical capillary pressure change? By what factor does the permeability change? *(5 marks)*

(iv) It takes 1,000 s for water to imbibe into a core of radius 1 cm. How long — all else being equal — does it take for water to imbibe into a matrix block in a reservoir that is around 1 m in radius? Explain your answer. *(6 marks)*

(v) For the reservoir-scale matrix block, how long will it take for imbibition if all the pore sizes are now half the size than in the core-scale experiment? *(6 marks)*

5. Capillary pressure and Leverett *J*-function. *(25 marks total)*
   You measure the following waterflood capillary pressure in the
   laboratory. The core has a permeability of 50 mD, a porosity of
   0.25, the interfacial tension is 50 mN/m.

Pressure (Pa)	Saturation
200,000	0.30
100,000	0.35
10,000	0.45
0	0.50
−10,000	0.55
−100,000	0.75
−200,000	0.85

(i) Write an equation that relates the capillary pressure to the
Leverett *J*-function. Define all the terms and give appropriate
units. *(5 marks)*

(ii) In the field the average permeability is 20 mD, the porosity
is 0.15 and the interfacial tension is 20 mN/m. Plot a graph
of capillary pressure as a function of water saturation for the
field. *(10 marks)*

(iii) Is the core sample water-wet, oil-wet or mixed-wet? Explain
your answer. What is the the the Amott wettability index for
water? *(10 marks)*

END OF EXAM

Index